Sustainable Green Technologies for Environmental Management

Shachi Shah • V. Venkatramanan • Ram Prasad
Editors

Sustainable Green Technologies for Environmental Management

Springer

Editors
Shachi Shah
School of Interdisciplinary
and Transdisciplinary Studies
Indira Gandhi National Open University
New Delhi, India

V. Venkatramanan
School of Interdisciplinary
and Transdisciplinary Studies
Indira Gandhi National Open University
New Delhi, India

Ram Prasad
School of Environmental Science
and Engineering
Sun Yat-Sen University
Guangzhou, China

ISBN 978-981-13-2771-1 ISBN 978-981-13-2772-8 (eBook)
https://doi.org/10.1007/978-981-13-2772-8

Library of Congress Control Number: 2018961570

This Springer imprint is published by the registered company Springer Nature Singapore Pte Ltd.
The registered company address is: 152 Beach Road, #21-01/04 Gateway East, Singapore 189721, Singapore

Preface

Albeit Earth has experienced many global changes of varying magnitudes, present-day environmental change is significant for two reasons. Firstly, the change is occurring in remarkably stable period of Holocene which has provided favourable environmental conditions for the genesis and growth of human civilization. Secondly, the global change occurring in the post-industrialization era is mainly due to human action. The rapid changes in the environment is a product of the human ability to dominate the Earth's ecosystem, to enable land use changes of great magnitude and to alter global carbon and nitrogen cycle. Humans in their quest to derive food, feed and other valuable resources modified the earth's landscape. Nevertheless, industrialization, urbanization, land use changes and unprecedented natural resource use driven by rising human population precipitated into a series of environmental crisis warranting a development trajectory with underpinnings and perspectives on sustainability, equity, innovations, green technologies, environment health, and human wellbeing.

Decades back, aspiring for healthful environment, humanity endeavoured for sustainable development that would boost the economy, protect the natural resources and ensure sustainable growth and social justice. At the turn of the last century, United Nations member states and international organizations committed to achieve Millennium Development Goals (MDGs) which included a set of social priorities factoring in prevailing concerns for food security, malnutrition, poverty, education, environmental degradation, and gender inequalities. As a follow-up to the MDGs after their 2015 deadline, Sustainable Development Goals have been adopted in the light of growing urgency for sustainable development, scientific understanding about new geological epoch "Anthropocene" and the dire need for the humanity not to transgress the planetary boundaries and operate well within the safe operating space. Greening the development trajectory through green technologies will ensure sustainable development and drive the future economy. As the green technologies aim to safeguard Earth's life support system, minimize the environmental degradation, mitigate climate change through low carbon emissions, conserve the natural resources for posterity and strive to maintain environmental health, it would be sine qua non for achieving sustainable development goals.

The book epitomizes the potential of green technologies for environmental management. It caters to the needs of environmentalist, microbiologists, agriculturalists and those who are interested in environmental stewardship and sustainability paradigms. We are honoured to receive chapters from leading scientists and professors with rich experience and expertise in the field of sustainable environmental management. Each chapter provides a detailed account of environmental management through sustainable green technologies that include sustainable landscaping, climate smart agriculture technologies, geospatial technologies, biofuels, phytoremediation, microbial technologies in waste management, green infrastructure, and biofortification.

Our heartfelt gratitude goes to the contributors for their reflections and perspectives on green technologies for sustainable environmental management. We sincerely thank Dr. Mamta Kapila, Senior Editor, Springer, Mr. N.S. Pandian, Ms. Raman Shukla and Mr. John Ram Kumar, Project Coordinator, for their generous assistance, constant support and patience in finalizing this book.

New Delhi, India — Shachi Shah
V. Venkatramanan
Guangzhou, China — Ram Prasad

Contents

Contributors

R. Anandham Department of Agricultural Microbiology, Agricultural College and Research Institute, Madurai, India

Department of Agricultural Microbiology, Tamil Nadu Agricultural University, Coimbatore, Tamil Nadu, India

Sagar Chhabra MicroGen Biotech Ltd, ERIC Centre, Institute of Technology Carlow, Carlow, Ireland

Department of Biotechnology, Invertis University, Bareilly, India

Charu Gupta Amity Institute of Herbal Research and Studies, Amity University-UP, Noida, Uttar Pradesh, India

Sneh Gupta Department of Zoology, RGPG College, Meerut, Uttar Pradesh, India

S. Kannojia Centre for Environment Science and Climate Resilient Agriculture, Indian Agricultural Research Institute, New Delhi, India

Manoj Kaushal International Institute of Tropical Agriculture, Ibadan, Oyo State, Nigeria

Amir Khan Department of Microbiology, College of Basic Sciences and Humanities, GBPUA&T, Pantnagar, Uttarakhand, India

R. Krishnamoorthy Department of Agricultural Microbiology, Agricultural College and Research Institute, Madurai, India

Madan Kumar School of Environmental Sciences, Jawaharlal Nehru University, New Delhi, India

Manoj Kumar Centre for Life Sciences, Central University of Jharkhand, Ranchi, Jharkhand, India

S. Kumar Centre for Environment Science and Climate Resilient Agriculture, Indian Agricultural Research Institute, New Delhi, India

Vivek Kumar Himalayan School of Biosciences, Swami Rama Himalayan University, Dehradun, India

K. Kumutha Department of Agricultural Microbiology, Agricultural College and Research Institute, Madurai, India

Piyush Malaviya Department of Environmental Science, University of Jammu, Jammu, Jammu and Kashmir, India

Krishna G. Misra Birbal Sahni Institute of Palaeosciences, Lucknow, India

Sandhya Misra Indian Institute of Technology Gandhinagar, Palaj, Gujarat, India

Birbal Sahni Institute of Palaeosciences, Lucknow, India

Monica Azucena Nazareno Laboratorio de Antioxidantes Procesos Oxidativos, Centro de Investigacin Transferencia de Santiago del Estero (CITSE-CONICET), Universidad Nacional de Santiago del Estero (UNSE), Santiago del Estero, Argentina

Mukund D. Patil Research Program – Asia, International Crops Research Institute for the Semi-Arid Tropics, Patancheru, India

K. V. Peter Kerala Agricultural University, Thrissur, Kerala, India

Dhan Prakash Amity Institute of Herbal Research and Studies, Amity University-UP, Noida, Uttar Pradesh, India

Ram Prasad School of Environmental Science and Engineering, Sun Yat-Sen University, Guangzhou, China

S. Prasad Centre for Environment Science and Climate Resilient Agriculture, Indian Agricultural Research Institute, New Delhi, India

Dinesh Rai Hydrological Investigation Division, National Institute of Hydrology, Roorkee, India

S. P. Rai Hydrological Investigation Division, National Institute of Hydrology, Roorkee, India

V. Venkatramanan School of Interdisciplinary and Transdisciplinary Studies, Indira Gandhi National Open University, New Delhi, India

Tongmin Sa Department of Environmental and Biological Chemistry, Chungbuk National University, Cheongju, Chungbuk, Republic of Korea

M. Senthilkumar Department of Agricultural Microbiology, Tamil Nadu Agricultural University, Coimbatore, Tamil Nadu, India

Shachi Shah School of Interdisciplinary and Transdisciplinary Studies, Indira Gandhi National Open University, New Delhi, India

Malleshaiah Kumar Sharath Horticulture and Product Physiology Group, Wageningen University and Research, Wageningen, The Netherlands

Pradeep Kumar Sharma Department of Environmental Science, Graphic Era University, Dehradun, Uttarakhand, India

Rozi Sharma Department of Environmental Science, University of Jammu, Jammu, Jammu and Kashmir, India

K. R. Sheetal Central Arid Zone Research Institute, Regional Research Station, Bikaner, Rajasthan, India

Ajay Veer Singh Department of Microbiology, College of Basic Sciences and Humanities, GBPUA&T, Pantnagar, Uttarakhand, India

Dharmaveer Singh Symbiosis Institute of Geo-informatics, Symbiosis International (Deemed University), Pune, India

GIS Cell, Motilal Nehru National Institute of Technology, Allahabad, India

Jyoti Singh Department of Microbiology, College of Basic Sciences and Humanities, GBPUA&T, Pantnagar, Uttarakhand, India

Shaili Srivastava Amity School of Earth and Environmental Sciences, Amity University, Gurugram, Haryana, India

Priyanku Teotia Department of Botany, CCS University, Meerut, India

Viabhav Kumar Upadhayay Department of Microbiology, College of Basic Sciences and Humanities, GBPUA&T, Pantnagar, Uttarakhand, India

Suhas P. Wani Research Program – Asia, International Crops Research Institute for the Semi-Arid Tropics, Patancheru, India

About the Editors

Dr. Shachi Shah is an environmentalist with nearly two decades of teaching and research experience in various reputed universities and institutes. She is Associate Professor (Environmental Studies) at the School of Interdisciplinary and Transdisciplinary Studies, IGNOU, New Delhi. A teacher of repute, she has more than 50 publications to her credit. Her research interests include green technologies for waste management and energy generation, bioremediation, waste valorization, plant growth-promoting organisms and biodiversity conservation.

Dr. V. Venkatramanan is Assistant Professor at the School of Interdisciplinary and Transdisciplinary Studies, Indira Gandhi National Open University, New Delhi. His area of interest includes climate change studies, biodegradation and green technologies for environmental management. He has more than 20 publications, including research papers in peer-reviewed journals and book chapters.

Dr. Ram Prasad is associated with Amity Institute of Microbial Technology, Amity University, Uttar Pradesh, India, since 2005. His research interest includes plant-microbe interactions, sustainable agriculture and microbial nanobiotechnology. Dr. Prasad has more than hundred publications to his credit, including research papers, review articles and book chapters and five patents issued or pending and edited or authored several books. Dr. Prasad has 12 years of teaching experience and he has been awarded the Young Scientist Award (2007) and Prof. J.S. Datta Munshi Gold Medal (2009) by the International Society for Ecological Communications; FSAB fellowship (2010) by the Society for Applied Biotechnology; the American Cancer Society UICC International Fellowship for Beginning Investigators, USA (2014); Outstanding Scientist Award (2015) in the field of microbiology by Venus International Foundation; BRICPL Science Investigator Award (ICAABT-2017); and Research Excellence Award (2018). Previously, Dr. Prasad served as Visiting Assistant Professor, Whiting School of Engineering, Department of Mechanical Engineering at Johns Hopkins University, USA, and, presently, works as Research Associate Professor at School of Environmental Sciences and Engineering, Sun Yat-Sen University, Guangzhou, China.

Chapter 1
Enviroscaping: An Environment Friendly Landscaping

Malleshaiah Kumar Sharath and K. V. Peter

Abstract The goal of developing ornamental landscapes that are safe, attractive, and functional for urban dwellers is pursued with great interest, and vast amounts of energy and material resources were used in this effort. However, direct and indirect energy consumption, the need for supplemental water, and the concerns about soil and ground water contamination raise serious questions regarding the long-term sustainability of urban landscapes. Sustainability in landscaping can be improved through a number of actions, such as planning and managing landscapes to function more like natural environments through cycling of resources and managing energy costs; integrating efforts to conserve water and energy, reduce green waste, improve soils, increase wildlife and reducing the demand for energy and material resources in other sectors of the urban environment through microclimate mitigation and habitat restoration. The objective of enviroscaping is to provide home gardeners and commercial landscapers with information that can help them to design and develop beautiful healthy, landscapes in an environmental friendly manner. The approach of enviroscaping is to manage landscapes as an interactive system by considering various components such as temperature, water/irrigation, fertilization, plants and trees, insect pest and pathogens control. Enviroscaping sets new dimension to landscape design and maintenance that can help us to conserve energy and water, recycle yard wastes on site and reduce inputs of fertilizers and pesticides into the environment.

Keywords Sustainability · Landscape design · Mitigation · Recycling · Green architecture

M. K. Sharath (✉)
Horticulture and Product Physiology Group, Wageningen University and Research, Wageningen, The Netherlands
e-mail: sharath.malleshaiah@wur.nl

K. V. Peter
Kerala Agricultural University, Thrissur, Kerala, India

S. Shah et al. (eds.), *Sustainable Green Technologies for Environmental Management*, https://doi.org/10.1007/978-981-13-2772-8_1

1.1 Introduction

Cities have a set of 'Wicked problems' to respond to including climate change, rising mountains of waste, loss of biodiversity, dramatically rising traffic congestion, air and water pollution are the age-old issues need to address and to create a liveable green city. In the era of rapid urbanization, the occupants of mega-cities have to deal with organised destruction of nature's resources like land, water, vegetation and the environment in cities to avail comfort lifestyle in the form of concrete-built infrastructure. In many developing countries, urbanization is taking place at rapid rate due to ample of opportunities in cities; this led to rose in urban population to secure employments and with the heedless perception of having better living standards in towns and cities compared to rural areas.

The face of city landscape is changing continuously due to multifaceted changes in employing the natural resources such as land, water, wildlife for deliberate and unintended economic activities by man. However, this global transformation of city landscapes poses global concern to rethink to build environment friendly and sustainable landscape to protect and conserve the environment for well-being of nature. Furthermore, increasing population in mega-cities is due to rapid rural to urban migration lead to complex socio-environment stresses. The most serious problems were environmental problems; all these were result of extensive human interference and exploitation with environment to build comfort lifestyles lead to experience a so-called ecological/ecosystem crisis.

The environmental catastrophe is a predicament of unplanned, inappropriate industrialisation and consequence of poorly designed cities while using ecological landscapes. In many ways, the environmental/ecological crisis is a result of 'design' crisis. Design plays a central role in connecting culture and nature in the terms of energy usage, exchange of materials and land use pattern or choices. Some of today's environmental problems have arisen due to failure of design process. Every new or altered design will have critical impact upon the nature because every decision on design is a decision on environment. Any landscape or city design is a consequence of how things are under consideration, and the designers have shaped or placed the available natural resources. The approach and manage of our landscapes while we designing will have a significant impact on natural ecosystem. Therefore, one should learn sound landscape design principles, practices and strategies to protect and manage our fragile environment to conserve our renewable and non-renewable energy resources.

In a constant world of economic and social developmental change in terms of intensifying metropolitan cities and environmental crisis, it is wise and imperative to re-define, re-think and to re-design the relationship between urban inhabitants and natural resources. In this twenty-first century it is expected that majority of humanity lives in mega-cities; by 2050, 65–66% of the global population is expected to be urban inhabitants. If the present trends of urban population continue, it will perpetrate irreversible damage on the environment (Eisenstein 2001). This massive urbanization, industrial revolution and the declining of environment in present days triggered and amplified awareness in the mind of public to restore and reconsider assets of nature in urban contexts. This led to innovative designing idea of landscapes under the broader brand of 'green urbanism' with single agenda to increase the quality of life in the towns and city while protecting surrounding environment.

This chapter endeavours to comprehend the best practices pertaining to eco-friendly sustainable landscapes. In the timeline to environmentally sustainable movement, '*enviroscaping*' emerged as a novel approach and indeed as a discipline that aims to further the interaction between the humanbeings and the environment. This gains importance from the perspective of minimising the environmental impact due to human action. The granular understanding existing between the natural landscapes and humans would aid the engineers and architects to foresee the consequence of their action. Incentives and support from the national governments along with scientific understanding about sustainable architecture can aid in sustainable environment friendly development and promotion of sustainable infrastructure. This chapter dwells extensively on eco-friendly practices including green walls, green roofs, green homes, and xeriscaping.

1.2 Designing the Environment Friendly Landscape

The design of landscapes vary according to location, subject to landscape usage, availability of resources and most importantly the consumers/customers desire. The decision on any of landscapes design depends on the desire of end users. Designers should meet the consumer needs by incorporating eco-friendly sustainable principles to build aesthetically pleasing and sound landscape.

Enviroscaping revolves on broader ideas of dynamic, holistic, intuitive and responsive approach (Fig. 1.1) to safeguard ecological landscape. Enviroscaping is dynamic approach, because it is a product of natural and cultural process with landscape as a mosaic of ecosystems and designs and management as ongoing processes. It is also holistic approach because it instantaneously reflects past and present as well as indigenous and provincial landscape processes and patterns. It serves as an intuitive approach by integrating emotions and imagination of creative art and nature of a user. Lastly, it is a very responsive approach, because it realizes the constraints and opportunities that exist in natural and cultural environment.

Enviroscaping attempts to conserve and safeguard clean water and air along with landfill spaces, the basic challenge of enviroscaping is to blend the aesthetical and environment friendly principles together to craft a landscape with environmentally comprehensive and looks aesthetically beautiful. An environment friendly landscape can prosper with minimal inputs of water, fertilizers, pesticides and labour. Designing environment friendly landscape is a thoughtful balance between usage of available resources and results achieved. We have to factor in environmental considerations in designing landscape, to create a pleasurable landscape with sound solutions to environmental problems.

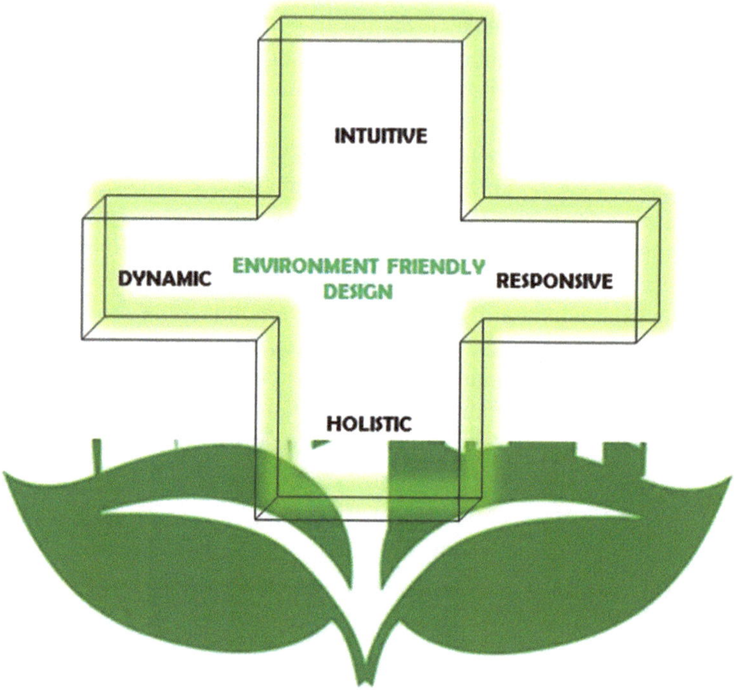

Fig. 1.1 Approaches for enviroscaping

Enviroscaping is to encourage the development of landscapes that do not pollute or waste natural resources. Building or developing environmental friendly landscape begins with a good well thought design based on rigorous assessments of location or site. In the process of designing/redesigning new or old landscapes, our primary concern is to achieve low maintenance landscapes because this saves money, time and conserves energy and water. Low maintenance landscape design is possible by selecting suitable plants and placing them on the right position on the site.

1.3 Principles of Enviroscaping

The built landscapes adopting sustainable practices can augment the health of the ecosystem, as the built landscapes are the part of the larger ecosystem. As regards the sustainable practices, mulching, composting, and water conservation play an important role in improving soil fertility, soil biodiversity and landscape management. Further, care must be taken to avoid invasive species and include plants that will benefit the ecosystem. The principles also include energy conservation and biological pest management.

1.4 Goals and Proposed Landscape Policies

Sustainable urban landscapes with ample of environmental benefits are possible by clearly identifying the demands for resources and energy for a long-term. The principal goal of enviroscaping is to provide space for the survival of all species within biosphere while not diminishing future options. Planning and management guidelines of environment friendly landscapes can increase the efficiency of our landscape practices and optimise environmental benefits. In essence, the planning goal is to conduct more of a cost-benefit analysis of urban landscapes with greater understanding of resources and environmental impacts, thus promoting a consistent and integrated framework of participation.

Natural environments can serve as models for sustainable systems. They are highly complex and interconnected, and demonstrate principles of energy flow and resource cycling. Planning for sustainable landscapes should begin with a program of policies and guidelines that can fit these principles of natural environments into the structure and functioning urban landscapes.

Efforts to achieve higher levels of environmental friendly landscapes require a range of policies, standards, and guidelines. Following are a number of key landscape goals proposed, requiring various levels of legislation to design, develop and to manage environmental friendly sustainable landscapes.

- Landscapes that meets 'CRR' idea of conserve, recycle, and reuse of available resources to achieve ideal levels of sustainability.
- Landscapes with increasing levels of preservation and efficiency in utilising energy.
- Landscapes based on principles of higher conservation and optimal use of water resources.
- Landscapes that produce optimum/higher levels of biomass, thus providing higher release of oxygen and storage of carbon.
- Landscapes built on the considerations of plants requirements such as soil, climate, water and maintenance needs.
- Landscapes with optimal levels of microclimate benefits to cut urban heat build-up effect to reduce energy usage for cooling and heating of built structures.
- Landscapes that integrate organic soil management practices and can accommodate composted landscape trimmings.
- Landscapes that devoid of toxic herbicides, pesticides and inorganic fertilizers that dilutes the principle of enviroscaping.
- Landscapes that support flora and fauna or any type of wildlife to build biodiversity.
- Events and programs to create awareness to educate public on the benefits from sustainable landscapes.

1.5 Energy-Efficient Landscapes

As we are in twenty-first century of the *Anno Domini* era, the continued heavy consumption of fossil fuels for energy looms as a major problem. Dwindling petroleum supplies and the effects of climate change can have a devastating impact on our environment and economy. A few degrees of warming may alter ocean currents, global air circulation patterns, rainfall regimes, and agricultural capabilities enough to cause mass starvation in some regions. I hope that we can avoid such worst-case scenarios, and in working toward that goal to achieve energy efficient landscapes are one means of conserving energy and creating a more sustainable society.

The demand for electricity for cooling in megacities increases about 2% for every 1 degree F rise in air temperature. Around 3–8% of the electricity used for cooling built structures in urban setup to compensate urban heat island effect (Akbari et al. 1992). The warmer temperatures in and around megacities compared to rural areas has other stressful effects like upsurges in carbon dioxide emission, increased water demand among urban inhabitants, unhealthy levels of ozone, dust pollution and depleted levels of oxygen and finally cause human discomfort. These grave problems of urban warming will deteriorate the standard of living in cities. The accelerating world trend towards rural-urban migration, especially in tropical regions, hastens the need for energy-efficient sustainable landscapes. Landscaping decisions can affect energy use through the decisions on choice of plants, plant forms, and their placement in the environment. Landscaping to an extent influences energy consumption. Albeit, estimation of energy savings from appropriate landscaping technology is difficult, a research conducted in California infers that landscaping with trees yields a marked reduction in energy use (McPherson and Simpson 2001).

Energy-efficient landscaping which aims to modify the microclimate of a home or, an urban region is construed as a component of *environmental landscaping* or *enviroscaping*. Energy savings in landscape architecture includes measures like landscaping around homes, commercial structures, office buildings, road dividers, parking lots, etc. Quite a number of landscaping techniques are available that can influence microclimates resulting in marked amount of energy savings. For instance, deciduous trees planted in the residential milieu help to shade buildings during the summer months and provide access to light and solar heat during the winter months. Planting of deciduous trees greatly influence the ambient temperature experienced inside a building. Hardscaping refers to the rocks, walls, and fences that can integrate with the plants, which can be use in landscaping for energy savings. In addition, diverse heights of shrubs, trees, and hardscape help to control wind speeds. Such windbreaks are important during winter months to reduce the effect of wind chill temperature. Plants that require water can be planted in a swale or rain garden. On the other hand, heat tolerant plants can be planted on a south facing slope, which will help in reducing water use.

An essential aspect of landscaping is to choose right plant for the right place to achieve energy conservation. As regards the plant choices, care is required to choose plants that require only few external inputs. Native and established plant flora are suitable for normal environments. During landscaping for extreme urban environments and harsh environments like parking lots and strips, road dividers, traffic circles, chosen plants need to be adapted to harsh environments like compacted soils, pollute environment, arid environment, high temperature, saline soil or water, etc.

1.5.1 Designing Energy-Efficiency Landscapes

Landscape with efficient energy optimisation and conservation requires a well-thought site-specific approach prior to landscape designing process. By upholding good landscape principles, plant components can be employed not only to achieve aesthetics but also for functional needs such as to save energy and water.

1.5.1.1 Sun Orientation

Landscape designers and gardening architects lack awareness on the movement and orientation of sun during the day, which is critical to consider during landscape design decisions. The sun orientation is the primary factor that influences energy consumption in four cardinal directions.

- Eastern direction: The eastern side of house/any-built structure that receive direct morning sun later in the mid noon the built structures will be shaded by the shadow of the building.
- Western direction: The western side of the house/any-built structure is shaded in the morning hours but fully exposed to the hot sun in the afternoon.
- Southern direction: The southern side of the house/any-built structure receives the most sunlight throughout the day but never as intense as the eastern or western direction.
- Northern direction: The northern facing side of the house/any-built structure is usually in shade.

1.5.1.2 Trees as Tools of Energy Conservation

Trees are employed as primary component in designing energy efficient landscapes. Trees can have a well-developed canopy in large dimension to shade house/any-built structures of landscape in helping to reduce cooling costs and increasing comfort of inhabitants. During selection of any form of plant components like trees, shrubs, and vines consideration should be given to determine the hardiness of plant species, soil and climatic requirements in line with location specific attributes of landscape. Plants species were chosen for their appearance as well as for their functional abilities to control climatic factors.

Fig. 1.2 A tree's shade potential depends upon its canopy shape and the density of the shadow it creates

Plants species harmonize with or enhance the built architecture of landscape. A wide range of plant species in the group of shrubs, trees and vines are available with rage of appropriate size, shapes and densities to achieve shading (Fig. 1.2).

Based on shading application, tree species are divided: deciduous (trees that lose their leaves during winter), evergreen, and broadleaf evergreens. Deciduous tree species help to block heat by sun in the summer days but let much of it in during the winter season, evergreen tree species, shrubs, and vines too provide year-round shade in hot climates.

While considering the plant species knowledge on leaf out period of species is essential because within deciduous group of trees and shrubs, few species leaf out early in the spring season and other much later in the season. The shadow pattern and coverage area of trees varies from species to species; some species cover larger area and some does lesser area due to large or smaller size of leaf/branches or due to slow/fast growing nature of species. In general, fast- growing tree species are not long lasting the landscape designer or horticulturist might consider an idea of interplanting the fast-growing tree species with slow-growing tree species. As a normal rule, for a tree species to become functional in few years to provide shade in any landscape, trees with height of 12–18 ft. tall are suitable for planting. The absorption and reflection of sun's rays depends on the tree canopy, leaf colour, texture, shape and size of tree leaves; tree species with a light or smooth leaf surface reflect more sunrays than those with dark and coarse surfaces. This attributes of trees decides the intensity of cooling. While selection and planting of shade tree species, the movement shade in the chosen location of landscape needs verification. Deciduous trees fits best in south and east sides of built structures. During winter months tree drops its leaves, sunlight can reach the built structures to build-up heat in built structures. Evergreen trees fits best on the north and western sides for best protection from the strong summer sun while sunset and also protection from strong cold winds during winter. Note: trees without leaves can able to block as much as 60% of the sun. Evergreen shrubs and small tree species can also serves as potential windbreak if we plant them under dense planting.

1.5.1.3 Adaptability and Hardiness

During selection of plant species, adaptability and hardiness should be major considerations. Species of hardy nature is of greater choice to survive under the extreme weather conditions. The climatic requirement of plants varies and different location/direction of landscape have varied climate conditions for example in northern side of the built structures, choose a shade-tolerant plant species and for northern location with extreme winter hardiness tree as to be evergreen and for southern and western sides plants need to be drought resistant.

1.5.1.4 Root Structure

Roots of plant species is the system to supply the required nutrients and water and provide anchoring to plants to stand straight in all weather conditions. While selection and planting of any shade tree species near to built structures, it is important to think of type of root systems that tree species own, different tree species have different root system and different root growth habits that is influenced by type and depth of soil. Few tree species have surface lateral root systems, while others have deeper lateral root systems, some have taproots, and some have a combination of taproots and lateral root systems. Selection should also on the location and depth of soil to prevent the foundation damage of built infrastructure.

1.5.1.5 Branch Structure

Trees species with well-structured will have branches spread and growing in all directions. Selection of tree species with an inherently strong branch attachment are suitable to plant near the house/buildings to avoid snap-off in heavy rains and windy situation.

1.5.1.6 Arrangement

Any identified trees species for energy conservation landscapes should be part of a much larger landscape. Arrangement in ecological landscapes is not just a matter of placing a single tree species in optimal solar directions. It is about creating an ecological landscape that structures and protects a wide range of plants species and fauna to make the whole landscape composition more beautiful and functional. Trees or plant species planted in groves are of perfect way to offer solutions that give you opportunities to use varying plant species to ensure biodiversity for ensuring wildlife in the ecological space.

1.5.1.7 Wind Control

During strong winter with powerful wind, plant species can help us protect the landscape. Instead of blocking strong winter winds, we can control them by diffusing with the help of plants species. Winds can be diverted, intercepted or channeled through the help of plant species. Combining evergreen and deciduous plant will help to diffuse the winter winds. The summer winds can be channeled by selectively combining shrubs and trees. Directing the wind around and over the plant species adjacent to any structures of landscape will boost natural ventilation.

Characteristics of an Effective Windbreak

- Sixty percent of foliage density on the windward side of landscape.
- Staggered 2–3 rows of evergreen tree species or deciduous tree species with 5–6 rows are ideal.
- The ideal length of a windbreak in the range of 10–12 times the mature width of standing tree species.
- Different heights of plant species within the chosen windbreak.

1.6 Water Smart Landscapes

Water is the most precious resource by nature; without water, no life survives on earth. Average person water use 100 gallons/day, that's sum up to 320 gallons/day by the average family. In the summer months, water usage will exceed for all the household purposes especially in hot and dry climates.

In dry climates, a household's outdoor water use is higher and about 50% is utilised for landscaping purposes to irrigate the plants in landscape. Ongoing population growth and recurring drought cycles have increased our concern for conserving water in urban landscapes. Questions about the availability of water and the costs associated with its use are becoming more important in the design and management decisions. It must be noted that the beautiful gardens and lawns are not because of extensive watering, fertilization, and pesticide application. Here is how best we can design, create and apply water-smart landscaping principles to create beautiful landscapes that save water resources.

1.6.1 Principles for Designing Water Smart Landscapes

The foundation for designing and managing water smart landscapes can be discovered by developing sensitivity to the regional environment and through application of specific design principles. This approach initially encourages to study the large-scale climate, topography, and soils in combination with native flora and fauna.

Such information provides context for understanding the issue of water conservation and many other aspects of ornamental landscapes, and brings attention to several key concepts of ecology. The ecological approach allows us to see the landscapes as functioning system of many parts well suited to its particular region and individual site. The regional climate, topography and soil factors provide the basic framework for landscape plants to adapt and flourish green. Patterns of rainfall, temperature extremities and edaphological properties are the prime factors of consideration. This macro scale views helps to evaluate the adaptability of certain plant species to arid or drought stress. The seasonal nature of moisture and temperature in a region indicates the nature cycles and rhythms that affect the plants. The availability of soil moisture, duration of long-day and short-day light and warmth bears directly on plant growth, dormancy and phonological habits. The design and management of a water smart landscape need to reflect these rhythms in order to achieve environment friendly landscapes by making appropriate decisions.

1.6.1.1 Selection of Native and Water Wise Plants

Plants evolve and adapt in groups or associations. Instead of treating plants as individual garden elements in a landscape, a sense of overall association is needed that enables plants to interact and complement each other while adapting to moisture, sun, soils, space and other considerations. Native and exotic plant species should not be introduced without some idea of their origin and natural habitat. Do they come from tropical, sub-tropical, Mediterranean, arid, or temperate climates? Do they grow in coastal, foothill or valley habitats? This background knowledge on plants enables to bring them together with greater levels of compatibility and suitability in the landscapes.

Preserving any existing plant species saves lot of resources because established plants species usually demands less nutrients, water and low-maintenance. After establishment, native plant species demands very little to no additional water beyond normal rainfall of the region. In addition, plants are adapted to local soils and climatic conditions and do not require any addition of nutrients and possess more resistant to insect-pests and pathogens. Needs more attention while selecting non-native species as some of them may become invasive and they can be water guzzler and may choke out native plant species of ecosystem.

There are many natural patterns and rhythms among plants. Physical characteristics include the size, shape, foliage colour, texture and fragrance; cultural adaptations include preferences and tolerances of sun, shade, moisture, and soil fertility. A successful landscape will develop plant associations that combine many layers or patterns of physical and cultural characteristics. A natural landscape demonstrates a strategy of species diversity that leads to flexibility and adaptability to many environmental conditions and changes. These observations provide ideas for use in urban landscapes.

1.6.1.2 Assemblage of Plant Species According to Water Requirement

Grouping plants species with similar water requirement into precise "hydrozones" will help us to save water. Turfgrasses or lawns and shrub planted areas needs separation due to varied water requirement. This approach will aid to cater to the water demands of the plants and thereby significantly conserving water. Hydrozones can be separated into different zones like hydrozone 0 (no irrigation), hydrozone 1 (irrigate monthly), hydrozone 2 (irrigate twice per month), hydrozone 3 (irrigate weekly) and hydrozone 4 (irrigate twice per week). Knowing the water requirements of the plants will help in efficiently using water by catering to the timely needs of the plants. The watering zone designs will depend on the amount of water available to use for the landscape to achieve aesthetic and environmental value.

1.6.1.3 Healthy Soils

Healthy soils will help to build strong water-smart landscape by helping plants to build strong and healthy root systems; landscape with healthy soil with good water holding capacity and rich with organic matters can effectively recycle nutrients, minimize runoff of nutrients and retain water. Healthy soils led to healthy and efficient plant root system that minimises the resources demanding in landscapes.

1.6.1.4 Reduce Turf Areas

Healthy green looking turfgrass is an aesthetic asset that act as backdrop and pulls the landscape design together to achieve harmony within the landscape. Turfgrasses demands more and considered as anti-component to have in designing water-smart landscaping but rather than designing the non-turf landscape its wise to use turf to significantly reduce the amount of irrigation water usage in landscape by choosing less water demanding turfgrasses and placing them in right position to improve the aesthetics of landscape. Less demand or water-conserving warm season turfgrass species, such as centipede, zoysia or bermudagrass can be chosen. Among the three, Bermudagrass cultivars are best at conserving water and they are drought-resistant.

1.6.1.5 Water Wisely

Right amount of watering at right time is an important part of conserving and using water in any landscape. Without a doubt, one of the easiest ways to reduce the amount of water use in the garden is by ensuring no wastage. It is common to see lawns or beds being watered with sprinklers that throw water beyond the intended area and onto nearby paths. Adjusting the pressure, or designing the watering system properly in the first place prevents this wasteful use of water.

1.6.1.6 Use Mulch

Mulching is most beneficial in water-smart landscapes to conserve water or moisture various mulches are available ranging from organic to inorganic and even plastic mulches. In environmentally friendly landscapes, organic mulches are of greater choice such as hardwood or bark chips, shredded bark, and compost leaves and turfgrass clippings are useful in retaining soil moisture by reducing soil evaporation. Other benefits of mulch are preventing water runoff, reduce weed growth, decrease soil temperature, and prevent soil erosion in the landscape.

The best water wise garden can be designed with all the above six principles in order to motivate people to help save water. The art and technique of water conserve landscaping that brings all the water conservation techniques together in terms of 'Xeriscaping' concept. The idea is to exploit natural rainfall with very little or no additional watering. Xeriscaping is an environmentally friendly landscaping and gardeners can reap the benefits beautiful landscape by employing low maintenance landscape practices.

1.7 Xeriscaping, the Art of Conserving Water

Xeriscaping emerged as new theory of landscaping under dry areas emerged during 1980s. The word xeriscaping originating from Greek "xeros" refers to "dry landscaping" (Iannotti 2017). The main drive of xeriscaping is to produce a good-looking landscape with vegetation that uses very little or no water (Beaulieu 2017c). The focus of xeriscaping is to reduce water usage in gardening to reduce the pressure on the water supply.

Fundamentally, xeriscaping is with seven key principles; planning and design, improvement of soil, plant selection, minimizing turf, efficient irrigation, use of mulch, and maintenance. Xeriscaping encourages reduction of turfgrass areas to adopt low maintenance ground covers as an alternative. Towards this end, plants with similar needs are grouped together. Appropriate planning and selection of plant species are important to lessen the maintenance and the amount of water usage (Friedman 2012). With regard to soil health, suitable nutrients must be provided in the soil through sustainable practices like mulching to maximise the plant growth. All these techniques helps to create beautiful garden while conserving water.

The best type of plant species to use while designing xeriscape garden are plants species with drought resistant or tolerant. Plants that originates in dry or desert habitat are hardy; require low care, and drought resistant. Few best examples that can grow in dry and sunny environments are Bird of Paradise, Bougainvillea, Fairy Duster, Lantana, Oleander, Pampas Grass, Purple Sage and Yellow Bells.

1.7.1 Planning and Design

Appropriate planning is a key step to design an operative xeriscape garden with right plants species based on the availability and requirement of water to synchronize with natural adjoining landscape. Environmental conditions like wind direction, location/site influence greatly the planning of xeriscape sites. Proper land grading is required so that the water instead of moving as runoff may soak the soil and help the plants to absorb water. Land grading includes slope management, which will aid water conservation. At times, obstructions and utilities lying underground may be challenging during land grading (Landscape America 2018). Naturally, raised beds dry out much quickly as compared to standard flat beds and in many cases; the standard flat beds are not encouraged. Drought tolerant plants species on the side of the dominant winds can shelter the less tolerant plants species. Plants species of similar water requirements can be in beds together to allow for optimum growth and development with limited water usage. (http://gardenline.usask.ca/yards/xeri1.html).

1.7.2 Soil Improvement

The growth and development of plant is influenced by physical, chemical and biological characteristics of soil. Soil samples can be analysed for its chemical composition, which will help in gauging the nutrient status of the soil. Soil management aids in growth of wide variety of vegetation. As regards the composition, sand, silt and clay content define the textural class. In case of sandy soils, the water drains quickly due to interconnected macropores. Sandy soils can be easily worked but needs a lot of vegetative cover to minimise the erosion losses. Due to its low moisture holding capacity and nutrient status, plants grown in sandy soils require regular irrigation and fertilization. Whereas, the clayey soils has high water holding capacity. Soil management with organic manures like compost, manure greatly improves the soil quality (http://gardenline.usask.ca/yards/xeri1.html).

1.7.3 Minimizing Turf

Lawn maintenance is a difficult proposition. Lawns require repeated watering, fertilizer application, and gasoline to keep them healthy. Therefore, in case of xeriscaping, the focus is on reducing the amount of lawn turf. This is achieved through either minimizing the actual grass cover or using alternative lawn coverings. Irrespective of the methods adopted for turf management, lawn grass need to be chosen based on the soil, and climatic condition. Interestingly, clover and mosses had emerged as a popular alternative to the lawn grass as clover and mosses require less maintenance and nutrients (Cacciatore et al. 2010; Beaulieu 2017b, 2018).

1.7.4 Appropriate Plant Selection and Efficient Irrigation Methods

Selection of plants for xeriscaping is important, as care must be taken to choose plants that are native and use less water their growth (Landscape Design Advisor 2018). Water conservation is paramount in xeriscaping. Irrigation demand varies with the plants. Plants that demand less water are preferred over other more water demanding plants (Landscape Design Advisor 2018). Efficient irrigation methods and practices are to be adopted for efficient landscaping. Efficient irrigation methods like drip irrigation system which provides irrigation water at regular interval, in small quantities cater to the irrigation demand of the plants most efficiently (Iannotti 2017). Through adoption of such methods, wastage of irrigation water and surface runoff are drastically reduced. Drip irrigation system operates at a little pressure to progressively supply right amount of water to the plant's root system using a network of laterals or small plastic tubes (Shock 2013).

1.7.5 Use of Mulch for Soil and Moisture Conservation

Mulches include leaves, debris, shredded bark, and compost. Mulches aid in reducing and moderating soil temperature, reduce weed growth and soil erosion. On decomposition, mulches add nutrients to soil and aid in improving physical, chemical and biological properties of soil. During the summer months, mulch moderates the soil temperature and protects the root system from soil temperature. Whereas, during winter months, mulch ensures a consistent ground temperature (Beaulieu 2017a). With respect to maintenance, xeriscaped gardens require trimming or pruning, weeding, and natural insect-pest management (Iannotti 2017).

1.7.6 Advantages of Xeriscape

Xeriscape gardens provides healthy environment and are economical as the cost involved in maintenance is less. As regards water conservation, xeriscape consumes less water. These gardens use less of fertilizers and pesticides and thereby the ground water contamination is averted. Urban xeriscaping has immense potential to reduce urban temperature, outdoor thermal discomfort and influence the urban water use. The detrimental effects of urban heat Island can be mitigated to some extent through sustainable approaches like growing low water intensive xerophytic plants within residential yards (Chow and Brazel 2012).

1.8 Constructed Wetlands

"Constructed wetland is an artificial wetland designed to exploit and elevate the natural biological processes that occur in natural wetlands to treat wastewater" (Rousseau et al. 2008). The idea of constructed wetlands was developed in Germany in late 1960s and refined in the 1970s (Vymazal 2005). It became popular because of lower capital costs. There are three main types constructed wetlands namely surface flow (SF) wetlands (free water surface wetlands), subsurface flow (SSF) wetlands, and vertical flow (VF) wetlands (Ghermandi et al. 2007).

Wastewater treatment using constructed wetlands prevent the discharge of harmful pollutants into the natural environment. Treated water derived from the constructed wetlands can be used for irrigation (Ghermandi et al. 2008; Rousseau et al. 2008). Constructed wetlands are piece of land covered with water or saturated with moisture permanently or on season basis. Wetland waters may be fresh, salty, or brackish but they serves the important function of environment by purifying water systems by removing the excess nutrients and other water pollutants in rain and storm water. Wetlands also helps in conserving biological biodiversity by supporting many species of amphibian, reptile, waterbird, fish, insects and other kind of animals. Wetlands can be home for various types of hydrophytic vegetation and selection of these aquatic plants needs a careful scrutiny of benefits that each plant can offer to its local ecosystem. In general, every hydrophytes with vigorous growth are able to uptake nutrients and absorb pollutants that can help us to improve the quality of water bodies. The most commonly used plant species for planting in constructed wetlands incudes Cattails (*Typha sp.)* and common reeds (*Phragmites australis*). In addition to their capacity to improve water quality, these two species also add the aesthetic value to wetlands and offer comfortable habitat for insects and small bird species. Maintenance of constructed wetlands is required in order to keep the wetland system functional and beautiful, regular trimming of wetlands vegetation, removal of dead and invasive species are essential. Some of examples of plants that are not recommended to use in constructed wetlands includes hydrilla (*Hydrilla verticillata*), and the water hyacinth (*Eichhornia crassipes*) due to invasiveness of these plant species that grow vigorously and multiply fast in wetlands rich with excess nutrients and their invasiveness is contrary to the principles of enviroscaping.

The advantages of adopting constructed wetlands include cost effectiveness, energy efficient, recreational facility, and reduction of pollution load (Cardoch et al. 2000). Though constructed wetlands for wastewater treatment have many advantages, the demerit with this technology is the land required for treatment. In many cases the constructed wetlands are preferred in the rural areas where the cost of the land may not be exorbitant (Cameron et al. 2003). Care must be taken to avoid overflow as that can lead to public health issues.

1.9 Water Management in Singapore: A Case Study

Singapore, a tropical island nation, receives abundant rainfall of about 2400 mm annually that unevenly distributed temporally. Two-thirds of the island in low-lying areas often suffers from widespread flooding (PUB 2014). The projected climate trends show increased uncertainty in frequency and intensity of heavy rainfall events, making the city more vulnerable to floods. An effective storm drainage system is therefore critical. In the past, conventional approaches were used to enlarge and line most natural waterways with concrete, including the Kallang River and Sungei Sembawang to reduce bank erosion. While concrete waterways are able to discharge higher flow, but they fail to build any habitat that support aquatic ecosystems for the uplift of environmental integrity. At the same time, Singapore experienced seasonal water shortages, which calls for a more integrated approach to inter-linked water issues. To address the flooding and other related water issues and improve the living environment of its people, Singapore envisioned becoming a 'City of Garden and Water' and developed a strategy of total water management. This strategy relies on precise control and management of the entire water cycle and different forms of water, to achieve the objectives of flood risk reduction and improve water environment and year-round availability.

These approaches involve precise assessment of water movement, from rainfall to water storage to runoff generation in order to determine the storage and infiltration capacity required to absorb the flood water from various sources. Such assessment and management of the water movement underpins the selection and implementation of a range of 'green' technologies and multifunctional infrastructures to reduce the flood hazards. Many of the green infrastructures such as constructed wetlands also serve to treat the storm water, store up water for dry seasons and improve the landscape.

The overall strategy and approach of Singapore water management are reflected in the very successful program called 'ABC' waters program that stands for Active, Beautiful, Clean waters program launched in 2006 by Singapore's Public Utility Board (PUB) as key program to implement the vision of 'City of Garden and Water'. This program is a deliberate effort to conserve and to improve the quality of water to harvest the full potential of the water management. ABC program integrated the reservoirs canals and drains all along the Singapore Island with the surrounding environment in an all-inclusive and eco-friendly manner (PUB 2014).

The major element of 2007 ABC Master Plan was to integrate Singapore city's rivers, canals, drains and reservoirs with the environment and to transform the city water bodies into lively, charming and hygienic flowing streams and lakes (PUB 2014).

The ABC Concept encapsulates Singapore's idea of harvesting the full potential of water resources to integrate them into the lifestyles and environment. The rationale is to use natural systems to retain and treat storm water on site before it flows into reservoirs and waterways, this will enhance Singapore biodiversity and the living environment in the country landscape (PUB 2014); ABC program also brings

Fig. 1.3 Singapore marina barrage reservoir

inter-agency collaboration. The ABC Waters design principles are also recognized and adopted by other sector agencies such as the Building & Construction Authority (BCA). For example, the BCA Green Mark Scheme is a benchmark scheme, which includes internationally recognized finest practices in building environment friendly design (PUB 2014). Two of the most important criteria in evaluating the building performance are water efficiency, and green features and innovations adopted such as green roof, balcony plantation, etc. The ABC Waters Program embraces a range of 'green' technologies for different functional levels of the water cycle. At the catchment management level, rooftop garden, sky/terrace/balcony garden, green planter box within open space and vertical greenery around the buildings are widely adopted in the city.

Innovatively designed water management infrastructure such as Marina Barrage reservoir (Fig. 1.3) is making considerable contributions to reduction in the water management of the city and other flood risks in the geographically lower areas of the city with other co-benefits. For example, the Barrage has also become a well-known landmark and lifestyle attraction of Singapore. Similarly, the floating wetlands are producing multiple benefits in flood risk management, water environment and shortage alleviation, and landscaping. The main green infrastructure at Marina Barrage is iconic green roof that built with environment friendly drainage cells and utilizes 100% recycled plastics this vast green roof acts as natural heat insulation for the built structure. Barrage as countries largest collection of solar panels that generates about 50% of the daytime electricity needed for indoor lighting and power supply at the barrage.

1.10 Green Roofs

Growing plants on rooftops is in practice for more than thousands of years. However, this idea took shape as green roofs, living roofs or eco-roofs only in last few decades. This urban greening is fairly a recent phenomenon advocated for the benefits these technologies provide for the building occupants, and ambient environment. Foliage rich green roofs symbolizes a range of benefits like energy reduction, providing clean air, ecological and aesthetic value, and water conservation.

The basic aim of the green roof is to have profuse foliage growth. Nevertheless, the generic green roof structure should not the damage the underlying structure. Generic green roof structure consists of different layers from bottom to top: roof decking (metal or concrete), insulation material, waterproof layer, drainage layer, plant growing medium, plant life (Banting et al. 2005) (Fig. 1.4). The waterproof membrane is essential to protect the underlying built structure and the drainage layer is constructed to drain the excess water. Soil or any lightweight substrate is used as a growing medium (Kosareo and Ries 2007). The substrate normally used in green roofs is rich with mineral nutrients and organic matter that will enable profuse plant growth and at the same time, it will exert more pressure on the built structure (Rogers 2013).

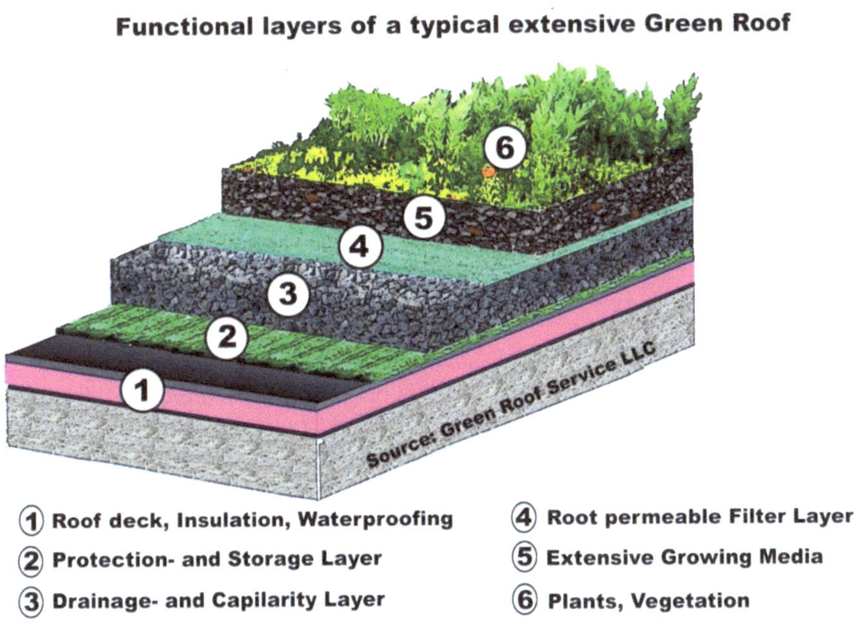

Fig. 1.4 Extensive green roof diagram. (Source: Green roof service LLC)

Table 1.1 Differences between extensive and intensive green roof

Details	Extensive green roof	Intensive green roof
Maintenance	Low	High
Irrigation	little or no irrigation	Often
Plant density	Low	High
Plant group	Grasses, Ground covers, Moss	Grasses, Perennials, Shrubs, Trees
System build up height	Up to 150 mm	Up to 1500 mm
System build up weight	50–150 kg per m^2	290–970 kg per m^2
Attractive	Less	Very attractive
Costs	Low	High

Green roof systems are broadly classified into extensive and intensive (Table 1.1) based on the desired function of the roof space.

1.10.1 Extensive Green Roofs

Extensive green roofs (Fig. 1.5) are not for public use and are designed and developed to achieve ecological and aesthetic benefits. The main attributes of them is involving low cost of establishment, lightweight (50–150 kg/m^2) and they are developed on the thin substrate medium of up to 100 mm. these type of green roofs have very minimal maintenance and inspection for the quality standards are only one to two times per year. Plants employed for extensive green roof are with low maintenance and most of them are self-generative.

1.10.2 Intensive Green Roofs

Intensive green roofs also called roof gardens (Fig. 1.6). The major attributes of intensive green roofs are:

- Greater load of about 150 kg/m^2 and more compared to extensive green roof.
- The substrates used as medium to grow plants will be more than 100 mm of thickness with a higher amount of mineral nutrient than extensive systems.
- Designed and developed to be accessible for people with well-incorporated areas of paving and seating and used as parks or building amenities.
- The can bear the added weight of people movement,
- Higher capital cost is involved in constructing and maintenance,
- Intensive planting strategies with low, medium and tall growing plants species were used and they have higher maintenance requirements.

Fig. 1.5 Extensive green roof on School of Art, Design & Media at NTU, Singapore

Fig. 1.6 Intensive green roof on parking structure, Singapore

- The plant selection for intensive green roofs ranges from ornamental lawn to shrubs and trees.
- Regular traditional garden maintenance such as watering, fertilizing, clipping of grass and weeding is mandatory.
- In addition to aesthetics and amenity spaces, these intensive green rooftops provide energy savings, aid in storm water management by absorbing huge bulks of storm water, and control air temperature.

1.10.3 Environmental Benefits of Green Roofs

Installing intensive or extensive green roofs is one option that can reduce the negative impact of urbanization and development while providing numerous environmental, economic, and social benefits such as

- Improve storm water management by reducing runoff by retaining 60–100% of the storm water and improving water quality,
- Green roofs will conserve energy by providing insulation to built infrastructure.
- Helps to mitigate the urban heat island in megacities,
- increase longevity of roofing membranes,
- helps to reduce noise and air pollution,
- Urban biodiversity improvisation by providing habitat for wildlife such as birds and butterflies,
- Opens up space for urban faming to grow fruits and vegetables,
- Offer a more visually pleasing and healthy environment to work and live for the urban dwellers,
- Furthermore the economic benefits includes; Construction and maintenance of green roofs deliver business opportunities for green industry members such as contractors of roofing or plant nurseries, landscape designers and contractors

1.10.3.1 Temperature Reduction

Typical hard and dark surfaced roofs pf buildings in the cities contribute to the urban heat island effect. The air temperature in the city landscape becomes 4–5 degrees warmer compared to rural areas termed as urban heat island effect. Higher urban temperatures will increase the level of pollution and increased use of cooling resources in the office spaces. Green roofs helps to mitigate the heat island effect by lowering temperature in city landscape by mixing of air vertically. Warm air above the hard roof surfaces rises due to lower density and it is substituted by denser cooler air from the green rooftop. Because of mixing of air, the overall temperature is reduced.

1.10.3.2 Improve Air Quality

Plant species under green roof will improve the quality of air directly by filtering airborne pollutants and releasing fresh oxygen by absorbing carbon dioxide and carbon monoxide by their leaves. They are also known to absorb gaseous pollutants through photosynthesis and sequester them in their leaves. Green roofs can also improve quality of air indirectly by reducing surface temperature. Green roofs can decrease such thermal air movements by reducing the energy available for heating. The increased air temperature above the roof surfaces also contributes to the chemical reaction responsible for low atmospheric ozone creation, which is the primary component of smog. Therefore, by monitoring the air temperature through the adoption of green roofs can avoid ozone formation thereby improving air quality.

1.10.3.3 Improve Rainwater Retention

Green roofs helps to retain storm water is very important in controlling runoff in megacities. Environmentally, this renders into benefits like reduced surface contaminants in the rainwater, limited soil erosion and improved wellbeing of urban biodiversity such as aquatic creatures, plants and animals. Studies in Berlin have shown that on the average size of green roofs is responsible for 75% of absorption of rainwater (Köhler 1989).

1.10.3.4 Habitat Preservation

Green roofs serves as an alternative habitat for urban flora and fauna. The presence of birds, butterflies and maintaining habitat for any type of wildlife may enrich the ecological quality and health of the environment and this will provide all kind of benefits to humans. Due to their height, green roofs may have less human intervention than gardens located at grade. This will provide a safer habitat for sensitive plants and certain animal species that is almost absent in the urban city. Studies in America show that butterflies will visit gardens as high as 20 stories. Bees and birds are able to inhabit beyond the 19th storey (Johnston and Newton 2004).

1.10.3.5 Lower Carbon Dioxide and Higher Oxygen

Green roofs greatly influence the air quality of the surrounding area. Quick growing green plant foliage has the potential to reduce the atmospheric concentration of carbon dioxide, nitrogen dioxide, ozone, and greatly reduce the smog, which is a product of primary and secondary pollutants.

1.10.3.6 Improve Water Quality

Green roofs are capable to filter out the heavy metals and mineral elements/nutrients present in rainwater (Johnston and Newton 2004). Green roofs in city landscape act as a natural filtration system in removal of any heavy metals and pollutants of rainwater before reaching water reservoirs.

1.10.3.7 Noise and Sound Insulation

The green roof systems consisting of soil/plant growth medium, plants and trapped air layers within the green roof structure can act as a sound insulation barrier. The plant growth medium has potential to block lower sound frequencies while the plants above the soil layer block higher sound frequencies. Sound insulation by green roofs is a function of the system and the substrate depth. Green roof with a 12 cm thickness can reduce sound by 40 dB.

1.11 Green Walls

Green walls are normally encouraged as beautifying feature in buildings. Green wall refers to the vegetated wall surfaces (Figs. 1.7 and 1.8). Green walls have huge potential as compared to green roofs. At the micro-level, green walls can be passive design solution with immense potential to improve the microclimate of building, which provide shade and provide evaporative cooling effect. The green walls are classified into green facade and living walls. Green facade include climbing plants or hanging plants that cover the vertical wall. Green facade are grouped into direct and indirect green facades. Under direct green facade, plants are attached directly to the wall of vertical surfaces. Whereas, in indirect green facades supporting structures were used for attaching vegetation.

Fig. 1.7 Interior green wall

Fig. 1.8 Exterior green wall

Living walls permit a rapid coverage of larger vertical surfaces with more of uniform growth (Manso and Castro-Gomes 2015). Green walls are becoming popular nowadays across the globe. They are quite widespread in European countries. World's largest living wall is found in II Fiordaliso Shopping Center in Milan, Italy, which is covered with more than 44,000 plants. Green walls are also gaining popularity in Singapore and Germany. As the green walls consist of a diverse group of plants, which attracts butterflies, pollinators, insects, etc., act as a nesting ground for a quite number of bird species adding value to urban biodiversity. In effect, green walls provide enormous environmental and social benefits.

1.12 Green Homes – An Environment-Friendly Home

Green homes evolved with the concept of bringing environmentally responsible process and resource-efficient strategies together. Every aspect of green home from interior to exterior design is as per the environment responsive touch. Green homes are constructed, operated and maintained in environment friendly manner. These homes are designed to exploit the available natural resources and to reduce carbon footprint. A green home comprises various passive design features like orientation of building for natural ventilation, natural insulation and solar shading in various seasons of the year, environmentally suitable use of building materials, fixtures and fittings. Green homes built in very cost-effective way by using raw materials that are good for the environment with distinct aesthetics that sets it apart from other conventional or non-environment friendly home designs. Various insights that architects, designers and horticulturists while designing green homes range from conditions of site to layout and orientation of home, from building materials to wall and floor materials, positioning of rooms and windows in to design, and use of solar panels to utilize the natural solar energy, an clean and renewable source of energy. Green homes and buildings at times are called as a sustainable building.

1.13 Conclusion

Enviroscaping is a shared sensible approach to design and maintain landscape, which conserves energy and water, recycles and reduce the inputs in environment friendly manner. Environment is our responsibility and we must do our part if we are to preserve natural beauty for generations to come. Enviroscaping is all about designing with nature and the environment in mind.

References

Akbari, H., Davis, S., Dorsano, S., & Huang, J (Eds.). (1992). *Cooling our communities. A guidebook on tree planting and light-colored surfacing* (Lawrence Berkeley Laboratory Report LBL-31587). Washington, DC: U.S. Environmental Protection Agency.

Banting, D., Doshi, H., & Li, J. et al. (2005). *Report on the environmental benefits and costs of green roof technology for the city of Toronto*. Ontario. https://commons.bcit.ca/greenroof/files/2012/01/banting_et_al.pdf

Beaulieu, D. (2017a). How to select a mulch for your yard. In *The Spruce*. https://www.thespruce.com/landscape-mulch-basics-2130792. Accessed 19 May 2018

Beaulieu, D. (2017b). Moss plants: Shade alternative to lawns. In *The Spruce*. https://www.thespruce.com/moss-plants-shade-alternative-to-lawns-2130879. Accessed 19 May 2018

Beaulieu, D. (2017c). Xeriscaping plants: A drought-resistant approach. In *The Spruce*. https://www.thespruce.com/xeriscaping-plants-a-drought-resistant-approach-2130777. Accessed 19 May 2018

Beaulieu, D. (2018). Top tips for low-maintenance landscaping. In *The Spruce*. https://www.thespruce.com/tips-for-low-maintenance-landscaping-2132466. Accessed 19 May 2018

Cacciatore, P., Hogan, J., Mattern, R., & Munoz, A. (2010). *Development of a sustainable landscape architecture best practices manual*. Degree of Bachelor of Science, Worcester Polytechnic Institute.

Cameron, K. C., Madramootoo, C., Crolla, A., & Kinsley, C. (2003). Pollutant removal from municipal sewage lagoon effluents with a free-surface wetland. *Water Research, 37*, 2803–2812.

Cardoch, L., Day, J., Rybczyk, J., & Kemp, G. (2000). An economic analysis of using wetlands for treatment of shrimp processing wastewater-a case study in Dulac, LA. *Ecological Economics, 33*, 93–101. https://doi.org/10.1016/s0921-8009(99)00130-5.

Chow, W., & Brazel, A. (2012). Assessing xeriscaping as a sustainable heat island mitigation approach for a desert city. *Building and Environment, 47*, 170–181. https://doi.org/10.1016/j.buildenv.2011.07.027.

Eisenstein, W. (2001). Ecological design, urban places, and the culture of sustainability. In *SPUR*. http://www.spur.org/publications/urbanist-article/2001-09-01/ecological-design-urban-places-and-culture-sustainability. Accessed 19 May 2018

Friedman, A. (2012). Edible landscaping and xeriscaping. In *Fundamentals of sustainable dwellings*. Washington, DC: Island Press/Center for Resource Economics.

Ghermandi, A., Bixio, D., & Thoeye, C. (2007). The role of free water surface constructed wetlands as polishing step in municipal wastewater reclamation and reuse. *Science of Total Environment, 380*, 247–258.
Ghermandi, A., van den Bergh, J., & Brander, L. et al. (2008) The economic value of wetland conservation and creation: A meta-analysis. *SSRN Electronic Journal*. doi: https://doi.org/10.2139/ssrn.1273002
http://gardenline.usask.ca/yards/xeri1.html
Iannotti, M. (2017). Xeriscape gardening – Planning for a water wise garden. In *The Spruce*. https://www.thespruce.com/xeriscape-gardening-1402341. Accessed 19 May 2018
Johnston, J., & Newton, J. (2004). Building green: A guide to using plants on roofs, walls and pavements. Published by Greater London Authority City Hall. ISBN 1 85261 637 7
Köhler, M. (1989). Ökologische Untersuchungen an extensiven Dachbegrünungen. *Verhandlungen der Gesellschaft für Ökologie (Essen 1988), 18*, 249–255
Kosareo, L., & Ries, R. (2007). Comparative environmental life cycle assessment of green roofs. *Building and Environment, 42*, 2606–2613. https://doi.org/10.1016/j.buildenv.2006.06.019.
Landscape America. (2018). Site analysis for developing a landscape plan. In Landscape-america.com. http://www.landscape-america.com/landscapes/design/site.html. Accessed 19 May 2018
Landscape Design Advisor. (2018). Natural outdoor living with xeriscape landscape design. In *Landscape design advisor*. http://www.landscape-design-advisor.com/ideas-tips/eco-friendly/water-wise. Accessed 19 May 2018
Manso, M., & Castro-Gomes, J. (2015). Green wall systems: A review of their characteristics. *Renewable and Sustainable Energy Reviews, 41*, 863–871. https://doi.org/10.1016/j.rser.2014.07.203.
McPherson, E. G., & Simpson, J. R. (2001). *Effects of California's urban forests on energy use and potential savings from large-scale tree planting*. USDA Forest Service, Pacific Southwest Research Station, Center for Urban Forest Research, Davis CA
PUB (Ed.). (2014). *Public utilities board: ABC waters design guidelines* (3rd ed.p. 112). Singapore: Public Utilities Board www.pub.gov.sg.
Rogers, J. (2013). Green, brown or grey: Green roofs as 'sustainable' infrastructure. *WIT Transactions on Ecology and the Environment, 173*, 323–333. https://doi.org/10.2495/sdp130271.
Rousseau, D. P. L., Lesage, E., Story, A., Vanrolleghem, P. A., & De Pauw, N. (2008). Constructed wetlands for water reclamation. *Desalination, 218*, 181–189.
Shock, C. (2013). *Drip irrigation: An introduction. Sustainable agriculture techniques* (EM 8782). Oregon State University Extension Service. https://www.cropinfo.net/pdf/extension/em8782-DripIrrigationIntroduction.pdf. Accessed 19 May 2018.
Vymazal, J. (2005). Horizontal sub-surface flow and hybrid constructed wetlands systems for wastewater treatment. *Ecological Engineering, 25*(1), 478–490.

Chapter 2
Climate Smart Agriculture Technologies for Environmental Management: The Intersection of Sustainability, Resilience, Wellbeing and Development

V. Venkatramanan and Shachi Shah

Abstract Agriculture sync with the existence of humanity. Converging demands from human population, food security, climate change mitigation and adaptation, agricultural resources, biofuel and oil prices, food prices, have engendered a new transformative, resilient and smart agricultural approach entitled "Climate Smart Agriculture". Rapid alteration by humans on agricultural landscape driven by aspirations to maximize production, productivity, and profit with scant regard to environmental concerns led to the degradation of agricultural lands, alteration in global carbon and nitrogen cycles, loss of soil fertility and biodiversity, pest and disease outbreaks. Under such circumstances, agriculture production system must be insured against impending danger of climate change; augmented with diversity of biological resources; enhanced with adaptive capacity and resilience; provided with site-specific sustainable management practices like integrated crop management, conservation agriculture, agriculture diversification and landscape management. Climate Smart Agriculture (CSA) is construed as a "comprehensive agricultural approach that aims at sustainable productivity enhancement, mitigation of and adaptation to climate change, and achieving global food security and other related sustainable development goals". CSA incorporates the virtues of "climate-smart food system", "climate-proof farms", and "climate-smart soils".

Keywords Climate smart agriculture · Food security · Resilience · Sustainable development goals · Environmental stewardship

V. Venkatramanan (✉) · S. Shah
School of Interdisciplinary and Transdisciplinary Studies, Indira Gandhi National Open University, New Delhi, India
e-mail: vvramanan@ignou.ac.in

S. Shah et al. (eds.), *Sustainable Green Technologies for Environmental Management*, https://doi.org/10.1007/978-981-13-2772-8_2

2.1 Introduction

Humans endeavour to transform the Earth's landscape fructified through agriculture. "Agriculture land expansion is construed as one of the important human modification to the global environment" (Matson et al. 1997). Nonetheless, land use changes associated with agriculture expansion was needed to sustain the growing population. Sooner the advent of new technology and innovations to use natural resources enabled humanity to dominate almost all the Earth's ecosystem (Vitousek et al. 1997). Agriculture is no exception. The green in "green revolution" brought into the agroecosystem a wide variety of chemical fertilizers, pesticides and other synthetic inputs. The alteration of global carbon and nitrogen cycle which is a splendid example of prowess of human action, resulted in increases of radiatively active gases like carbon dioxide, nitrous oxide, methane, etc. (Ciais et al. 2013). Agriculture growth driven by synthetic fertilizers led to major changes in nitrogen cycle. The nitrous oxide concentration drastically increased after 1950s corroborating the indiscriminate use of fertilizers. Further, agricultural activity particularly lowland paddy cultivation and rearing of ruminants are largely responsible for the increases in atmospheric methane concentration. In effect, agriculture too is responsible for releasing GHGs into the atmosphere. Agriculture, forestry and other land use (AFOLU) sector is responsible for GHG emissions of 10–12 gigatonne of CO_2-equivalent per year (Lipper et al. 2014; Smith et al. 2014).

Nonetheless, the ecosystem services rendered by agriculture are invaluable. A food system encompasses all the processes and infrastructure aiming to cater to food demands of the society. It includes food gathering/fishing; growing, harvesting and storing of food products; processing, packaging, transporting, marketing, and consumption of food; and disposing of food waste (Porter et al. 2014). Global food production has to be increased by 60% by 2050 to meet the needs of 10 billion population. Presently about two billion people suffer from micronutrient deficiencies (Wheeler and von Braun 2013). Achieving food security is an uphill task as it is very complex and changing climate mounts further pressure on all the dimensions of food security (Porter et al. 2014). Further, through the negative impacts on food production, food prices and accessibility, consumption and utilization, climate change has pervasive effect on global food security (Porter et al. 2014). Food production is found to be the most focused among all the dimensions of food security. Indeed, studies gear up to augment food production through innovative and sustainable green technologies and approaches (Venkatramanan 2017). Sustainable agricultural production is challenged by a complex web of concerns including climate change, resource constraints, biodiversity loss, food price fluctuations, structural diet shift (Venkatramanan 2017). Agriculture being a climate sensitive sector, is influenced greatly by climate change and variability, though the effects vary with regions, crops, degree of adaptiveness and resilience in agriculture production system. However, the developing countries and low latitude countries are more vulnerable to climate change due to poverty, income instability and lesser adaptive capacity (Lipper et al. 2014). This chapter synthesizes and endeavours to reflect on the broad contours of relationship and interaction between agriculture and climate change including the impacts and GHGs emissions from agriculture. This indeed provide a stable plinth and growing need for structural transformations, resilient growth pathways, environmental stewardship, stakeholder integration, and strategies to align agriculture growth with sustainable development goals. To maximise the synergy existing between the mitigation and adaptation to climate change and to achieve global food security in the times of climate change and other multiple stresses, FAO and World Bank put forward an approach called "Climate Smart Agriculture" to transform agriculture. The approach is gaining currency for its innate positive potential and goals to establish a smart agriculture system which is inclusive of "climate-smart food system" (Wheeler and von Braun 2013), "climate-proof farms" (Schiemeier 2015) and "climate-smart soils" (Paustian et al. 2016).

2.2 World Agriculture at Crossroads

Land-use changes by humans have modified the face of the Earth surface (Foley et al. 2005). The bottom line of land-use changes are exploiting the natural resources for meeting the human needs and such changes most often destabilize the environmental conditions which is indicated by changes in atmospheric composition, global carbon and nitrogen cycle, ecological crisis and catastrophes. Nevertheless, land-use practices are essential for the service of the humanity as they provide ecosystem services like food, feed, fodder, shelter, and freshwater. As regards the land-use change and food production, agriculture occupies almost 40% of the land surface. Incidentally, the increase in crop land area combined with technological development had enabled global agriculture to augment food grain production (Foley et al. 2005). Indeed, the modern agriculture is squarely based on "(i) the varietal improvements in targeted crops, achieved through either conventional plant breeding or through genetic engineering, and (ii) increased utilization of purchased inputs, such as inorganic fertilizers, agrochemical crop protection, and usually petroleum-derived energy to support large-scale production with extensive mechanization" (Uphoff 2015).

Sustainable development of agriculture system is challenged by environmental concerns like soil security, water security, climate change, food and nutritional security, energy security, agro-biodiversity loss, etc. These challenges have in common certain characteristics like global occurrence, complex and interconnected nature, and great difficulty to resolve (McBratney et al. 2014). Further, the pervasive influence of climate change on agriculture sector is being observed through reduction in food production, food insecurity, food crisis and food riots.

2.3 A Primer on Climate Change

Climate change is not uncommon in the earth's history. Availability of weather records is sine qua non for climate change analysis. Nevertheless, weather data is available only for few hundred years in developed countries but in less developed countries, the weather data exists for few decades only. However, the proxy indicators like ice cores, pollen analysis, tree rings, sea sediments have enabled the past climate reconstruction. Fifth assessment report of IPCC stated that "warming of the climate system is unequivocal, and since the 1950s, many of the observed changes are unprecedented over decades to millennia" (IPCC 2013). Climate change refers to the variation in either mean state of the climate or in its variables persisting for a long and an extended period (decades or longer). Climate change includes rising temperature, changes in precipitation amount and frequency, rise in sea level rise and more importantly increased occurrence of extreme weather events. While on the contrary, climate variability refers to sudden and discontinuous seasonal or monthly or periodic changes in climate or its components without showing any specific trend of temporal change (IPCC 2013).

Natural and anthropogenic processes alter the global energy budget and drives climate change. "The strength of drivers is quantified as Radiative Forcing (RF) in unit's watts per square metre (Wm^{-2}). Radiative forcing refers to the change in energy flux caused by a driver, and is calculated at the tropopause or at the top of the atmosphere" (IPCC 2013). While total anthropogenic RF for 2011 relative to 1750 is 2.29 Wm^{-2}. the RF from emissions of well-mixed greenhouse gases that includes CO_2, CH_4, N_2O, and halocarbons for 2011 relative to 1750 is 3.00 Wm^{-2}. The largest contribution to total radiative forcing is caused by the increase in the atmospheric concentration of CO_2 since 1750 (IPCC 2013).

Global surface temperature is the most common indicator of global climate. Global warming led to a decline in snow cover and sea ice, rise in sea level and cause many more visible impacts on the biosphere. Climate change and variability markedly influence all the components of biosphere. The globally averaged greenhouse gas (GHGs) concentrations (Fig. 2.1) are found to increase since the onset of industrialization. The increases in GHGs are responsible for increase in surface temperature and the Fig. 2.2 throws light on globally averaged combined land and ocean surface temperature anomaly. One of the visible indicator of warming climate is sea level rise as evidenced from Fig. 2.3. Fifth assessment report of IPCC stated that "globally averaged combined land and ocean surface temperature data as calculated by a linear trend, show a warming of 0.85 °C (Fig. 2.2), over the period 1880-2012" (IPCC 2014a). Between 1906 and 2005, global climate system observed an increase of 0.74 °C.

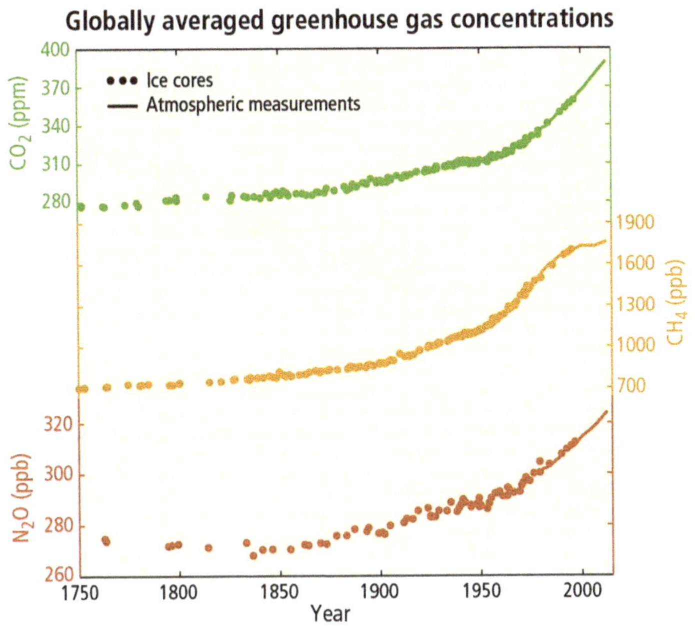

Fig. 2.1 Observed changes in atmospheric greenhouse gas concentrations. Atmospheric concentrations of carbon dioxide (CO_2, green), methane (CH_4, orange), and nitrous oxide (N_2O, red). Data from ice cores (symbols) and direct atmospheric measurements (lines) are overlaid. (With permission from *Climate Change 2014: Synthesis Report. Contribution of Working Groups I, II and III to the Fifth Assessment Report of the Intergovernmental Panel on Climate Change* [Core Writing Team, Pachauri, R.K. and Meyer, L. (eds.)]. IPCC, Geneva, Switzerland)

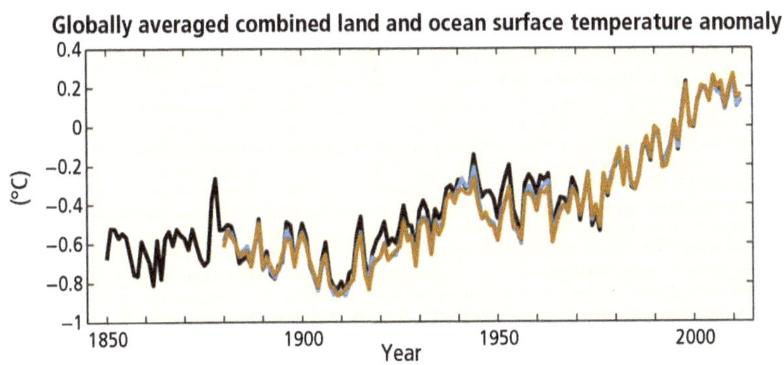

Fig. 2.2 Annually and globally averaged combined land and ocean surface temperature anomalies relative to the average over the period 1986 to 2005. Colours indicate different data sets. (With permission from *Climate Change 2014: Synthesis Report. Contribution of Working Groups I, II and III to the Fifth Assessment Report of the Intergovernmental Panel on Climate Change* [Core Writing Team, Pachauri, R.K. and Meyer, L. (eds.)]. IPCC, Geneva, Switzerland)

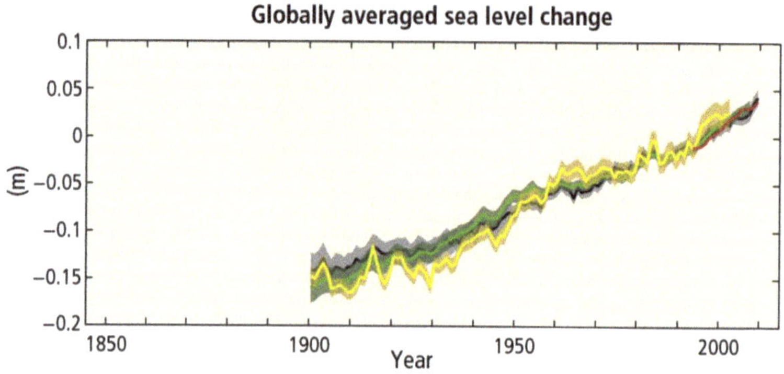

Fig. 2.3 Annually and globally averaged sea level change relative to the average over the period 1986 to 2005 in the longest-running dataset. Colours indicate different data sets. (With permission from *Climate Change 2014: Synthesis Report. Contribution of Working Groups I, II and III to the Fifth Assessment Report of the Intergovernmental Panel on Climate Change* [Core Writing Team, Pachauri, R.K. and Meyer, L. (eds.)]. IPCC, Geneva, Switzerland)

2.4 Interrelationship Between Climate Change and Agriculture

World agriculture including croplands and pastures occupies about 38% of Earth's terrestrial surface (about 5 billion hectares) and it is one of the largest terrestrial biomes on our planet (Foley et al. 2005; Ramankutty et al. 2008). Climate change and variability has potential to influence crop geography, crop production and productivity, and exacerbate the risks associated with crop farming activities (Scherr et al. 2012). Assessment reports of IPCC upheld the gravity of climate change impacts on agricultural production and productivity in several agricultural regions of the world, and firmly expressed the vulnerability of developing countries and island and low-lying countries to negative impacts of climate change (IPCC 2014a). Impacts from extreme weather events like droughts and floods, heat and cold waves, must be reckon with in the coming decades through devising appropriate climate resilient pathways. Granular understanding of the intersection between agriculture and climate change expose the net negative impacts of climate change on agricultural landscape, greenhouse gas emission from agriculture and uncovers the immense potential of agricultural sector to reduce the anthropogenic emissions.

2.4.1 Greenhouse Gas Emissions from Agriculture

Human actions continue to change the atmosphere composition and increase the concentration of radiatively active gases. Given the fact that the human population is increasing, the kindred issues with population growth would pump more

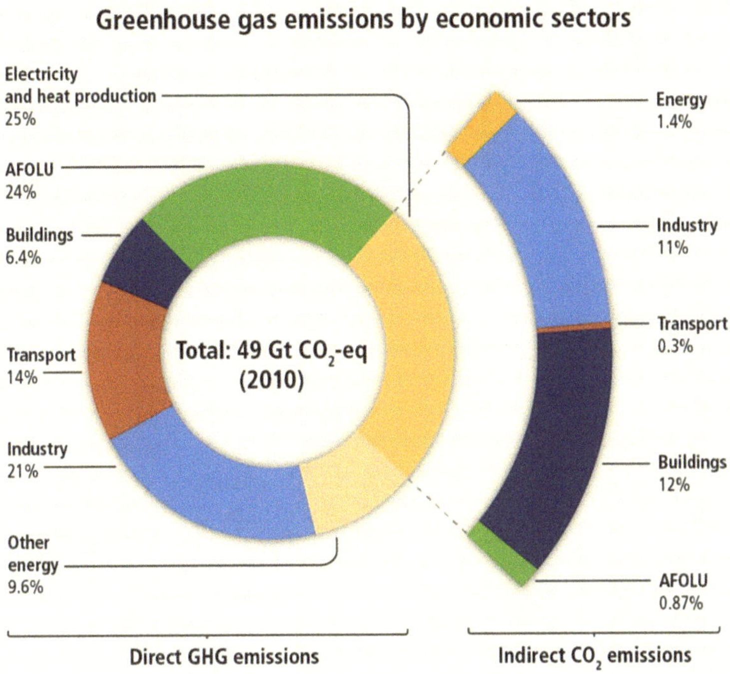

Fig. 2.4 Total anthropogenic greenhouse gas (GHG) emissions (gigatonne of CO_2equivalent per year, $GtCO_2$-eq/yr) from economic sectors in 2010. The circle shows the shares of direct GHG emissions (in % of total anthropogenic GHG emissions) from five economic sectors in 2010. The emission data on agriculture, forestry and other land use (AFOLU) includes land-based CO_2 emissions from forest fires, peat fires and peat decay that approximate to net CO_2 flux from the sub-sectors of forestry and other land use (FOLU). (With permission from *Climate Change 2014: Synthesis Report. Contribution of Working Groups I, II and III to the Fifth Assessment Report of the Intergovernmental Panel on Climate Change* [Core Writing Team, Pachauri, R.K. and Meyer, L. (eds.)]. IPCC, Geneva, Switzerland)

greenhouse gases into the atmosphere and consequently increasing the earth's surface temperature. Even if the GHG concentration is kept at the present day level, the Earth's surface temperature is bound to increase by 0.6 °C by the end of this century. The total GHG emissions due to human action (Fig. 2.4) is about 49 gigatonne of CO_2-equivalent per year ($GtCO_2$-eq/yr) (IPCC 2014a). Greenhouse gases emissions data are presented in terms of CO_2-equivalents based on 100-year Global Warming Potential (GWP100). Equivalent carbon dioxide emission is defined as "the amount of carbon dioxide emission that would cause the same integrated radiative forcing, over a given time horizon, as an emitted amount of a greenhouse gas or a mixture of greenhouse gases" (IPCC 2013).

The well-mixed greenhouse gases have very long lifetimes so that they are relatively homogeneously mixed in the troposphere layer of the atmosphere. For instance, while CH_4 has a lifetime of 10 years, N_2O has a lifetime of 100 years.

Importantly, the three GHGs CO_2, CH_4, and N_2O together amount to 80% of the total radiative forcing and also their concentration found to increase in the post-industrial era. Increase in the atmospheric concentration of these three gases are primarily due to the fossil fuel combustion, land use changes and agricultural activities. These three GHGs are important from the agriculture perspective as agricultural activities are responsible for the GHGs emissions (IPCC 2014a) (Fig. 2.5). IPCC (2013) reports that "CO_2 increased by 40% from 278 ppm about 1750 to 390.5 ppm in 2011. During the same time interval, CH_4 increased by 150% from 722 ppb to 1803 ppb, and N_2O by 20% from 271 ppb to 324.2 ppb in 2011" (Ciais et al. 2013). Half of the CO_2 emissions was sequestered in the ocean and vegetation biomass, and undisturbed soil ecosystem. As regards the increase in CH_4 concentration in the recent past, the causes were natural wetlands emissions (177–284 Tg (CH_4) yr^{-1}), agriculture and waste (187–224 Tg (CH_4) yr^{-1}), fossil fuel related emissions (85–105 Tg (CH_4) yr^{-1}), other natural emissions (61–200 Tg (CH_4) yr^{-1}), and biomass and biofuel burning (32–39 Tg (CH_4) yr^{-1}). "N_2O emissions are caused mainly by microbial mediated nitrification and de-nitrification reactions occurring in soils and in the ocean" (Ciais et al. 2013). While application of synthetic N fertilizer is responsible for the emission of 1.7–4.8 TgN N_2O yr^{-1}, fossil fuel use and industrial processes emit 0.2 to 1.8 TgN N_2O yr^{-1} and biomass and biofuel burning emit 0.2–1.0 TgN N_2O yr^{-1} (IPCC 2014a).

The agriculture, forestry and other land use (AFOLU) sector contributes about 10–12 gigatonne of CO_2-equivalent per year. The GHGs from agriculture as shown in the Figs. 2.5 and 2.6 are mainly due to land use and land use changes and forestry related activities, enteric fermentation in ruminants, biomass and biofuel burning, lowland paddy cultivation, and use of synthetic nitrogen fertilizers (Lipper et al. 2014; Smith et al. 2014).

On account of land use and land use changes like deforestation and degradation, CO_2 is emitted into the atmosphere and the atmospheric CO_2 is sequestered by land use activities such as afforestation, and reforestation. Global net CO_2 emissions due to land use change from 2000 to 2009 is estimated at 1.1 ± 0.8 Pg C yr^{-1} (Ciais et al. 2013). Increase in atmospheric methane is mainly due to anthropogenic emissions. Nevertheless, the sources of CH_4 can be either thermogenic or biogenic. The thermogenic sources include mainly the natural emissions from geological sources. On the other hand, the biogenic sources include natural and anthropogenic biogenic sources. Anthropogenic biogenic emissions of methane is important as it is increasing due to the human activities like low land paddy cultivation, rearing of ruminants, man-made lakes and waste management including the emissions from landfills. Ruminants like cattle, sheep, goats, etc. produce CH_4 due to food fermentation occurring in their anoxic rumen environment. Increase in atmospheric concentration of N_2O after 1950s is mainly due to agricultural intensification which involves extensive use of synthetic N fertilizers and manure application (Matson et al. 1997). "Human alterations of the nitrogen cycle have increased concentration of N_2O" (Vitousek et al. 1997). Soil microbial processes like nitrification and denitrification are squarely responsible for increased atmospheric N_2O concentration. N_2O emissions from soil processes may increase on account of growing food demand and

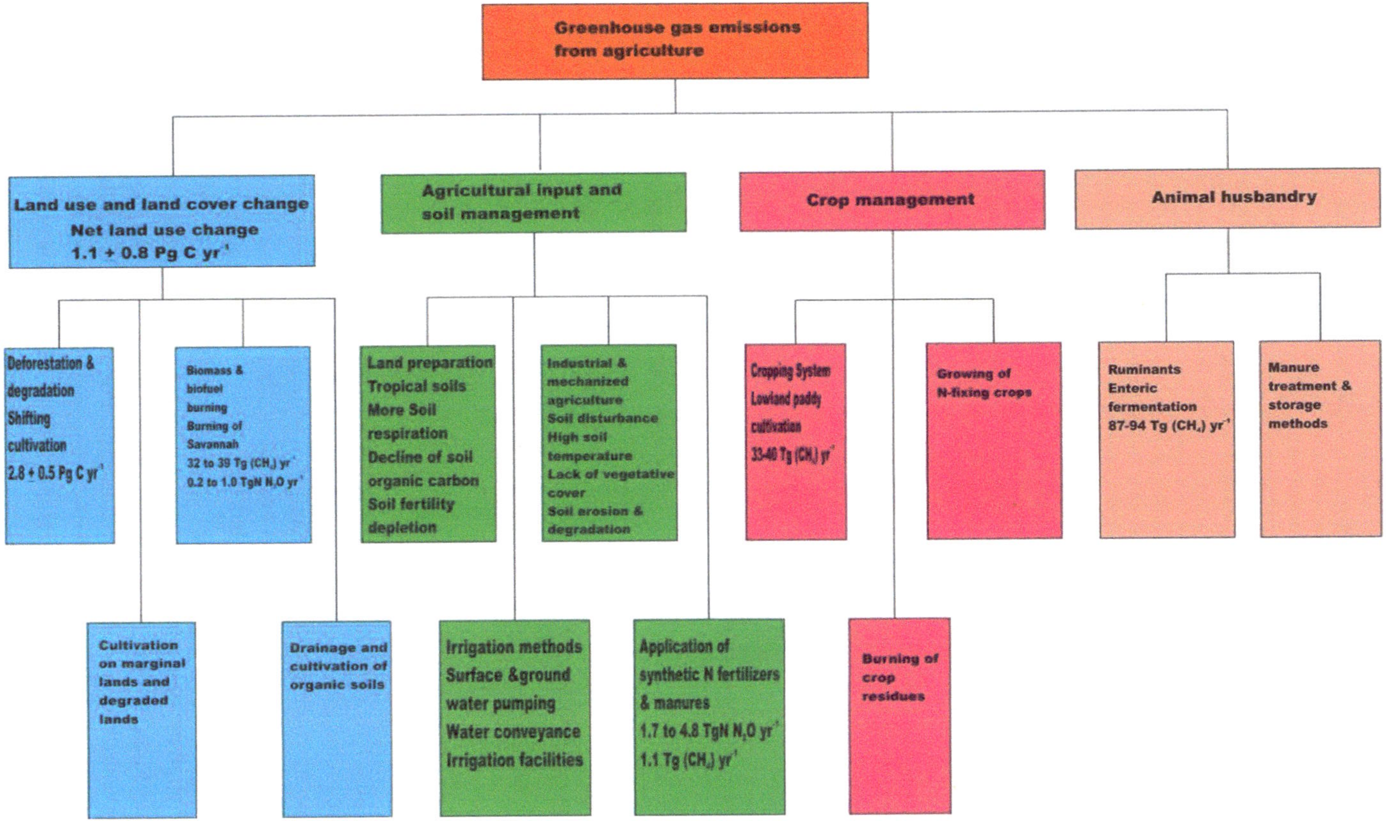

Fig. 2.5 Greenhouse gas emissions from agriculture. The GHG values mentioned represent the annual anthropogenic fluxes in Tg/year averaged over 2000–2009 time period. (Modified from Lal (2002); GHG values from Ciais et al. (2013))

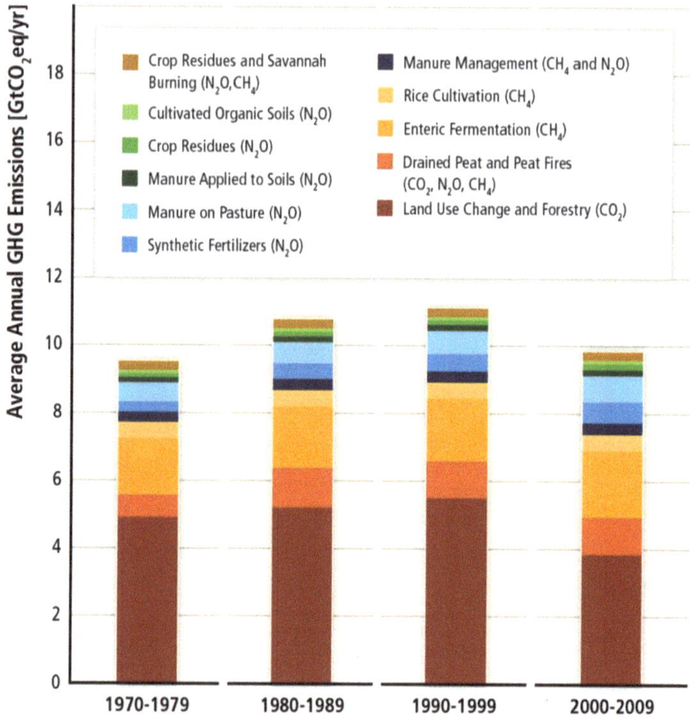

Fig. 2.6 AFOLU emissions for the last four decades. (With permission from Smith et al. (2014))

dependency of modern agriculture on external inputs like nitrogenous chemical fertilizers (IPCC 2014a, b, c).

2.4.2 *Climate Change Impacts on Agriculture*

Agriculture is a climate sensitive sector. Climate change and variability affects agriculture production and productivity in several regions of the world. Incidentally, the developing countries where agriculture is a major economic sector is reported to be vulnerable to climate change on account of poverty, income instability and lesser adaptive capacity (IPCC 2014a, b; Lipper et al. 2014). Research has shown that crop yields reduce in response to extreme daytime temperatures particularly around 30°C. High daytime and night time temperature was reported to reduce the growth, yield and quality of rice and wheat crops which are the staple food crops of South Asia (Venkatramanan and Singh 2009a, b). Porter et al. (2014) also stated that the global warming reduces yield levels of important cereal crops like rice, wheat, and maize, which, although, differ spatially. "Estimated impacts of both historical and future climate change on cereal crop yields show that yield loss can be up to 35%

for rice, 20% for wheat, 50% for sorghum, 13% for barley, and 60% for maize depending on the geographic location, climate scenarios and projected year" (Porter et al. 2014; Khatri-Chhetri et al. 2017). Further, increased tropospheric ozone shows negative response on crop yields. While the CO_2 fertilization effect stimulates the growth of crop plants with C_3 photosynthetic pathway, climate change and increasing atmospheric CO_2 concentration engenders weed menace as the weed plants become competitive in the changing climatic conditions (Aggarwal et al. 2009; Pathak et al. 2012; Porter et al. 2014). Increase in CO_2 concentration in the atmosphere have positive effect in C_3 plants as compared to C_4 plants as the C_4 photosynthetic rates are less responsive to increases in atmospheric CO_2 concentration. As regards the fertilization effect of CO_2, positive and beneficial responses were observed in case of tuber crops wherein the extra carbohydrates are stored in the underground storage organs (Porter et al. 2014).

2.4.3 Climate Change Impacts on Food Security

Food system encompasses activities involving agricultural production, processing, transport, and food consumption. Global food system is a complex system as their sustainability is driven by its innate potential, biophysical environment, and socio-economic conditions (Venkatramanan 2017). The global food chain is under constant pressure from pest and disease outbreaks, life cycle GHG emissions, food safety issues and other demand and supply side constraints. The integrity of food systems faces challenges from increasing human population, structural changes in diet, food price volatility, and mounting competition for land, water, and energy. Food security can be construed as a product of functioning food system and inherent potential of the food systems. As regards the challenges facing food security, human population growth is foremost. World human population could reach 9–10 billion by the year 2050. Interestingly, world average per capita food availability for direct human consumption improved to 2770 kilo calories per person per day in 2007 providing a rosy picture that global food production can meet the food demand of growing population (Alexandratos and Bruinsma 2012). Nevertheless, due to varying socio-economic conditions and biophysical environmental constraints, there is a huge disparity in agricultural production in the global agricultural landscape. It is reported that "some 2.3 billion people live in countries with under 2,500 kcal, and some 0.5 billion in countries with less than 2,000 kcal, while at the other extreme some 1.9 billion are in countries consuming more than 3,000 kcal" (Alexandratos and Bruinsma 2012). Demographic changes primarily in the food-insecure countries like in sub-Saharan Africa which is scourged with low per capita food consumption and high population growth can modify the development trajectory in world food security.

Food security as a comprehensive term was used to describe whether a country had access to enough food to meet dietary energy requirements (Pinstrup-Andersen 2009). The concept of food security emerged with complete clarity at the World

Food Summit in 1996, wherein it was categorically stated that "food security exists when all people, at all times, have physical, social and economic access to sufficient, safe and nutritious food to meet their dietary needs and food preferences for an active and healthy life" (FAO 1996, 2009; Barrett 2010). The four cornerstones of food security are availability, access, utilization and stability. The nutritional perspective is an intrinsic part of food security (FAO 2009).

Availability It relates to "the availability of sufficient quantities of food of appropriate quality, supplied through domestic production or imports" (Wheeler and von Braun 2013). This dimension is concerned with the ability of the agricultural system to cater to the rising food demand on account of increasing population.

Access It includes "access by individuals to adequate resources to acquire appropriate foods for a nutritious diet" (Wheeler and von Braun 2013). This dimension is greatly influenced by the changes in the real income and food prices, transport of food grains, and purchasing power of consumers.

Stability It refers to the food accessibility to individuals, household, population at all times.

Utilization This dimension reflects on adequate food, clean water, sanitation, health care, and food safety. This component of food security aims to achieve nutritional wellbeing (FAO 2008).

Nutritional Security An essential ingredient of food security is nutritional security. Incorporation of pulse crops and nutri-cereals in the crop basket will enable to achieve nutritional security and these crops are climate-resilient. Further, pulse crops enrich soil fertility and augment the rural livelihood security of farmers and cultivators. Crop and farm diversification enable the resilience and adaptive capacity of agricultural system.

Climate change and food systems including food security have risk and uncertainities for human societies and ecologies (Wheeler and von Braun 2013). Through the negative impacts on food production, food prices and accessibility, consumption and utilization, climate change has pervasive effect on all the dimensions of food security (Porter et al. 2014). In effect, climate change will affect the integrity and stability of food systems. Climate change indirectly impacts food security through its effects on access to drinking water, income, health, sanitation, income and food supply chain (Fig. 2.7). Vulnerable, disadvantaged particularly smallholder farmers and food insecure are likely to be the first affected from climate change (FAO 2009). Climate change affects food markets and enhances the consumer risks to food supply (Lipper et al. 2014). Agriculture-dependent and rural livelihoods that are vulnerable to food crises face potential threats under the emerging realities of climate change (Lipper et al. 2014; Venkatramanan 2017). Perusal of impacts of climate change warrants appropriate mitigation and adaptation strategies involving green technologies, conservation management and wise use of agricultural inputs. Climate

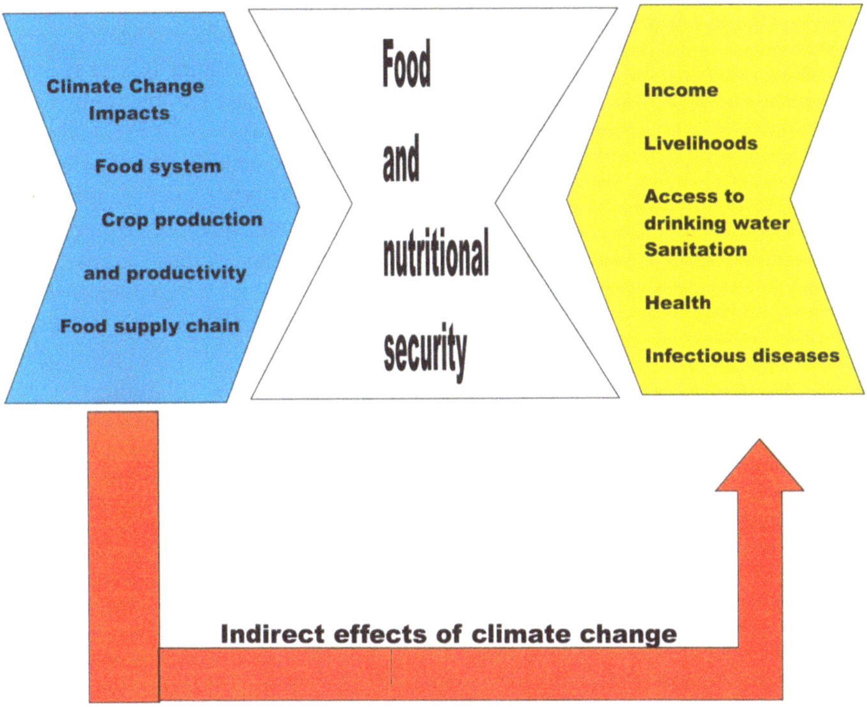

Fig. 2.7 Direct and indirect effects of climate change on food security

smart agriculture is sine qua non to achieve global food security, to empower the rural livelihoods and to fortify and enhance the resilience of agriculture system and stakeholders (Venkatramanan 2017).

2.5 Climate Smart Agriculture: The Way Forward

Agriculture is at crossroads. Mounting pressure from growing food demand, declining soil fertility, competing uses for agriculture inputs like land, water, global change including climate change demand greening of agricultural practices, growth and development. Greening of agriculture growth warrants a paradigm shift in agricultural planning, research and development, innovations in food systems, biotic and abiotic risk reduction, hazard management and climate management (Venkatramanan 2017). In effect, agriculture production system requires resilient development pathways to augment global food grain production, reduce GHG emissions from agricultural activity and adapt agriculture to climate change and variability (Lipper et al. 2014; Venkatramanan 2017). Due to growing concerns for hunger and malnutrition (Grebmer et al. 2012), potential threats of climate change and extreme weather events on agricultural food systems (Lipper et al. 2014), declining marginal productivity of agriculture inputs, FAO and World Bank conceptualised "Climate Smart Agriculture". Climate Smart Agriculture (CSA) is aptly defined as an "agricultural approach that sustainably increases productivity, resilience (adaptation), reduces/removes greenhouse gases (mitigation) and enhances achievement of national food security and development goals" (FAO 2013). The goals of CSA (Fig. 2.8) focusses on food security and poverty reduction; maintaining and augmenting the productivity and resilience of natural and agricultural ecosystem functions (Steenwerth et al. 2014). Climate change threats to agriculture system can be minimised by adopting CSA approach (Fig. 2.9) which aims at increasing the resilience, the adaptive capacity of the farmers and resource use efficiency (Lipper et al. 2014). Adaptation is defined as the "decision-making process and the set of actions undertaken to maintain the capacity to deal with future changes or perturbations to socio-ecological system without undergoing significant changes in function, structural identity, while maintaining the option to develop" (Nelson et al. 2007). Relationship existing between climate change and food production to a large extent depends on which, when and how the adaptation strategies and actions are taken. Multi-level adaptation strategies including agronomic adaptation greatly

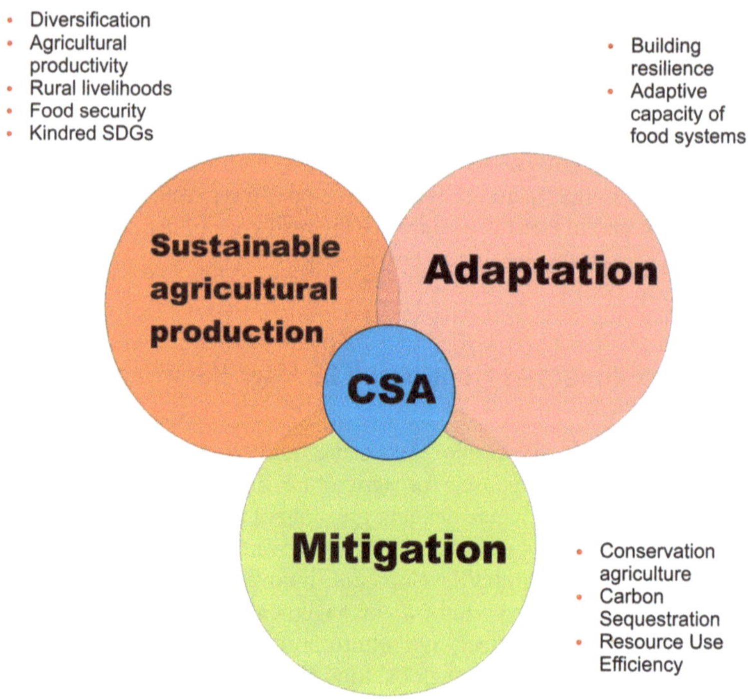

Fig. 2.8 Goals of Climate Smart Agriculture

minimize the negative impacts (Porter et al. 2014). For instance, transformative policies and proactive management practices aiming to build resilience of the agro-ecosystem would greatly empower and insure the stakeholders and marginal and small farmers from climate change impacts.

Climate smart agriculture provide a pathway to achieve sustainable development goals which focuses on poverty reduction, food security, environmental health. SDGs are integrated into the development agenda as a follow-up to the MDGs. There are in total 17 SDGs with 169 targets and the agenda is set for 2030 (Griggs et al. 2013; Lu et al. 2015). The SDGs are interconnected goals having a global reach (Glaser 2012) and they squarely address the challenges and barrier to sustainable development like poverty, food and nutritional security, environmental degradation, health and education (ICSU 2015). While SDG 2 deals with measures to end hunger, and achieve food security through sustainable agriculture and improved nutrition, the SDG 13 aims to combat climate change and its impacts. SDGs are well connected as evidenced in the connection of SDG 2 with SDG 13 (climate security), SDG 6 (water security), SDG 15 (soil security), SDG 7 (energy security), and SDG 5 (gender equality).

2.5.1 Essentiality of Resilience for Climate Smart Agriculture

Resilience as a concept factor in a range of issues like disaster risk reduction, adaptation, food and nutritional security, etc. (Tanner and Horn-Phathanothai 2014). Resilience refers to the "amount of change a system can undergo and still retain the same function and structure while maintaining options to develop" (Nelson et al. 2007). The key characteristics of the resilience concept are shown in Fig. 2.10 (Bahadur et al. 2013).

High diversity greatly augments resilience capacity, ecosystem stability, and ecosystem functioning. For instance, rich agro-biodiversity insulates the agricultural system from stresses like climate change. Further, the diversity concept underlines the need for more options for rural livelihood, looking beyond the horizon, and factoring in multiple technologies, and dovetailing the non-farm livelihood opportunities with on-farm livelihoods. As regards the agro-ecosystem resilience, the system should have the ability to change. Inability to change can be construed as lack of resilience. Further, the transient behavior of ecosystem should be upheld. The natural ecosystem are stated to be resilient as they are continually in a transient state. Further, resilience thinking enable the adaptation planners' and policy makers to plan not only to the predicted climate change but also to all possible scenarios including the failure scenario (Bahadur et al. 2013; Tanner and Horn-Phathanothai 2014).

Agricultural sustainability as enunciated by Jordan (2013) includes recognition of feedback interaction in ecosystems that enable the system to be controlled and

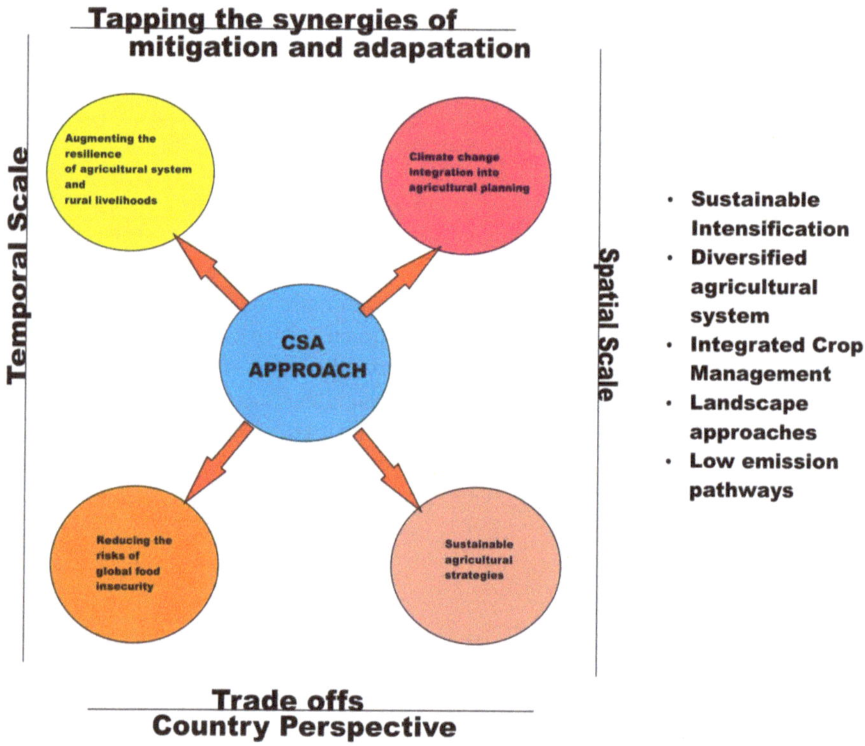

Fig. 2.9 Dissection of Climate Smart Agriculture approach

self regulated; maintaining the stability and sustainability of the ecosystem through use of free services of nature and increasing the species and landscape diversity. Environmental sustainability is essential to meet the food security goals. Agriculture on its part must strive to achieve certain environmental goals like reducing GHG emissions, enhancing resource use efficiency, agrobiodiversity conservation. In order to live sustainably well within the planetary limits, agricultural productivity must be enhanced through producing more from same piece of land, and increasing the efficiency of resource use (Fedoroff 2015). Further, to achieve the food security, smart agriculture must enable enhanced agricultural production, ensure integrity and stability of food system, improve accessibility and availability of food (Fig. 2.11) (Foley et al. 2011).

The community participation is one of the hall marks of resilient system. Community participation in tandem with traditional and indigenous knowledge enables community resilience as the representatives from all the sections of the community involve in decision making. Also, the individuals can better assess their capability and vulnerability, identify the coping strategies, and earnestly prepare and plan for the potential disturbances. Redundancy is another important characteristic of resilient system. It refers to the ability of the response pathway, elements of the system, processes and capacities of ecosystem to allow for a partial failure of the

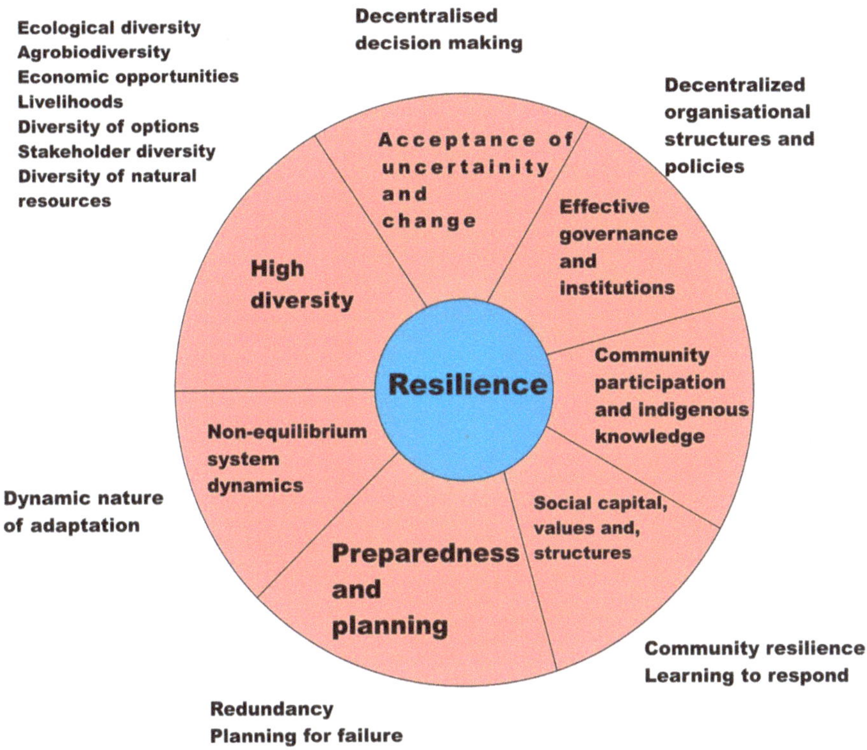

Fig. 2.10 Virtues of resilience – An adaptation to climate smart agriculture approach. (Modified from Bahadur et al. 2013).

system without a complete collapse of the system. Resilience and sustainable development intersect on the idea and essentiality of equity in achieving climate resilient growth and development. Justice and equitable distribution of assets, wealth, livelihood opportunities, and equitable economy will safeguard the socio-ecological system from hazards and stresses (Bahadur et al. 2013). In effect, while resilience approach deals with non-linear dynamics, uncertainties, and surprises, adaptation aim at reducing the expected damage from climate scenarios (Tanner and Horn-Phathanothai 2014).

2.5.2 Climate Resilient Pathways in Agriculture

As agriculture sector encounters many biotic and abiotic stresses, there exists a multitude of development pathways. However, the CSA pathway leads to higher resilience and lower risks to food security (Fig. 2.12) (Lipper et al. 2014). The drivers of change greatly enhances the vulnerabilities. However, the CSA approach aims at reducing the impact of drivers of change through adaptation and mitigation measures.

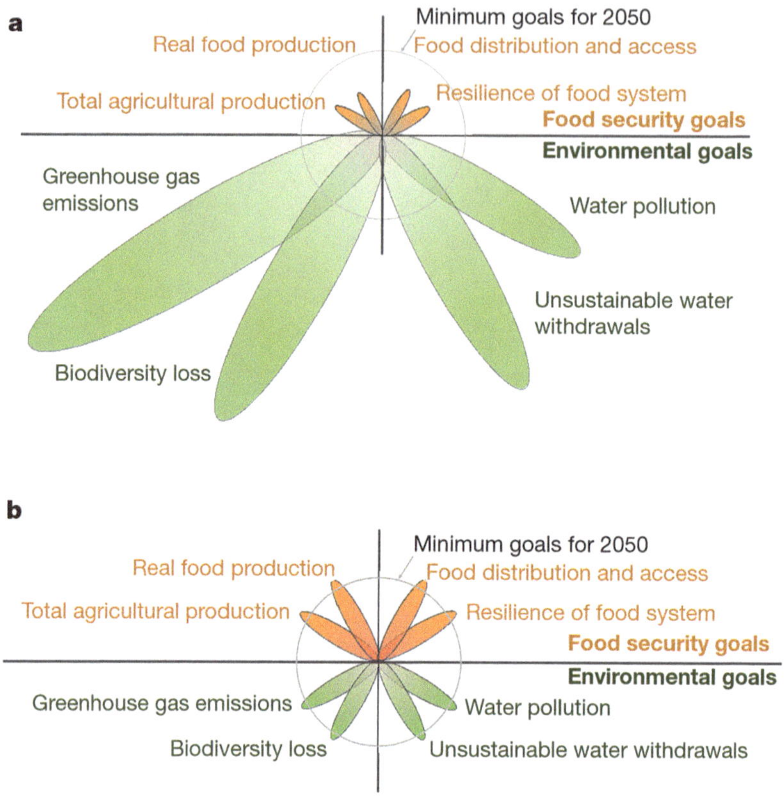

Fig. 2.11 Meeting goals for food security and environmental sustainability. (Reprinted by permission from Springer Nature, Solutions for a cultivated planet, Foley et al. 2011. Nature 478: 337–342)

Climate smart agriculture practices include integrated crop management, integrated farming system, conservation agriculture, agro-forestry, management of crop residues, agronomic adaptation, water conservation and irrigation management. As regards the spatial scale of CSA adoption, CSA practices are applied from farm to global level through food supply chain and landscape. The driving force of CSA is ecosystem approach which is based on the systems approach. The cornerstone of CSA is adoption of mitigation and adaptation strategies and dovetailing the same into development trajectory of agriculture growth. Mitigation to climate change refers to human mediated interventions aiming to reduce or remove the GHGs through enhancing the sequestration of GHGs. The mitigation potential and strategies for mitigation of climate change are function of crop and soil environment. To cite an example, lowland paddy cultivation under waterlogged and anaerobic condition, releases methane due to the action of methanogens upon the soil organic matter. The Methane emission can be reduced through adoption of appropriate mitigation strategies. Irrigation management is crop and region specific. Mid-season aeration greatly aids in mitigating methane emissions. Methane emission from

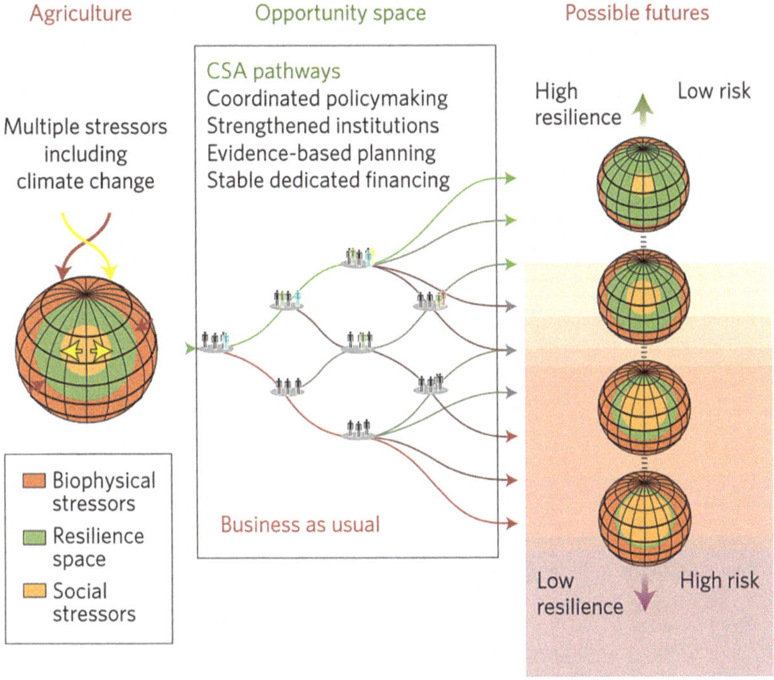

Fig. 2.12 Climate-resilient transformation pathways for agriculture. (Reprinted by permission from Springer Nature, Climate–smart agriculture for food security. Lipper et al. 2014. Nature Climate Change. Vol.4: 1068–1072)

ruminants can be reduced by adopting protein rich diet (Aggarwal et al. 2009; Pathak et al. 2012). Adaptation to climate change is the process of adjustment to actual or expected climate and its effects. Adaptation strategies include crop diversification, use of indigenous knowledge, development of tolerant crop varieties, etc.

2.5.3 Climate Smart Agriculture Technologies

The core of CSA is climate smart agriculture technologies. "Adaptation options that sustainably enhance productivity, enhance resilience to climatic stresses, and reduce greenhouse gas emissions are known as climate-smart agricultural (CSA) technologies, practices and services" (FAO 2010; Khatri-Chhetri et al. 2017). These technologies though are needed in all the agroecosystem, it is pertinent for the developing countries as these practices can bring in agricultural transformation; insure them against climate change and achieve food security. Albeit, several studies have reported diverse technologies as part of CSA, the technologies can be stated as smart technologies provided they are able to achieve either of the goals of CSA (Fig. 2.13). CSA technology adoption is mainly dependent on the bio-physical

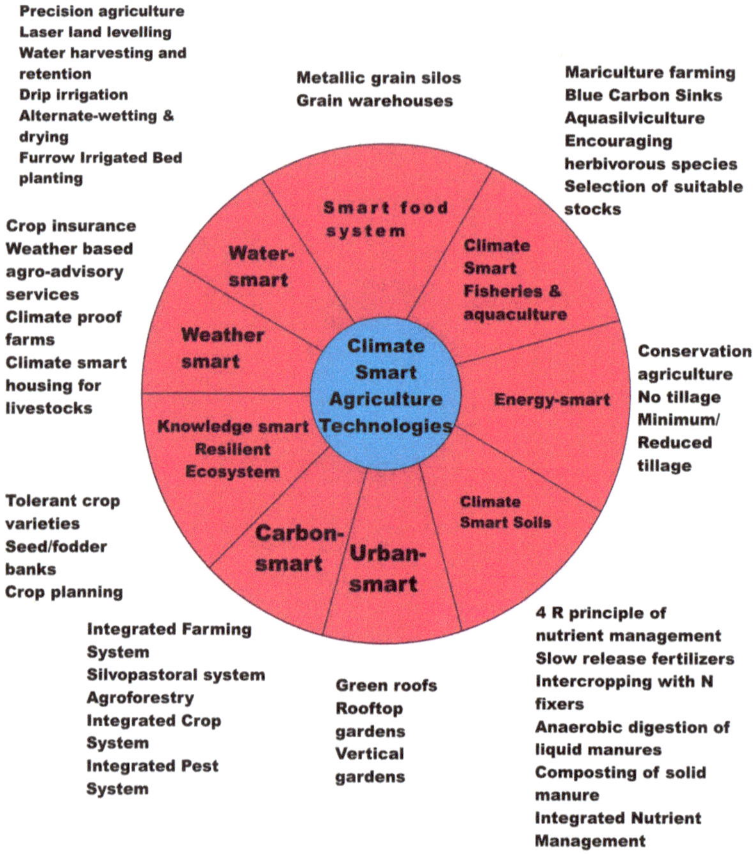

Fig. 2.13 Matrix of climate smart agriculture technologies. (Modified from FAO (2010) and Khatri-Chhetri et al. 2017)

environment, socio-economic characteristics of farmers, and virtues of CSA technologies (Khatri-Chhetri et al. 2017).

2.6 Conclusion

Agriculture is at crossroads. Perhaps, the perform or perish paradigm is stifling global agriculture sector. Agriculture being one of the major GHG emitter, has to factor in many transformative policies to mitigate climate change. On the other hand, being an ecosystem service provider, agriculture strives hard to upscale agriculture production and achieve global food security. So, in order to achieve increased agricultural production and decreased GHG emissions, the climate smart agriculture approach came into existence. This CSA approach enables along with food security, mitigation of and adaptation to climate change as it is embedded with sustainability principles of resilience and adaptation. Further, the approach as both temporal and spatial dimensions. This CSA approach is applicable from local to global scale and it has short term and long term goals. Inherently, the approach recognises the fact that climate change mitigation and adaptation measures are country specific and it also possess enough elbow space to factor in trade offs. Additionally, stability, integrity, and resilience of agricultural system is sine qua non for human existence.

References

Aggarwal, P. K., Joshi, H. C., Singh, S. D., Bhatia, A., Jain, N., Prasad, S., Chaudhary, A., Gupta, N., & Pathak, H. (2009). Agriculture and environment. In *Hand book of agriculture* (pp. 62–92). New Delhi: Directorate of Information and Publication Agriculture, ICAR.

Alexandratos, N., & Bruinsma, J. (2012). World agriculture towards 2030/2050: the 2012 revision. ESA Working Paper No. 12-03. Rome: FAO.

Bahadur, A., Ibrahim, M., & Tanner, T. (2013). Characterising resilience: Unpacking the concept for tackling climate change and development. *Climate and Development, 5*, 55–65. https://doi.org/10.1080/17565529.2012.762334.

Barrett, C. (2010). Measuring food insecurity. *Science, 327*, 825–828. https://doi.org/10.1126/science.1182768.

Ciais, P., Sabine, C., Bala, G., et al. (2013). Carbon and other biogeochemical cycles. In T. F. Stocker, D. Qin, G.-K. Plattner, M. Tignor, S. K. Allen, J. Boschung, A. Nauels, Y. Xia, V. Bex, & P. M. Midgley (Eds.), *Climate change 2013: The physical science basis. Contribution of working group I to the fifth assessment report of the intergovernmental panel on climate change*. Cambridge/New York: Cambridge University Press.

FAO. (1996). Rome declaration on world food security and world food summit plan of action, World food summit, Rome, 13 to 17 November 1996 http://www.fao.org/docrep/003/w3613e/w3613e00.htm. Accessed 27 May 2018

FAO. (2008). Climate change and food security: A framework document. In Fao.org. http://www.fao.org/forestry/15538-079b31d45081fe9c3dbc6ff34de4807e4.pdf. Accessed 26 May 2018

FAO. (2009). Declaration of the world summit on food security, World summit on food security, Rome, 16 to 18 November 2009 http://www.fao.org/fileadmin/templates/wsfs/Summit/Docs/Final_Declaration/WSFS09_Declaration.pdf. Accessed 27 May 2018

FAO. (2010). Climate smart agriculture: Policies, practices and financing for food security, adaptation and mitigation. Food and Agriculture Organization of the United Nations (FAO), Rome, Italy. http://www.fao.org/docrep/013/i1881e/i1881e00.pdf. Accessed 31 May 2018.

FAO. (2013). Climate-smart agriculture: Source book, pp. 557.

Fedoroff, N. (2015). Food in a future of 10 billion. *Agriculture & Food Security*. doi: https://doi.org/10.1186/s40066-015-0031-7

Foley, J., DeFries, R., & Asner, G. (2005). Global consequences of land use. *Science, 309*, 570–574. https://doi.org/10.1126/science.1111772.

Foley, J., Ramankutty, N., Brauman, K., et al. (2011). Solutions for a cultivated planet. *Nature, 478*, 337–342. https://doi.org/10.1038/nature10452.

Glaser, G. (2012). Base sustainable development goals on science. *Nature, 491*, 35–35. https://doi.org/10.1038/491035a.

Grebmer, K., Ringler, C., & Rosegrant, M. et al. (2012) Global hunger index. The challenge of hunger: Ensuring sustainable food security under land, water, and energy stresses. Washington. https://doi.org/10.2499/9780896299429

Griggs, D., Stafford-Smith, M., Gaffney, O., et al. (2013). Sustainable development goals for people and planet. *Nature, 495*, 305–307. https://doi.org/10.1038/495305a.

ICSU, ISSC. (2015). Review of the sustainable development goals: The science perspective. International Council for Science (ICSU), Paris, pp. 88.

IPCC. (2013). Summary for policymakers. In T. F. Stocker, D. Qin, G.-K. Plattner, M. Tignor, S. K. Allen, J. Boschung, A. Nauels, Y. Xia, V. Bex, & P. M. Midgley (Eds.), *Climate Change 2013: The physical science basis. Contribution of working group I to the fifth assessment report of the intergovernmental panel on climate change*. Cambridge/New York: Cambridge University Press.

IPCC. (2014a). In Core Writing Team, R. K. Pachauri, & L. A. Meyer (Eds.), *Climate change 2014: Synthesis report. Contribution of working groups I, II and III to the fifth assessment report of the intergovernmental panel on climate change* (p. 151). Geneva: IPCC.

IPCC. (2014b). In C. B. Field, V. R. Barros, D. J. Dokken, K. J. Mach, M. D. Mastrandrea, T. E. Bilir, M. Chatterjee, K. L. Ebi, Y. O. Estrada, R. C. Genova, B. Girma, E. S. Kissel, A. N. Levy, S. MacCracken, P. R. Mastrandrea, & L. L. White (Eds.), *Climate change 2014: Impacts, adaptation, and vulnerability. Part A: Global and sectoral aspects. Contribution of working group II to the fifth assessment report of the intergovernmental panel on climate change* (p. 1132). Cambridge/New York: Cambridge University Press.

IPCC. (2014c). In O. Edenhofer, R. Pichs-Madruga, Y. Sokona, E. Farahani, S. Kadner, K. Seyboth, A. Adler, I. Baum, S. Brunner, P. Eickemeier, B. Kriemann, J. Savolainen, S. Schlömer, C. von Stechow, T. Zwickel, & J. C. Minx (Eds.), *Climate change 2014: Mitigation of climate change. Contribution of working group III to the fifth assessment report of the intergovernmental panel on climate change*. Cambridge/New York: Cambridge University Press.

Jordan, C. (2013). *An ecosystem approach to sustainable agriculture* (pp. 245–246). Dordrecht: Springer Netherlands.

Khatri-Chhetri, A., Aggarwal, P., Joshi, P., & Vyas, S. (2017). Farmers' prioritization of climate-smart agriculture (CSA) technologies. *Agricultural Systems, 151*, 184–191. https://doi.org/10.1016/j.agsy.2016.10.005.

Lal, R. (2002). Greenhouse effect, world soils and. In *Encyclopedia of soil science* (pp. 621–625). New York: Marcel Dekker, Inc..

Lipper, L., Thornton, P., Campbell, B., et al. (2014). Climate-smart agriculture for food security. *Nature Climate Change, 4*, 1068–1072. https://doi.org/10.1038/nclimate2437.

Lu, Y., Nakicenovic, N., Visbeck, M., & Stevance, A. (2015). Policy: Five priorities for the UN sustainable development goals. *Nature, 520*, 432–433. https://doi.org/10.1038/520432a.

Matson, P., Parton, W., Power, A., & Swift, M. (1997). Agricultural intensification and ecosystem properties. *Science, 277*, 504–509. https://doi.org/10.1126/science.277.5325.504.

McBratney, A., Field, D., & Koch, A. (2014). The dimensions of soil security. *Geoderma, 213*, 203–213. https://doi.org/10.1016/j.geoderma.2013.08.013.

Nelson, D., Adger, W., & Brown, K. (2007). Adaptation to environmental change: Contributions of a resilience framework. *Annual Review of Environment and Resources, 32*, 395–419. https://doi.org/10.1146/annurev.energy.32.051807.090348.

Pathak, H., Aggarwal, P. K., & Singh, S. D. (2012). *Climate change impact, adaptation and mitigation in agriculture: Methodology for assessment and application* (p. 302). New Delhi: IARI.

Paustian, K., Lehmann, J., Ogle, S., et al. (2016). Climate-smart soils. *Nature, 532*, 49–57. https://doi.org/10.1038/nature17174.

Pinstrup-Andersen, P. (2009). Food security: Definition and measurement. *Food Security, 1*, 5–7. https://doi.org/10.1007/s12571-008-0002-y.

Porter, J. R., Xie, L., Challinor, A. J., Cochrane, K., et al. (2014). Food security and food production systems. In C. B. Field, V. R. Barros, D. J. Dokken, K. J. Mach, M. D. Mastrandrea, T. E. Bilir, M. Chatterjee, K. L. Ebi, Y. O. Estrada, R. C. Genova, B. Girma, E. S. Kissel, A. N. Levy, S. MacCracken, P. R. Mastrandrea, & L. L. White (Eds.), *Climate change 2014: Impacts, adaptation, and vulnerability. Part A: Global and sectoral aspects. Contribution of working group II*

to the fifth assessment report of the intergovernmental panel on climate change (pp. 485–533). Cambridge/New York: Cambridge University Press.
Ramankutty, N., Evan, A., Monfreda, C., & Foley, J. (2008). Farming the planet: 1. Geographic distribution of global agricultural lands in the year 2000. *Global Biogeochemical Cycles, 22*, n/a–n/a. https://doi.org/10.1029/2007gb002952.
Scherr, S., Shames, S., & Friedman, R. (2012). From climate-smart agriculture to climate-smart landscapes. *Agriculture & Food Security, 1*, 12. https://doi.org/10.1186/2048-7010-1-12.
Schiermeier, Q. (2015). Quest for climate-proof farms. *Nature, 523*, 396–397. https://doi.org/10.1038/523396a.
Smith, P., Bustamante, M., Ahammad, H., et al. (2014). Agriculture, forestry and other land use (AFOLU). In O. Edenhofer, R. Pichs-Madruga, Y. Sokona, E. Farahani, S. Kadner, K. Seyboth, A. Adler, I. Baum, S. Brunner, P. Eickemeier, B. Kriemann, J. Savolainen, S. Schlömer, C. von Stechow, T. Zwickel, & J. C. Minx (Eds.), *Climate change 2014: Mitigation of climate change. Contribution of working group III to the fifth assessment report of the intergovernmental panel on climate change*. Cambridge, New York: Cambridge University Press.
Steenwerth, K., Hodson, A., Bloom, A., et al. (2014). Climate-smart agriculture global research agenda: Scientific basis for action. *Agriculture & Food Security, 3*, 11. https://doi.org/10.1186/2048-7010-3-11.
Tanner, T., & Horn-Phathanothai, L. (2014). *Climate change and development* (1st ed.). New York: Routledge.
Uphoff, N. (2015). Alternate paths to food security. In R. J. Herring (Ed.), *The oxford handbook of food, politics, and society* (p. 202). New York: Oxford university press.
Venkatramanan, V. (2017). Climate change and food security. UGC ePG Pathshala. http://epgp.inflibnet.ac.in/epgpdata/uploads/epgp_content/S000017GE/P001781/M025271/ET/1512451985Module36,VVRamanan,ClimateChangeandFoodSecurity.pdf. Accessed 27 May 2018
Venkatramanan, V., & Singh, S. D. (2009a). Differential effects of day and night temperature on the growth of rice crop. *Pusa Agri Science, 32*, 57–62.
Venkatramanan, V., & Singh, S. D. (2009b). Differential effects of day and night temperature on the growth of wheat crop. *Annals of Agricultural Research, 30*(1&2), 49–52.
Vitousek, P., Aber, J., Howarth, R., et al. (1997). Human alteration of the Global nitrogen cycle: Sources and consequences. *Ecological Applications, 7*, 737–750. https://doi.org/10.1890/1051-0761(1997)007[0737:haotgn]2.0.co;2.
Wheeler, T., & von Braun, J. (2013). Climate change impacts on global food security. *Science, 341*, 508–513. https://doi.org/10.1126/science.1239402.

Web Links

http://www.ipcc.ch/report/ar5/syr/
http://www.ipcc.ch/report/ar5/wg1/
http://www.ipcc.ch/report/ar5/wg2/
http://www.ipcc.ch/report/ar5/wg3/

Chapter 3
Phosphorus Management in Agroecosytems and Role and Relevance of Microbes in Environmental Sustainability

Sagar Chhabra

Abstract Phosphorus (P) is an important macronutrient source for plant growth. However, it is a limiting mineral resource based on its availability in the environment and the form it is available to the plants. High amount of phosphorus use in soil is often considered to be non-productive to agriculture and can lead to mineral and heavy metal accumulation, soil leaching, surface run-off, and eutrophication in water bodies. This chapter reviews literature concerning P management practices in agriculture, the importance, role and relevance of microorganisms in P availability, environmental sustainability and the perspective of these microbes is discussed.

Keywords Fungi · Bacteria · Phosphorus · Environmental sustainability

3.1 Introduction

Agriculture is at crossroads. Modern intensive agriculture is under immense pressure to cater to the needs of growing population. The bedrock of modern agriculture is its dependence on chemical fertilizers, and pesticides. As regards the soil nutrients, phosphorus is an essential macronutrient source in crop production and plays a vital role in increasing plant productivity as it helps in plant development, photosynthesis, transfer of genetic traits and is component of metabolic pathways, etc. (Armstrong 1988). The high P fertilization in agro-ecosystems is a common practice around the world today due to low availability of available form of P from either organic or chemical P fertilizer sources (Rodríguez and Fraga 1999). The organic form of P constitute 20-80% of P pool in soil and is the major portion of immobilized P reservoir in soils whereas the chemical form of P fertilizer added as organic P substitute such as from mineral rock P resource do not exceeds more than 30% as

S. Chhabra (✉)
MicroGen Biotech Ltd, ERIC Centre, Institute of Technology Carlow, Carlow, Ireland

Department of Biotechnology, Invertis University, Bareilly, India

S. Shah et al. (eds.), *Sustainable Green Technologies for Environmental Management*, https://doi.org/10.1007/978-981-13-2772-8_3

available P in soil for plant uptake on an estimate (Norrish and Rosser 1983; Richardson 1994).

The low availability of P in soils is due to low plant available form of P as orthophosphate anions (HPO_4^{2-}, $H_2PO_4^-$) in soil which results primarily by weathering of rock phosphate in environment or due to mineralization, desorption or dissolution of organic or inorganic P sources or pool present in soil (Compton et al. 2000; Hinsinger 2001; Arcand and Schneider 2006).

The plant available form of P (orthophosphate anions) is usually highly reactive in soil and very often form mineral complexes with calcium, iron and aluminium present in soil depending on the soil pH (Shen et al. 2011). The excessive addition of P to overcome nutrient P deficiency in soil is therefore considered important and can lead to nutrient accumulation, nutrient imbalance, heavy metal accumulation (such as Cadmium (Cd), lead (Pb) and arsenic (As), etc. that are naturally present in resource P as in mineral rock phosphate, soil acidification, soil leaching and surface runoff resulting in eutrophication and pollution of water bodies worldwide (Giuffréde López Carnelo et al. 1997; Keller and Schulin 2003; Gupta et al. 2014; Wang et al. 2015b).

The current P management practices in agriculture therefore undermine the future food security and also has impact on environment and human health. This chapter focuses on sustainability measures to P management with the major emphasis on the role and relevance of microbes in P availability and environmental sustainability.

3.2 Approaches to P Management and Relevance of Microorganisms in Soil

There are number of sustainability measure that are suggested for the improvement of phosphorus usage and P management in agriculture. The 4R Nutrient management approach advocated by the IFA (2009) and IPNI (2014) are useful measure on the selection of P fertilizer and its utilization on land such as with the focus on the right sources, the right rate, the right time and the right placement to limit excessive usage of P resource on land. The changes in fertilization practices with the adoption of land management practices that protect soil quality and limit soil erosion (Sharpley 2016). The suitable land use planning and the suitable plant breeding program and effective nutrient recycling and controls are other important measures and approaches to P management in soil (McDowell et al. 2001; Sharpley 2016).

The microbes play vital role in nutrient cycling and P availability in agriculture. The use of microorganisms is an important step to promote the sustainable use of phosphorus resources (Sharpley 2016). Figure 3.1 depicts the nutrient P resource and microbial mediated nutrient P cycling and the control measures to sustainable P management.

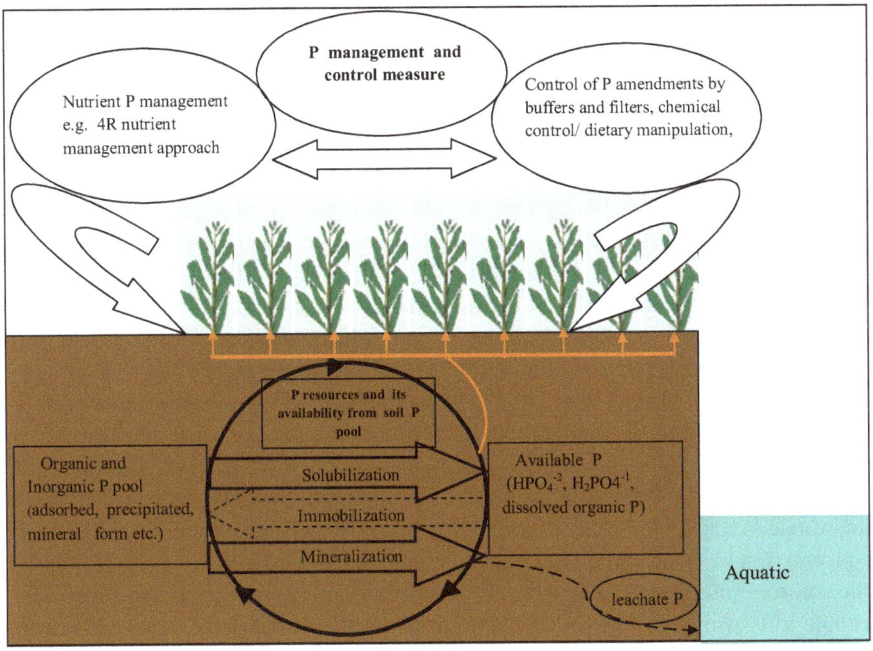

Fig. 3.1 Strategies for phosphorus management and its controls measures in agriculture. The *4R* management refers to Right rate of P to match crop needs, in the Right source, at the Right time, and in the Right place. (From IFA (2009) and IPNI (2014)

Table 3.1 Examples of phosphate functional microbiome

Microorganisms	Diversity
Bacteria	*Achromobacter, Agrobacterium, Bacillus, Bradyrhizobium, Burkholderia, Citrobacter, Enterobacter, Erwinia, Gordonia, Klebsiella, Micrococcus, Nitrobacter. Nitrosomonas, Pantoe, Proteus, Pseudomonas, Rhizobium, Serratia, Thiobacillus, Xanthomonas*
Fungi	*Aspergillus, Candida, Chaetomium, Cladosprium, Cunnighamella, Curvularia,Humicola, Helminthosporium, Paecilomyces, Penicillium, Populospora, Phoma, Pythium, Trichoderma*
AMF	*Glomus*

3.2.1 Microbial Mediated P Availability in Soil

The soil contain highly diverse group of microorganisms such as bacteria, fungi or AMF and the P availability is long being associated with microbial functions (Table 3.1) (Richardson and Simpson 2011). Among the well-known associations, AMF association with land plants is the most primitive and is known to be associated with 80% of land plants (Kim et al. 2017). AMF improve P availability in land plants such as by indirectly increasing root soil interface and improves the P availability to plants from the soil (Johri et al. 2015). AMF are also known to release hydrolytic enzyme such as acid phosphatase and contribute to P availability directly by organic P mineralization in soil (Sato et al. 2015). The AMF function has been reported to significantly improve P availability in soils especially with low P content and in soil having high P fixing capacity (i.e. soil having higher immobilized form of P) (Balemi and Negisho 2012). AMF are also associated with other important plant growth promotion functions and are associated with heavy metal remediation potential (Khan et al. 2014a). Besides AMF, a number of other group of fungi and fungal endophyte that reside asymptomatically (with no visible symptoms of infection) in plants are associated in improving P availability (Johri et al. 2015).

The microbial P availability is also associated with number of bacteria present in soil which are involved in P solubilization and mineralization activity to increase P availability in soil in form of orthophosphate anions for the plant uptake (Rodríguez and Fraga 1999; Chhabra et al. 2013). An estimated P-solubilizing bacterial population in soil constitute in a range of 1–50% of total population whereas the fungi in soil are in range of 0.5%–0.1% of the total population in soil (Kucey 1983; Gyaneshwar et al. 2002). The bacterial contribution to P availability to land plants are also associated with other or indirect function such as phytostimulation of plant root and root hairs which can help increase P uptake by plant from soil by increasing the soil root interface (Chhabra and Dowling 2017). The microorganisms also contribute to P availability in soil by acting as P resource themselves, as a significant amount of P is cycled through the microbial biomass on land on an annual basis which is estimated to be typically 10 to 50 kg P/ha or may be as high as 100 kg P/ha in soil (Richardson 2007).

3.2.2 Microbial Phosphate Solubilization and Mineralization Activity

Microorganisms such as bacteria and fungi isolated from plants have been shown under *in* vitro conditions to be involved in mineral phosphate solubilization (MPS) activity by production of extracellular organic acid anions such as gluconic acid (Otieno et al. 2015), malic acid (Jog et al. 2014), citric acid, salicylic acid, benzene acetic acid (Chen et al. 2014), tartaric acid, and acetic acid (Khan et al. 2014b), etc. which help to release P linked to minerals and increase P availability in soil (Kepert et al. 1979). The mineral P availability in soil is also associated with other functions such as production of siderophores, protons, hydroxyl ions, CO_2 and are linked to the release free orthophosphate anions in the extracellular medium (Rodríguez and Fraga 1999; Sharma et al. 2013).

The Organic phosphate mineralization activity and its availability occurs largely at the expense of plant or animal remains or in the form such as phytic acid (inositol hexakisphosphate), mono- and diesters, phospholipids, nucleotides, sugar phosphates, and phosphoproteins. Also phosphonate (a class of organophosphonates) degradation is associated with mineralization activity (Tate 1984; Rodríguez and

Fraga 1999). The mineralization mechanism of organic P involves hydrolysis of the phosphoester or phosphoanhydride bonds carried out by phosphatases (or phosphohydrolases) or cleavage of the C-P bond in the case of phosphonates. There are several different phosphatases (or phosphohydrolases) that are involved in hydrolysis of phosphoester or phosphoanhydride bonds and these have been classified based on their activity under optimal pH conditions or their specificity such as acidic or alkaline phosphatase etc. (Rodríguez and Fraga 1999).

3.2.3 Functional Aspects of Phosphate Solubilizing Microbes and Its Role in Heavy Metal Remediation

The phosphate solubilizing or mineralizing microorganisms are recognized with many other functional aspects or plant growth promoting traits such as plant hormone production, biocontrol function, nutrient cycling, etc. and some are detailed in Tables 3.2 and 3.3 and Fig. 3.2. The bioremediation potential of these microbes are recognized to be useful to treat heavy metals polluted soil and can be beneficial for soils having high immobilized form of P or high heavy metal contamination such as those that are also sourced by higher P fertilizers management in soil such cadmium (Cd), arsenic (As), chromium (Cr), lead (Pb) etc (Gupta et al. 2014) (Table 3.4).

The phosphate solubilizing microorganisms can help in heavy metal remediation by its immobilization or stabilization in soil also known as phytostabilization (Ahemad 2015). The phytoextraction of heavy metals by microbial mediated mobilization in soil and its uptake by plant is another mechanism for heavy metal remediation. The phytostabilization of heavy metals by P microbes are proposed to involve biosorption or precipitation mechanism or microbial chelation activity (Ahemad 2015). The phytoextraction of heavy metal or its mobilization may involve microbial functions such as organic acid production, secretion of siderophores, etc. (Table 3.4), which may help in heavy metal mobilization or its phytoextraction in plants (Ahemad 2015). Some of these functions from literature are detailed in Table 3.4.

3.2.4 Phosphate Solubilizing or Mineralizing Bioinoculant Potential for Sustainable Environment Management

The current strategies that underlie with sustainable P management in agriculture is to reduce reliance on fertilized P resource and to better exploit P accumulated in soil. The use of microorganisms can be useful as a biofertilizer resource for increasing the mobilization of P in soil or to enhance the P availability to crops plants. The phosphate solubilizing microorganisms themselves are useful P resource (with available P in their biomass) or it can help in increasing the P availability to plant

Table 3.2 Examples of P mobilizing microorganisms with recognized functional aspects

Organism	Isolation source	Functional aspects	References
Acinetobacter calcoaceticus	Roots of Brassica sp. (*Brassica oleracea*)	Mineral phosphate solubilization, Phosphatase activity, heavy metals tolerance	Ghoreishi and Etemadifar (2017)
Aspergillus sp., *Emmericella* sp. and *Penicillium* sp.	Arid and semi-arid soil	Phytase and phosphatases	Yadav and Tarafdar (2003)
Bacillus species PSB10	Conventional (cultivated) fields soil	Mineral phosphate solubilization IAA, siderophores, HCN, ammonia	Wani and Khan (2010)
Bacillus thuringiensis	Agricultural fields soil	Mineral phosphate solubilization, IAA, antibiotic resistant, heavy metal resistant	Sandip et al. (2011)
Bacillus sp., *Azotobacter* sp. and *Pseudomonas* sp.	Contaminated soil	Mineral phosphate solubilization heavy metals, IAA	Mohamed and Almaroai (2017)
Gluconacetobacter Diazotrophicus	Roots of sugar cane (*Saccharum officinarun)*	Mineral phosphate solubilization, Biological nitrogen fixation	Crespo et al. (2011)
Pantoea agglomerans	Root nodules of peanut (*Arachis hypogaea L.*)	Mineral phosphate solubilization, siderophore production	Taurian et al. (2013)
Penicillium oxalicum and *Aspergillus niger*	Rhizosphere of Asparagus, *Asparagus officinalis L.*)	Mineral phosphate Solubilization, biocontrol activity	Ruangsanka (2014)
Pseudomonas aeruginosa strain OSG41	Rhizosphere of mustard *(Brassica campestris)*	Mineral phosphate Solubilization, IAA, siderophores	Oves et al. (2013)
Pseudomonas. fluorescens strain Psd	Rhizosphere of Black gram (*Vigna mungo*)	Mineral phosphate Solubilization, IAA, siderophores, HCN, antibiotics, biocontrol activity	Upadhyay and Srivastava (2010)
Pseudomonas sp. TLC 6-6.5-4	Lake sediment	Mineral phosphate solubilization, IAA, siderophore	Li and Ramakrishna (2011)
Rhizobium endophyticum sp. Nov	Seed of Common Bean/French bean (*Phaseolus vulgaris*)	Phytate	López-López et al. (2010)
Streptomyces tricolor mhce0811	Wheat (*Triticum aestivum*)	Mineralphosphate solubilization phytase, siderophores, IAA, chitinase	Jog et al. (2014)
Trichoderma spirale	Seedling of Meranti (*Shorea leprosula)* and *(Shorea selanica*)	Mineral phosphate solubilization and inhibition of fungal pathogen (*Fusarium*)	Hakim et al. (2015)

Table 3.3 Examples of P mobilizing microorganisms and plant benefits

Organism	Benefitted plant	Benefit	References
Aspergillus. niger	Subterraneum Clover (*Trifolium subterraneum* L.)	Increased in seedling growth	Hayes et al. (2000)
Aspergillus niger	Wheat (*Triticum aestivum*)	Improved growth	Xiao et al. (2013)
Aspergillus niger, Penicillium aculeatum	Chinese cabbage (*Brassicasp.*)	Increased growth	Wang et al. (2015a)
Bacillus amyloliquefaciens FZB45	Maize (*Zea mays* cv. Elita)	Improved root growth	Idriss et al. (2002)
Bacillus sp. and *Pseudomonas* sp.	Sesame (*Sesamum indicum*)	Increased seed yield	Jahan et al. (2013)
Bacillus thuringiensis	Rice (*Oryza sativa*)	Increased shoot length	Paul Raj et al. (2014)
Burkholderia cepacia	Maize (*Zea mays*)	Improved growth	Zhao et al. (2014)
Burkholderia sp.	Chili (*Capsicum frutescens* L. cv. Hua Rua)	Improved growth	Surapat et al. (2013)
Burkholderia sp.	*Lotus* (*Lotus japonicus* B-129 'Gifu')	Increased in plant dry weight and shoot length	Unno et al. (2005)
Pantoea agglomerans PSB-1 and *Burkholderia anthina* (PSB-2)	Mung bean (*Vigna radiata* L.)	Improved growth	Charana Walpola and Yoon (2013)
Piriformospora indica	Maize (*Zea mays*)	Increase in biomass and growth	Kumar et al. (2011); Gill et al. (2016)
Pseudomonas fluorescens L321	Pea (*Pisum sativum* L.)	Improved growth	Otieno et al. (2015)
Pseudomonas sp. *and Rhizobium* sp.	Faba bean (*Vicia faba* L.)	Increased plant height	Demissie et al. (2013)
Serratia sp.	Wheat (*Triticum aestivum*)	Improved growth	Swarnalakshmi et al. (2013)
Sporotrichum thermophile	Wheat (*Triticum aestivum*)	Increased growth of wheat	Singh and Satyanarayana (2010)

for its uptake and often concomitantly perform many other plant growth promotion functions (Chhabra and Dowling 2017).

The bioremediation potential of these microorganisms can be useful in heavy metal remediation such as to minimize the heavy metal mobilization in food crops or it can be useful in improving the phytoremediation potential for heavy metals remediation (such as by use of selective microbial P inoculant source in soil) (Ahemad 2015)

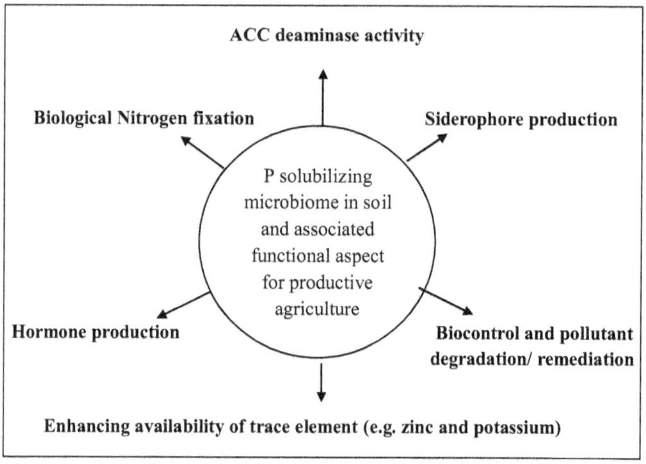

Fig. 3.2 Functional aspects of P microorganisms or microbiome for more productive agriculture

Table 3.4 Examples of phosphate mobilizing microorganisms and mediated metal remediation

Organism	Plant	Heavy metal	Mechanism	References
Acinetobacter haemolyticus RP19	Pearl millet (*Pennisetum glaucum*)	Zn	Phytostabilization	Misra et al. (2012)
Achromobacter xylosoxidans strain Ax10	Indian mustard (*Brassica juncea*)	Cu	Phytoextraction	Ma et al. (2009a)
Enterobacter aerogenes NBRI K24, *Rahnella aquatilis* NBRI K3	Indian mustard (*Brassica juncea*)	Ni, Cr	Phytostabilization	Kumar et al. (2009)
Pseudomonas aeruginosa strain MKRh3	Black gram (*Vigna mungo*)	Cd	Phytostabilization	Ganesan (2008)
Pseudomonas sp., *Bacillus* sp.	Indian mustard (*Brassica juncea*)	Cr	Phytostabilization	Rajkumar et al. (2006)
Pseudomonas fluorescens	Soybean (Glycine max PK 564)	Hg	Phytostabilization	Gupta et al. (2005)
Pseudomonas sp.	Chickpea (*Cicer arietinum*)	Ni	Phytostabilization	Tank and Saraf (2009)
Pseudomonas sp. TLC 6-6.5-4	Maize (Zea mays) and Sunflower (*Helianthus annuus*)	Cu	Phytoextraction	Li and Ramakrishna (2011)
Pseudomonas sp. SRI2, *Psychrobacter* sp. SRS8, *Bacillus* sp. SN9	Indian mustard (*Brassica juncea*), *Brassica oxyrrhina*	Ni	Phytoextraction	Ma et al. (2009b)

Adapted from Ahemad (2015)

A number of P microorganisms are being recognized with P function or tested with functional potential in field, green house or in vitro studies and the influence of these microorganisms in soil or on plant nutrient uptake can be variable among strains and is considered to be dependent on host species/cultivars, microbial taxa, and environmental conditions (Kageyama et al. 2008). The selection of an appropriate microorganisms in agriculture production is an important step for its commercialization and its utilization on land and mostly dependent on its field performance or its function.

The AM fungi belonging to the phylum Glomeromycota have been shown to increase P uptake in diverse crop plants (Berruti et al. 2015; Igiehon and Babalola 2017). However, the AMF application as a commercial bioinoculant source can be limited due to its unculturable characteristics and difficultly to produce at a large scale at a competitive costs (as AMF is a obligate biotroph and requires host for growth) (Berruti et al. 2015). The plant root or soil associated fungi are another group of microorganisms which can mobilize inorganic P such as rock phosphate to increase inorganic P availability in soil. Also, these microorganisms can increase P availability from organic P sources such as by phosphatase (phosphohydrolases) activity (Rodríguez and Fraga 1999). A number of soil and plant associated fungi are isolated and described in literature (Tables 3.1, 3.2 and 3.3). Among the commercially available examples, the fungal bioinoculant *Penicillium bilaiae* marketed by Mosanto BioAg as JumpStart® are considered to be useful for P availability (Mosanto BioAg, Canada). The *Penicillium bilaiae* colonize the plant's root of many plants and increases the availability of bound mineral P forms in soil and fertilizer phosphate. Besides role of other P fungi isolated from soil and rhizosphere are recognized (Johri et al. 2015). The role of fungal endophytes (that reside asympotmatically in plants) are also described in literature on P availability, for instance *Piriformospora indica* (phylum Basidiomycota; order Sebacinales) which is considered to be a potential candidate as a bio-inoculant as it is known to colonize a wide range of monocots and dicots plants (Prasad et al. 2005, 2008, 2013; Gill et al. 2016).

The potential of bacteria are also been recognized as a commercial inoculants in market such as N-Fix® (Azotic Technologies, UK) or NITROFIX™- AD (AgriLife, India) formulated using bacteria *Gluconacetobacter diazotrophicus.* The *G. diazotrophicus* recognized with biological nitrogen fixation in plants are also recognized with functional trait such as phosphate solubilization activity (Intorne et al. 2009). The bacterial P microorganism with heavy metal tolerance and remediation are also being successfully tested by number of studies (Table 3.4). A number of phosphatase mineralization and phytase microorgansims are recognized and have been tested for increasing plant yield (Singh and Satyanarayana 2011).

3.3 Conclusion

Phosphorus solubilizers and mobilizers can be utilized for improving agricultural productivity, and soil productivity through sustainable use of P resources and toxic pollutant remediation. The P microbial functional potential can be tapped as a useful bioresource which can help in increasing agricultural production; help to reduce mineral P inputs; aid in bioremediation, and perform many plant growth promoting functions.

Acknowledgement The author acknowledges the support provided by MicroGen Biotech Ltd, Ireland and its seed funding agency and research support team.

References

Ahemad, M. (2015). Phosphate-solubilizing bacteria-assisted phytoremediation of metalliferous soils: A review. *Biotech, 5*, 111–121.

Arcand, M. M., & Schneider, K. D. (2006). Plant- and microbial-based mechanisms to improve the agronomic effectiveness of phosphate rock: A review. *Anais da Academia Brasileira de Ciências, 78*, 791–807.

Armstrong, D. L. (1988). *Role of phosphorus in plants: In better crops with plant Food pp 4-DI Armstrong*. Atlanta: Potash and Phosphate Institute.

Balemi, T., & Negisho, K. (2012). Management of soil phosphorus and plant adaptation mechanisms to phosphorus stress for sustainable crop production: A review. *Journal of Soil Science and Plant Nutrition, 12*, 547–562.

Berruti, A., Lumini, E., Balestrini, R., & Bianciotto, V. (2015). Arbuscular mycorrhizal fungi as natural biofertilizers: Let's benefit from past successes. *Frontiers in Microbiology, 6*, 1559.

Charana Walpola, B., & Yoon, M. H. (2013). Phosphate solubilizing bacteria: Assessment of their effect on growth promotion and phosphorus uptake of mung bean (*Vigna radiata* [L.] R. Wilczek). *Chilean Journal of Agricultural Research, 73*(3), 275–281.

Chen, Y., Fan, J.-B., Du, L., Xu, H., Zhang, Q.-H., & He, Y.-Q. (2014). The application of phosphate solubilizing endophyte *Pantoea dispersa* triggers the microbial community in red acidic soil. *Applied Soil Ecology, 84*, 235–244.

Chhabra, S., & Dowling, D. N. (2017). Endophyte-promoted nutrient acquisition: Phosphorus and iron. In *Functional importance of the plant microbiome* (pp. 21–42). Cham: Springer.

Chhabra, S., Brazil, D., Morrissey, J., Burke, J. I., O'Gara, F., & N Dowling, D. (2013). Characterization of mineral phosphate solubilization traits from a barley rhizosphere soil functional metagenome. *Microbiology Open, 2*, 717–724.

Compton, J., Mallinson, D., Glenn, C. R., Filippelli, G., Föllmi, K., Shields, G., & Zanin, Y. (2000). Variations in the global phosphorus cycle. In *Marine Authigenesis: From global to microbial* (Vol. 66, pp. 21–33). United States: SEPM.

Crespo, J. M., Boiardi, J. L., & Luna, M. F. (2011). Mineral phosphate solubilization activity of gluconacetobacter diazotrophicus under P-limitation and plant root environment. *Agricultural Sciences, 2*, 16–22.

Demissie, S., Muleta, D., & Berecha, G. (2013). Effect of phosphate solubilizing bacteria on seed germination and seedling growth of faba bean (*Vicia faba* L.). *International Journal of Agricultural Research, 8*, 123–136.

Ganesan, V. (2008). Rhizoremediation of cadmium soil using a cadmium-resistant plant growth-promoting rhizopseudomonad. *Current Microbiology, 56*, 403–407.

Ghoreishi, F., & Etemadifar, Z. (2017). Heavy metal removal by phosphate solubilizing *Acinetobacter calcoaceticus* isolated from rhizosphere. *Journal of Biology, 6*, 230–239.
Gill, S. S., Gill, R., Trivedi, D. K., Anjum, N. A., Sharma, K. K., Ansari, M. W., Ansari, A. A., Johri, A. K., Prasad, R., Pereira, E., Varma, A., & Tuteja, N. (2016). *Piriformospora indica*: Potential and significance in plant stress tolerance. *Frontiers in Microbiology, 7*(332). https://doi.org/10.3389/fmicb.2016.00332.
Giuffréde López Carnelo, L., de Miguez, S. R., & Marbán, L. (1997). Heavy metals input with phosphate fertilizers used in Argentina. *Science of the Total Environment, 204*, 245–250.
Gupta, A., Rai, V., Bagdwal, N., & Goel, R. (2005). In situ characterization of mercury-resistant growth-promoting fluorescent pseudomonads. *Microbiological Research, 160*, 385–388.
Gupta, D. K., Chatterjee, S., Datta, S., Veer, V., & Walther, C. (2014). Role of phosphate fertilizers in heavy metal uptake and detoxification of toxic metals. *Chemosphere, 108*, 134–144.
Gyaneshwar, P., Naresh Kumar, G., Parekh, L. J., & Poole, P. S. (2002). Role of soil microorganisms in improving P nutrition of plants. *Plant and Soil, 245*, 83–93.
Hakim, S. S., Budi, S. W., & Turjaman, M. (2015). Phosphate solubilizing and antifungal activity of root endophyte isolated from *Shorea leprosula* Miq. and *Shoreal selanica* (DC) Blume. *Jurnal manajemen hutan tropika, 21*, 138–146.
Hayes, J. E., Simpson, R. J., & Richardson, A. E. (2000). The growth and phosphorus utilization of plants in sterile media when supplied with inositol hexaphosphate, glucose 1-phosphate or inorganic phosphate. *Plant and Soil, 220*, 165–174.
Hinsinger, P. (2001). Bioavailability of trace elements as related to root-induced chemical changes in the rhizosphere. In G. R. Gobran, W. W. Wenzel, & E. Lombi (Eds.), *Trace elements in the rhizosphere* (pp. 163–182). United States: CRC Press.
Idriss, E. E., Makarewicz, O., Farouk, A., Rosner, K., Greiner, R., Bochow, H., Richter, T., & Borriss, R. (2002). Extracellular phytase activity of *Bacillus amyloliquefaciens* FZB45 contributes to its plant-growth-promoting effect. *Microbiology, 148*, 2097–2109.
Igiehon, N. O., & Babalola, O. O. (2017). Biofertilizers and sustainable agriculture: Exploring arbuscular mycorrhizal fungi. *Applied Microbiology and Biotechnology, 101*, 4871–4881.
International Fertilizer Association IFA. (2009). The global '4R' nutrient stewardship framework: Developing fertilizer best management practices for delivering economic, social and environmental benefits, Paris, France. http://www.fertilizer.org/ifa/Home-Page/LIBRARY/Publication-database.html/The-Global-4R-Nutrient-Stewardship-Framework-for-Developing-and-Delivering-Fertilizer-Best-Management-Practices.html2
International Plant Nutrient Institute IPNI. (2014). 4R Nutrient stewardship portal. Available at: http://www.ipni.net/4R. Accessed Feb 9 2018
Intorne, A. C., de Oliveira, M. V. V., Lima, M. L., da Silva, J. F., Olivares, F. L., & de Souza Filho, G. A. (2009). Identification and characterization of *Gluconacetobacter diazotrophicus* mutants defective in the solubilization of phosphorus and zinc. *Archives of Microbiology, 191*, 477–483.
Jahan, M., Nassiri Mahallati, M., Amiri, M. B., & Ehyayi, H. R. (2013). Radiation absorption and use efficiency of sesame as affected by biofertilizers inoculation in a low input cropping system. *Industrial Crops and Products, 43*, 606–611.
Jog, R., Pandya, M., Nareshkumar, G., & Rajkumar, S. (2014). Mechanism of phosphate solubilization and antifungal activity of Streptomyces spp. isolated from wheat roots and rhizosphere and their application in improving plant growth. *Microbiology, 160*, 778–788.
Johri, A. K., Oelmüller, R., Dua, M., Yadav, V., Kumar, M., Tuteja, N., Varma, A., Bonfante, P., Persson, B. L., & Stroud, R. M. (2015). Fungal association and utilization of phosphate by plants: Success, limitations, and future prospects. *Frontiers in Microbiology, 6*, 984.
Kageyama, S. A., Mandyam, K. G., & Jumpponen, A. (2008). Diversity, function and potential applications of the root-associated endophytes. In P. D. A. Varma (Ed.), *Mycorrhiza* (pp. 29–57). Berlin/Heidelberg: Springer.
Keller, A., & Schulin, R. (2003). Modelling heavy metal and phosphorus balances for farming systems. *Nutrient Cycling in Agroecosystems, 66*, 271–284.

Kepert, D. G., Robson, A. D., & Posner, A. M. (1979). The effect of organic root products on the availability of phosphorus to plants. In J. L. Harley & R. S. Russell (Eds.), *The soil–root interface* (pp. 115–124). London: Academic.

Khan, A., Sharif, M., Ali, A., Shah, S. N. M., Mian, I. A., Fazli Wahid, B. J., Adnan, M., Nawaz, S., & Ali, N. (2014a). Potential of AM fungi in phytoremediation of heavy metals and effect on yield of wheat crop. *American Journal of Plant Sciences, 5*, 1578.

Khan, M. S., Zaidi, A., & Ahmad, E. (2014b). Mechanism of phosphate solubilization and physiological functions of phosphate-solubilizing microorganisms. In *Phosphate solubilizing microorganisms* (pp. 31–62). Cham: Springer.

Kim, S. J., Eo, J.-K., Lee, E.-H., Park, H., & Eom, A.-H. (2017). Effects of arbuscular mycorrhizal fungi and soil conditions on crop plant growth. *Mycobiology, 45*, 20–24.

Kucey, R. M. N. (1983). Phosphate-solubilizing bacteria and fungi in various cultivated and virgin alberta soils. *Canadian Journal of Soil Science, 63*, 671–678.

Kumar, K. V., Srivastava, S., Singh, N., & Behl, H. M. (2009). Role of metal resistant plant growth promoting bacteria in ameliorating fly ash to the growth of *Brassica juncea*. *Journal of Hazardous Materials, 170*, 51–57.

Kumar, M., Yadav, V., Kumar, H., Sharma, R., Singh, A., Tuteja, N., & Johri, A. K. (2011). *Piriformospora indica* enhances plant growth by transferring phosphate. *Plant Signaling & Behavior, 6*, 723–725.

Li, K., & Ramakrishna, W. (2011). Effect of multiple metal resistant bacteria from contaminated lake sediments on metal accumulation and plant growth. *Journal of Hazardous Materials, 189*, 531–539.

López-López, A., Rogel, M. A., Ormeño-Orrillo, E., Martínez-Romero, J., & Martínez-Romero, E. (2010). *Phaseolus vulgaris* seed-borne endophytic community with novel bacterial species such as Rhizobium endophyticum sp. nov. *Systematic and Applied Microbiology, 33*, 322–327.

Ma, Y., Rajkumar, M., & Freitas, H. (2009a). Isolation and characterization of Ni mobilizing PGPB from serpentine soils and their potential in promoting plant growth and Ni accumulation by Brassica spp. *Chemosphere, 75*, 719–725.

Ma, Y., Rajkumar, M., & Freitas, H. (2009b). Inoculation of plant growth promoting bacterium *Achromobacter xylosoxidans* strain Ax10 for the improvement of copper phytoextraction by *Brassica juncea*. *Journal of Environmental Management, 90*, 831–837.

McDowell, R. W., Sharpley, A. N., Condron, L. M., Haygarth, P. M., & Brookes, P. C. (2001). Processes controlling soil phosphorus release to runoff and implications for agricultural management. *Nutrient Cycling in Agroecosystems, 59*, 269–284.

Misra, N., Gupta, G., & Jha, P. N. (2012). Assessment of mineral phosphate-solubilizing properties and molecular characterization of zinc-tolerant bacteria. *Journal of Basic Microbiology, 52*, 549–558.

Mohamed, H. M., & Almaroai, Y. A. (2017). Effect of phosphate solubilizing bacteria on the uptake of heavy metals by corn plants in a long-term sewage wastewater treated soil. *International Journal of Environmental Science and Development, 8*, 366.

Norrish, K., & Rosser, H. (1983). *Soils: An Australian viewpoint*. Melbourne/London: CSIRO/Academic Mineral phosphate; pp. 335–361.

Otieno, N., Lally, R., Kiwanuka, S., Lloyd, A., Ryan, D., Germaine, K., & Dowling, D. (2015). Plant growth promotion induced by phosphate solubilizing endophytic *Pseudomonas* isolates. *Frontiers in Microbiology, 6*, 745.

Oves, M., Khan, M. S., & Zaidi, A. (2013). Chromium reducing and plant growth promoting novel strain Pseudomonas aeruginosa OSG41 enhance chickpea growth in chromium amended soils. *European Journal of Soil Biology, 56*, 72–83.

Paul Raj, D., Rhema SB Linda, R., & Babyson, S. (2014). Molecular characterization of phosphate solubilizing bacteria (PSB) and plant growth promoting rhizobacteria (PGPR) from pristine soils. *International Journal of Transportation Science and Technology, 1*, 317–324.

Prasad, R., Pham, G. H., Kumari, R., Singh, A., Yadav, V., Sachdev, M., PeskanT, H. S., Oelmuller, R., Garg, A. P., & Varma, A. (2005). Sebacinaceae: Culturable mycorrhiza–like endosymbiotic fungi and their interaction with non-transformed and transformed roots. In S. Declerck, D. G. Strullu, & J. A. Fortin (Eds.), *In vitro culture of Mycorrhizas* (Vol. 4, pp. 291–312). Berlin/Heidelberg: Springer-Verlag.

Prasad, R., Bagde, U. S., Pushpangdan, P., & Varma, A. (2008). *Bacopa monniera* L.: Pharmacological aspects and case study involving *Piriformospora indica. International Journal of Integrative Biology, 3*, 100–110.

Prasad, R., Kamal, S., Sharma, P. K., Oelmueller, R., & Varma, A. (2013). Root endophyte *Piriformospora indica* DSM 11827 alters plant morphology, enhances biomass and antioxidant activity of medicinal plant *Bacopa monniera. Journal of Basic Microbiology, 53*(12), 1016–1024.

Rajkumar, M., Nagendran, R., Lee, K. J., Lee, W. H., & Kim, S. Z. (2006). Influence of plant growth promoting bacteria and Cr6+ on the growth of Indian mustard. *Chemosphere, 62*, 741–748.

Richardson, A. E. (1994). Soil microorganisms and phosphorus availability. In C. E. Pankhurst, B. M. Doube, G. VVSR, & P. R. Grace (Eds.), *Soil biota: Management in sustainable farming systems* (pp. 50–62). Melbourne: CSIRO.

Richardson, A. E. (2007). Making microorganisms mobilize soil phosphorus. In *First international meeting on microbial phosphate solubilization* (pp. 85–90). Dordrecht: Springer.

Richardson, A. E., & Simpson, R. J. (2011). Soil microorganisms mediating phosphorus availability update on microbial phosphorus. *Plant Physiology, 156*, 989–996.

Rodríguez, H., & Fraga, R. (1999). Phosphate solubilizing bacteria and their role in plant growth promotion. *Biotechnology Advances, 17*, 319–339.

Ruangsanka, S. (2014). Identification of phosphate-solubilizing fungi from the asparagus rhizosphere as antagonists of the root and crown rot pathogen *Fusarium oxysporum. Science Asia, 40*, 16–20.

Sandip, B., Subrata, P., & Swati, R. G. (2011). Isolation and characterization of plant growth promoting Bacillus Thuringiensis from agricultural soil of West Bengal. *Research Journal of Biotechnology, 6*, 9–13.

Sato, T., Ezawa, T., Cheng, W., & Tawaraya, K. (2015). Release of acid phosphatase from extraradical hyphae of arbuscular mycorrhizal fungus *Rhizophagus clarus. Soil Science & Plant Nutrition, 61*, 269–274.

Sharma, S. B., Sayyed, R. Z., Trivedi, M. H., & Gobi, T. A. (2013). Phosphate solubilizing microbes: Sustainable approach for managing phosphorus deficiency in agricultural soils. *SpringerPlus, 2*, 587.

Sharpley, A. (2016). Managing agricultural phosphorus to minimize water quality impacts. *Science in Agriculture, 73*, 1–8.

Shen, J., Yuan, L., Zhang, J., Li, H., Bai, Z., Chen, X., Zhang, W., & Zhang, F. (2011). Phosphorus dynamics: From soil to plant. *Plant Physiology, 156*, 997–1005.

Singh, B., & Satyanarayana, T. (2010). Plant growth promotion by an extracellular HAP-phytase of a thermophilic mold *Sporotrichum thermophile. Applied Biochemistry and Biotechnology, 160*, 1267–1276.

Singh, B., & Satyanarayana, T. (2011). Microbial phytases in phosphorus acquisition and plant growth promotion. *Physiology and Molecular Biology of Plants, 17*, 93–103.

Surapat, W., Pukahuta, C., Rattanachaikunsopon, P., Aimi, T., & Boonlue, S. (2013). Characteristics of phosphate solubilization by phosphate-solubilizing bacteria isolated from agricultural chili soil and their efficiency on the growth of chili (*Capsicum frutescens* L. cv. Hua Rua). *Chiang Mai Journal of Science, 40*, 11–25.

Swarnalakshmi, K., Prasanna, R., Kumar, A., Pattnaik, S., Chakravarty, K., Shivay, Y. S., Singh, R., & Saxena, A. K. (2013). Evaluating the influence of novel cyanobacterial biofilmed biofertilizers on soil fertility and plant nutrition in wheat. *European Journal of Soil Biology, 55*, 107–116.

Tank, N., & Saraf, M. (2009). Enhancement of plant growth and decontamination of nickel-spiked soil using PGPR. *Journal of Basic Microbiology, 49*, 195–204.

Tate, K. R. (1984). The biological transformation of P in soil. In *Biological processes and soil fertility* (pp. 245–256). Berlin: Springer.
Taurian, T., Anzuay, M. S., Ludueña, L. M., Angelini, J. G., Muñoz, V., Valetti, L., & Fabra, A. (2013). Effects of single and co-inoculation with native phosphate solubilising strain Pantoea sp J49 and the symbiotic nitrogen fixing bacterium *Bradyrhizobium* sp SEMIA 6144 on peanut (*Arachis hypogea* L.) growth. *Symbiosis, 59*, 77–85.
Unno, Y., Okubo, K., Wasaki, J., Shinano, T., & Osaki, M. (2005). Plant growth promotion abilities and microscale bacterial dynamics in the rhizosphere of Lupin analysed by phytate utilization ability. *Environmental Microbiology, 7*, 396–404.
Upadhyay, A., & Srivastava, S. (2010). Evaluation of multiple plant growth promoting traits of an isolate of *Pseudomonas fluorescens* strain Psd. *Indian Journal of Experimental Biology, 48*, 601–609.
Wang, H.-Y., Liu, S., Zhai, L.-M., Zhang, J.-Z., Ren, T.-Z., Fan, B.-Q., & Liu, H.-B. (2015a). Preparation and utilization of phosphate biofertilizers using agricultural waste. *Journal of Integrative Agriculture, 14*, 158–167.
Wang, R., Guo, S., Li, N., Li, R., Zhang, Y., Jiang, J., Wang, Z., Liu, Q., Wu, D., Sun, Q., Du, L., & Zhao, M. (2015b). Phosphorus accumulation and sorption in calcareous soil under long-term fertilization. *PLoS One, 10*, e0135160.
Wani, P. A., & Khan, M. S. (2010). *Bacillus* species enhance growth parameters of chickpea (*Cicer arietinum* L.) in chromium stressed soils. *Food and Chemical Toxicology, 48*, 3262–3267.
Xiao, C., Zhang, H., Fang, Y., & Chi, R. (2013). Evaluation for rock phosphate solubilization in fermentation and soil–plant system using a stress-tolerant phosphate-solubilizing *Aspergillus niger* WHAK1. *Applied Microbiology and Biotechnology, 169*, 123–133.
Yadav, R. S., & Tarafdar, J. C. (2003). Phytase and phosphatase producing fungi in arid and semi-arid soils and their efficiency in hydrolyzing different organic P compounds. *Soil Biology and Biochemistry, 35*, 745–751.
Zhao, K., Penttinen, P., Zhang, X., Ao, X., Liu, M., Yu, X., & Chen, Q. (2014). Maize rhizosphere in Sichuan, China, hosts plant growth promoting *Burkholderia cepacia* with phosphate solubilizing and antifungal abilities. *Microbiological Research, 169*, 76–82.

Chapter 4
Management of Heavy Metal Polluted Soils: Perspective of Arbuscular Mycorrhizal Fungi

R. Krishnamoorthy, V. Venkatramanan, M. Senthilkumar, R. Anandham, K. Kumutha, and Tongmin Sa

Abstract In recent years, intensive research have been initiated on remediation of metal polluted soil due to the public concerns on ecosystem deterioration. Plants are used as an effective tool in remediation of metal polluted soil. In natural ecosystem, plants are associated with soil microorganisms which plays an important role in enhancing plant growth in metal contaminated site and phytoremediation process. Among the microorganisms, arbuscular mycorrhizal fungi (AMF) contributes markedly in the phytoremediation process in metal contaminated site by enhancing plant stress tolerance and metal extraction from soil (phytoextraction) and immobilization of metals in soil (phytostabilization). This chapter deals with our study on the effect of heavy metal on AMF root colonization and diversity in heavy metal and metalloid contaminated sites. In addition, this chapter summarizes the mechanisms involved in AMF mediated phytoremediation of metal polluted

R. Krishnamoorthy · K. Kumutha
Department of Agricultural Microbiology, Agricultural College and Research Institute, Madurai, India

V. Venkatramanan
School of Interdisciplinary and Transdisciplinary Studies, Indira Gandhi National Open University, New Delhi, India

M. Senthilkumar (✉)
Department of Agricultural Microbiology, Tamil Nadu Agricultural University, Coimbatore, Tamil Nadu, India
e-mail: msenthilkumar@tnau.ac.in

R. Anandham
Department of Agricultural Microbiology, Agricultural College and Research Institute, Madurai, India

Department of Agricultural Microbiology, Tamil Nadu Agricultural University, Coimbatore, Tamil Nadu, India

T. Sa
Department of Environmental and Biological Chemistry, Chungbuk National University, Cheongju, Chungbuk, Republic of Korea

S. Shah et al. (eds.), *Sustainable Green Technologies for Environmental Management*, https://doi.org/10.1007/978-981-13-2772-8_4

soil. Potential prospects lies in revealing the mechanisms behind the tripartite interaction among plant species, AMF species and heavy metals for effective management of polluted soils.

Keywords Arbuscular mycorrhizal fungi · Heavy metals · Phytoextraction · Phytostabilization

4.1 Introduction

Plant root symbiosis with fungi occurs in different forms, which are referred as mycorrhiza. Mycorrhizal associations are emblematic of a symbiosis between plants and fungi. The term mycorrhiza was coined by Frank (1885) which was derived from two Greek words '*mycos*' (meaning fungus) and '*rhiza*' (meaning root). Different type of mycorrhiza are described (arbuscular, ecto, ectendo-, arbutoid, monotropoid, ericoid and orchidaceous mycorrhizae) so far, in which arbuscular mycorrhizae and ectomycorrhizae are the most abundant and widespread (Allen et al. 2003). Mycorrhizal associations are important from the perspective of plant growth, plant nutrition, plant diversity, nutrient cycling, ecosystem processes, and ecosystem functioning (Sanders and Croll 2010). Arbuscular mycorrhizal fungi (AMF) are able to establish symbiotic association with more than 80% of the plant families. AMF takes about 20% of the carbon from plant and in return help the plant to uptake more water, phosphorus and other nutrients from soil. Arbuscular mycorrhizal fungi are grouped into the phylum Glomeromycota (Smith and Read 2008).

Arbuscular mycorrhizal fungi (AMF) are important soil microorganisms that play a key role in facilitating nutrient uptake by crops in a variety of agroecosystems, particularly in low-input farming systems, and in revegetation and rhizomerediation processes. Studies in glasshouse and fields have assessed the positive effects of AMF on nutrient uptake, plant growth and yield. Enhancing the mycorrhizal system of a low-fertility or degraded soil helps the root system acquire more nutrients. It is widely recognized that AMF play an important role in improving the uptake of low mobile ions, in phosphate (PO_4^{3-}) and in ammonium (NH_4^+) phases (Martin et al. 2007). AMF not only increase the rate of nutrient transfer from the roots to the host plant, but they also increase resistance to biotic and abiotic stresses (Singh 2006; Martin et al. 2007). The key effects of AM symbiosis can be summarized as follows: (1) enhancing uptake of low mobile ions, (2) improving quality of soil structure, (3) enhancing plant community diversity, (4) improving rooting and plant establishment, (5) improving soil nutrient cycling, and (6) enhancing plant tolerance to biotic and abiotic stress (Smith and Read 2008).

Glomalin is a glycoprotein secreted by AMF hypha and the soil organic matter fraction called glomalin related soil protein (GRSP) (Rillig 2004). GRSP appears to be a component of the hyphae and spore wall of AMF, likely released into the soil by mycelium turnover. The role of glomalin in the ecosystem is still unclear; however, the hypothetical role has evolved from that of an active secretion to enhance soil aggregation (Rillig and Mummey 2006) and heavy metal sequestration.

4.2 Heavy Metals

Chemically, heavy metals refer to transition metals which have the atomic mass over 20 and specific gravity more than five (Rascio and Navari-Izzo 2011). Nevertheless, from the biological perspective, the term heavy metal refers to metals and metalloids that cause toxic effect to plant and animals. Some metals like Co, Cu, Fe, Mn, Mo, Ni and Zn are essential for normal growth and development of plants (Gohre and Paszkowski 2006) at low concentration. Among the metals, Cu and Fe act as a cofactor in many metalloenzymes and Zn play a vital role in function and stabilization of proteins. These metals are called trace elements, required for enzyme activity and have structural functions in nucleic acid metabolism (Zenk 1996). However, some heavy metals and metalloids (from here heavy metal and metalloids will be grouped as heavy metals) like As, Cd, Hg, Pb or Se, are not essential, since they do not perform any known physiological function in plants. Although heavy metals are natural constituents of soil and occur naturally in the environment, nowadays these toxic metals are major concern globally as it has deleterious effects on plants and human health.

4.2.1 Effect of Heavy Metal on AMF Colonization and Diversity

Mycorrhizal colonization and diversity in the vicinity of the heavy metal contaminated Janghang smelter (36° 00′ N; 126° 39′ E) located in South Chungcheong province, South Korea was analyzed (Krishnamoorthy et al. 2015). Soil samples were collected from the three sites which have varying concentration of heavy metal and metalloids and the details are given in Table 4.1.

Based on soil heavy metal concentration (except Ni) site 1, 2 and 3 may be considered as highly, moderately and less contaminated sites, respectively. Heavy metal concentration of site 1 was significantly higher than site 3. Frequency of mycorrhizal colonization (F%) was found higher in site 1 (85%) followed by site 2 (83.3%) and site 3 (80.8%) (Fig. 4.1). Root colonization of AMF was considered to be an important factor for the successful establishment of plants in metal contaminated soils (Gonzalez-Chavez et al. 2009). In line with this, we have observed higher root colonization in the plants grown in highly and moderately contaminated soils compared to less contaminated soil. An earlier study showed that root colonization of maize was found to be higher in high contaminated soil compared to non-contaminated soil (Weissenhorn et al. 1995). Higher root colonization in heavy metal contaminated sites reduces heavy metal and metalloid stress by phytoextraction or phytostabilization process.

Table 4.1 Soil chemical properties of samples collected from heavy metal contaminated sites

Samples	pH (1:5)	EC_{se} dS m^{-1}	Av.P_2O_5	As	Cd	Cu	Ni	Pb	Zn
			mg kg^{-1} of soil						
Site 1	7.0 ± 0.02^{a}	0.2 ± 0.02^{a}	1436 ± 198^{a}	17.9 ± 1.8^{a}	3.1 ± 0.2^{a}	122.9 ± 4.2^{a}	4.4 ± 0.2^{ab}	219.9 ± 4.1^{a}	67.3 ± 8.1^{a}
Site 2	6.2 ± 0.15^{b}	0.3 ± 0.06^{a}	943 ± 482^{a}	9.3 ± 4.2^{ab}	2.2 ± 0.2^{b}	118.5 ± 4.8^{a}	4.0 ± 0.0^{b}	179.1 ± 34.7^{a}	25.9 ± 5.3^{b}
Site 3	6.6 ± 0.05^{ab}	0.2 ± 0.04^{a}	967 ± 27^{a}	4.9 ± 0.1^{b}	1.1 ± 0.0^{c}	21.8 ± 2.8^{b}	4.6 ± 0.1^{a}	14.9 ± 2.1^{b}	17.3 ± 3.4^{b}

Values presented here are the mean of four replicas ± SE (standard error). Same letters in the column are not significantly different between sites at $P < 0.05$
Adopted from Krishnamoorthy et al. (2015)

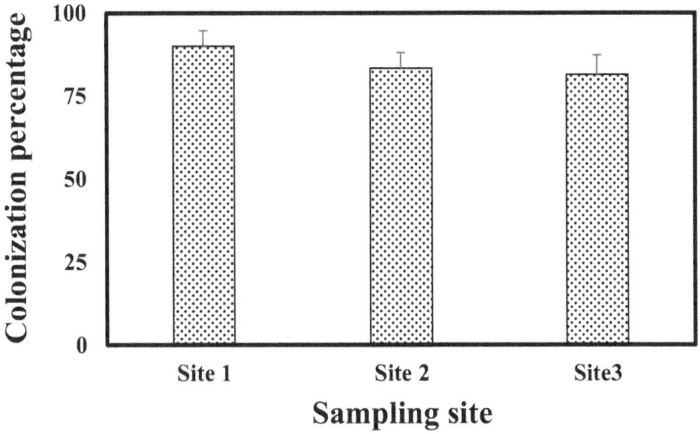

Fig. 4.1 Intensity of frequency of mycorrhizal colonization (F%) in total root system of the samples collected from heavy metal contaminated soils. Data are presented as mean ± SE from 4 replications; letters shows significant differences between sites according to t- test ($P < 0.05$). (From Krishnamoorthy 2015)

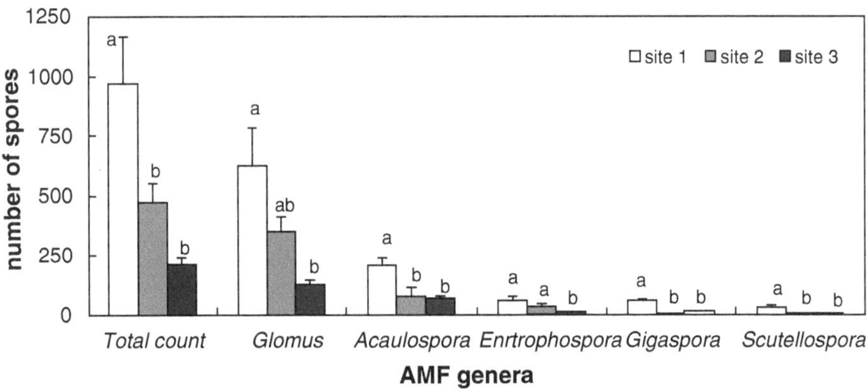

Fig. 4.2 Total number of spores (in 100 g soil) and genus level identification of AMF spores in heavy metal contaminated sites. Data are presented as mean ± SE from 4 replications; letters shows significant differences between sites according to t- test ($P < 0.05$). (From Krishnamoorthy 2015)

Spore density (SD) in heavy metal contaminated sites ranges from 201 to 970 per 100 g of soil, and it was significantly higher in site 1 compared to site 2 and site 3 (Fig. 4.2). According to the recent classification of AMF (Sieverding and Oehl 2006), out of 11 genera, 5 genera were found in all sampled areas. *Glomus* was the dominant genus in all three sites, followed by *Acaulospora, Entrophospora, Gigaspora* and *Scutellospora*. Spore density of all genera were found to be significantly higher in site 1 compared to site 3. Genus richness was higher in site 1 compared to site 2 and site 3. Frequency of occurrence for *Glomus* and *Acaulospora* genus was 100%, which means both the genus were found in all the samples.

Entrophospora, Gigaspora and *Scutellospora* genus frequency was 66.7, 86.7 and 60% respectively. Genus relative abundance was higher for *Glomus* (66%) followed by *Acaulospora* (23%) whereas other genera contributed for the remaining 11% (Fig. 4.2). *Glomus* genus RA was higher in site 2 followed by site 1 and site 3. In contrast RA for *Acaulospora* was found higher in site 3 followed by site 1 and site 2. High root colonization may lead to produce more extraradical hyphae (EH) in soil for efficient uptake of nutrient under stress conditions. Higher root colonization in high and moderately contaminated soils may lead to production of large amount of EH in soil. In this cause, EH might be exposed to heavy metal and metalloids stress and resorted to spore production as a survival strategy. Helgason and Fitter (2009) reported that rapid production of spores was a likely adaptation mechanism of AMF under stress condition. In agreement with this, our results also showed higher spore density in highly contaminated soil compared to less contaminated soils. *In vitro* study with *Glomus intraradices* revealed that EH and spore density in root compartment was increased as Pb, Zn and Cd concentrations increased (Pawlowska and Charvat 2004).

Soil heavy metal concentration exhibited positive correlation to mycorrhizal spore count and species richness. Soil arsenic concentration showed significantly positive correlation to total spore count ($R^2 = 1$; $P < 0.05$) (Fig. 4.3), *Glomus* ($R^2 = 0.98$), *Acaulospora* ($R^2 = 0.92$), *Entrophospora* ($R^2 = 0.93$), *Gigaspora* ($R^2 = 0.82$) and *Scutellospora* spore count ($R^2 = 0.81$). Similarly soil Zn concentration also showed significant positive correlation for AMF spore count and richness. Positive correlation of spore count with soil Zn concentration was observed by Zarei et al. (2008). Significant positive correlation for total, *Glomus* and *Entrophospora* spore count and species richness was observed for soil cadmium. Lead showed significant positive correlation to *Glomus* and *Entrophospora* spore count. Soil copper and nickel were not significantly improved species richness.

Diversity could vary due to environmental stress with succession of new/different AMF species colonization over decades but the number of AMF species may not change significantly. However, interestingly in our studied soils As ($R^2 = 0.99$; $P < 0.001$), Cd ($R^2 = 0.926$; $P = 0.09$) and Zn ($R^2 = 0.97$; $P < 0.05$) concentration significantly increased richness of AMF and other metals not influences much on species richness. Simpson's diversity index and evenness were not significantly reduced by soil heavy metals and metalloids. Similar to this, no significant reduction in AMF species was observed between high and less contaminated soil (Hassan et al. 2011). Other soil characters like organic matter and available phosphorus were exhibiting the negative correlation for AMF diversity. Plants may not prefer mycorrhizal symbiosis when available phosphorus is higher in the soil which may lead to the suppression of AMF species (Nogueira et al. 2007). Interestingly, *Glomus* genus richness exhibited positive correlation for soil As ($R^2 = 0.86$) and Zn ($R^2 = 0.96$) concentration. In line with this, *Glomus* genus dominance in heavy metal contaminated site was observed by many authors (Renker et al. 2005; Vallino et al. 2006; Zarei et al. 2010).

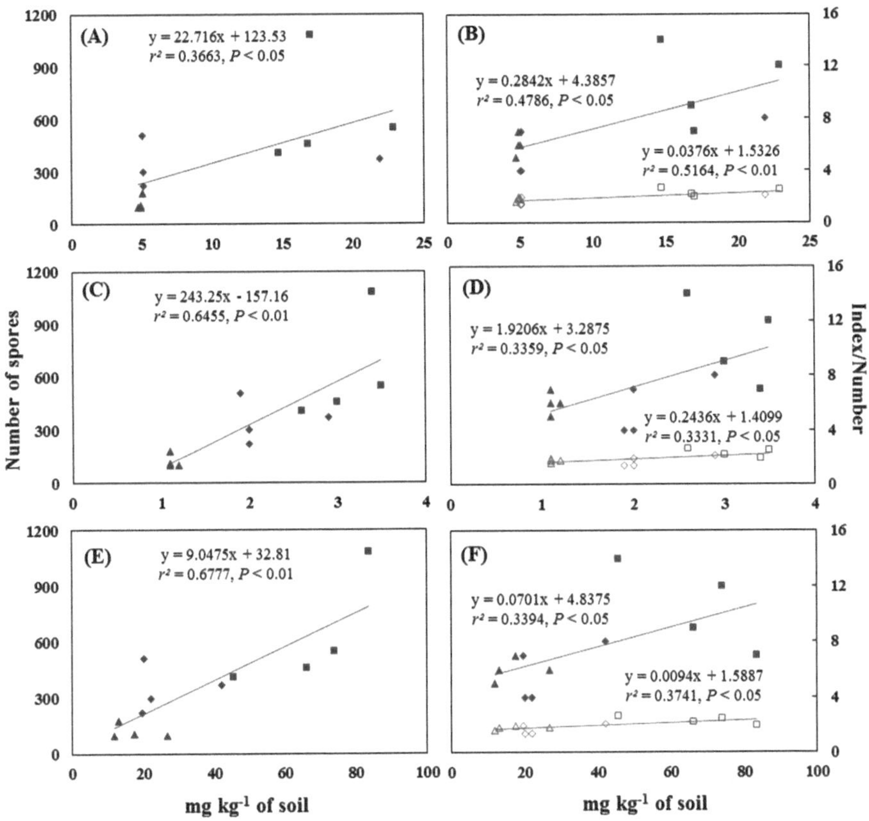

Fig. 4.3 Regression analysis of soil As (**a**, **b**), Cd (**c**, **d**) and Zn (**e**, **f**) against *Glomeraceae* spore count (**a**, **c**, **e**), richness and diversity (**b**, **d**, **f**). Square, diamond and triangle represents site 1, site 2 and site 3 respectively. In Fig. 4.3 **b**, **d** and **f** the open data series represents the diversity index and the closed ones are represents the richness. (adopted from Krishnamoorthy et al. 2015)

4.2.2 *Heavy Metal Stress on Plant*

Depending on the oxidation status of the heavy metals, it causes different damages to plant cells. At cellular and molecular level, it alters different plant physiological process like inactivation, denaturing of enzymes, proteins and DNA. It blocks the functional group in metabolically important molecules and substitute important metal ion from biomolecules and functional units, conformational changes and damaging cell membranes, which reduce the plant metabolism, photosynthesis, respiration and key enzymes in cellular process (Villiers et al. 2011). Further, it also disturbs redox homeostasis by forming free radicals and reactive oxygen species (ROS) (Hossain et al. 2012). Figure 4.4 illustrates the important negative impact of heavy metals on plant cells. Naturally, some plants have developed tolerance mechanism against heavy metals, which is governed by interrelated plant physiological

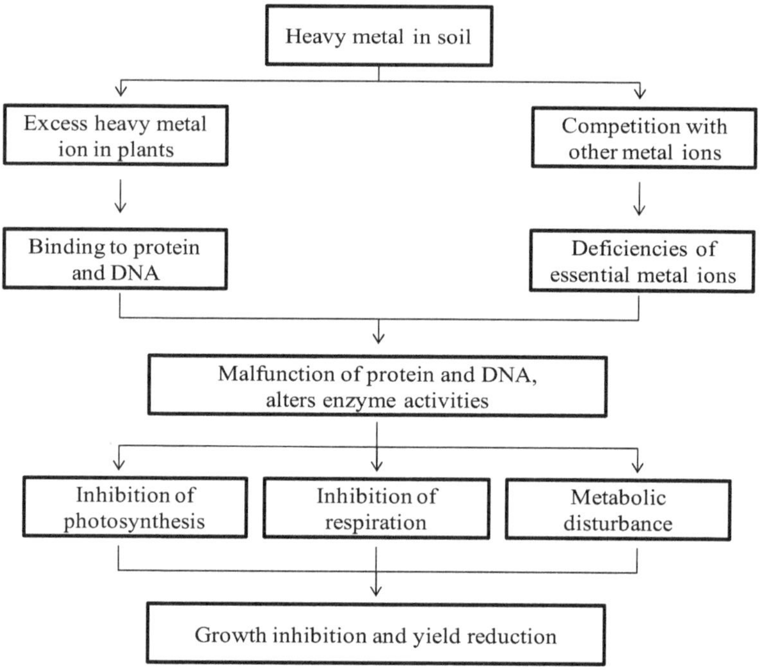

Fig. 4.4 Effects of heavy metals on plant growth. (modified from Hossain et al. 2012)

and molecular mechanisms. Some of the mechanisms are restricted uptake of HM inside the cell (Zhu et al. 2011), immobilization in cell walls, production of antioxidants (Srivastava et al. 2007), and sequestration in the cell and by phytochelation mechanisms (Zhang et al. 2005).

These heavy metal and metalloids cannot be chemically degraded in soil and it need to be physically removed or immobilized (Kroopnick 1994). Traditionally the heavy metal contaminated soils are excavated and disposed to the landfill sites. However, this kind of remediation disposed heavy metal from one place to another place and to the adjacent environments. Alternate to this, soil washing was followed to remove the contaminant, but this method is costly and need further treatments to remove contaminant. Use of physio-chemical methods in soil remediation affects the soil biological activities, which lead to the land not suitable for plant growth (Gaur and Adholeya 2004).

4.3 AMF Mediated Phytoremediation

AMF establish symbiotic association with plants grown in most ecosystem, including metal contaminated soils. By enhancing the uptake of phosphate, water and micronutrient, it improves plant growth and stress tolerance. Similarly, heavy metals are also taken up by the fungal hypha and transported to the plants. Hence, in some cases mycorrhizal plants can exhibit higher heavy metal uptake and root to shoot transport (phytoextraction) while in other cases AMF immobilize the metals in soil (phytostabilization). Mycorrhizal mediated phytoextraction and phytostabilization depends on the plant-fungus and heavy metal combination in contaminated sites.

Bacterial mediated detoxification of heavy metal in soil was recently illustrated by Krishnamoorthy et al. (2017). This chapter mainly focuses on the AMF mediated phytoremediation in metal contaminated sites. Heavy metal present in soil affects AMF spore germination and hyphal growth. However, spores isolated from heavy metal contaminated sites are more tolerant to elevated concentration of metal compared to the isolate from non-polluted soil (Gohre and Paszkowski 2006). Indigenous AMF handles heavy metal pollutants in two different ways. One is phytostabilization, in which the heavy metal is immobilized with in the soil. On the other hand, AMF enhance heavy metal uptake and transport to shoots, which is called as phytoextraction (Fig. 4.5).

Immobilization of heavy metal in soil is performed by seven different process. The first process involves chelating agent (Glomalin) secreted by AMF which precipitate heavy metal in polyphosphate granules in the soil. The second and third processes are binding of precipitated heavy metal to the cell wall of fungus and selectively prevent uptake in fungal plasma membrane. Specific and non-specific metal transporters will transport it to the cytosol (fourth) and chelates by metallothioneins produced by fungus (fifth). Sixth and seventh processes are exporting the transformed heavy metal to the soil again and arrest in vacuoles. In addition to that, it also transports heavy metal through fungal hypha and transport to the plant cells (phytoextraction). Recently, Gonzalez-Chávez et al. (2011) reported that, AMF hypha uptake arsenate (highly toxic form) and convert it into arsenite (less toxic form) in their cytosol, then it is secreted to the soil by arsenite efflux pump. Lower level of heavy metal was observed in the maize plant under mycorrhizal inoculation (Kaldorf et al. 1999), but controversially Tonin et al. (2001) reported that mycorrhizal inoculated clover roots showed higher accumulation of heavy metals. These phytostabilization and phytoextraction process mainly depends on the mycorrhizal fungi and plant combination.

4.3.1 *Phytostabilization*

Phytostabilization is an *in situ* method involving metal tolerant plant and soil amendment which can reduce the migration of metals to air, surface and groundwater and reduce the soil toxicity. AMF contribute to the immobilization of heavy metal in soil beyond the rhizosphere and thereby improve phytostabilization. AMF secrete specific compounds to precipitate heavy metal in polyphosphate granules in soil, adsorbs to fungal cell wall and chelate inside the fungus (Gaur and Adholeya 2004).

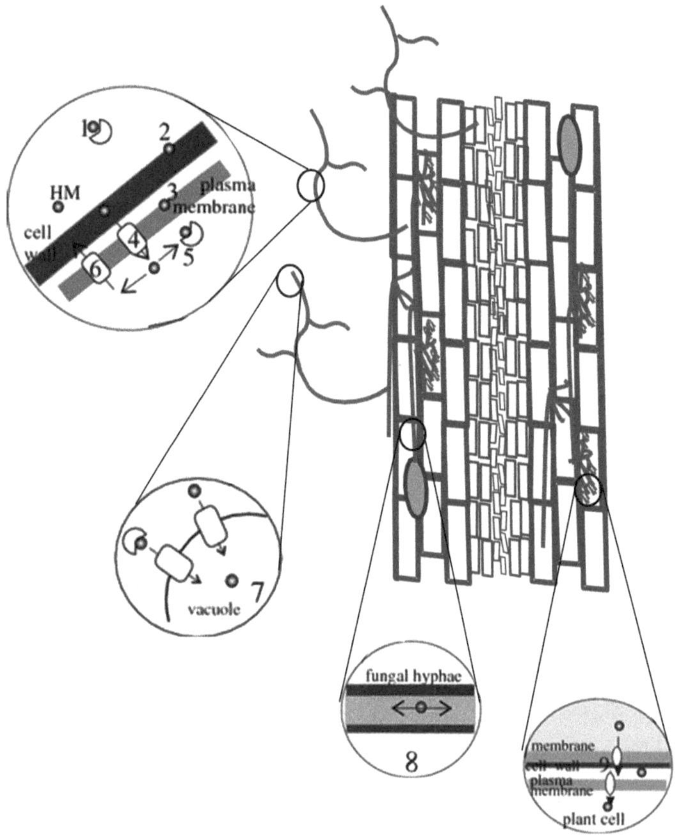

Fig. 4.5 Heavy metal and metalloid detoxification mechanism of AMF; 1- Chelating agents bind metals and metalloids in soil; 2- Bind the heavy metal to the cell wall of fungus; 3- Plasma membrane of the fungi act as a selective barrier to heavy metal; 4- Active and passive heavy metal transporters in the cell wall; 5- Chelates heavy metal in cytosol (Metallothioneins); 6- Export from cell to another cell via specific or nonspecific transporters; 7- Sequestration of heavy metal in vacuole; 8- Transport of heavy metal in the hypha of the fungus; 9- heavy metal are exporting from fungus to plant cell through arbuscules. (modified from Gohre and Paszkowski 2006)

Some AMF species reduce the metal uptake by reducing the metal bioavailability via precipitation, alkalinization and complexation processes. AMF exhibit the ability to absorb metals through substratum alkalinization activity by release of OH^-, therefore affecting the metal stability in soil (Budel et al. 2004). Glomalin secreted by the mycorrhizal hypha reduce metal mobility and sequesters in the glomalin protein called complexation of metal (Vodnik et al. 2008). Binding of heavy metal to the fungal cell wall chitin reduces the bioavailability of metal in soil. Study conducted by Joner et al. (2000) proved that AMF hypha absorb Cd upto 0.5 mg per mg dry biomass. Interestingly, the metal tolerant AMF can absorb 2–4 times higher heavy metals in hypha compared to the metal susceptible AMF. Similar to plant vacuoles, AMF vesicles may store the toxic compounds and thereby, could provide an additional detoxification mechanisms for metal tolerance.

Plants growing in metal contaminated sites uses excluder strategy by which the metals are either sequestered in the vacuoles of the root cortical cells or bound to the cell wall (Barceló and Poschenrieder 2003). It is proven that the symbiosis between AMF and plants improve nutrient and water up take from soil and protect the plant from metal toxicity (Audet and Charest 2007; Rajkumar et al. 2012), which may be crucial for the survival of plants in metal contaminated soil. Generally plants grown in soil with high metal concentration has higher mycorrhizal root colonization compared to the plants grown in soil with less metal concentration (Krishnamoorthy et al. 2015; Zhang et al. 2018a).

Recently *Funneliformis mosseae* and *Funneliformis caledonium* were used for enhancing sunflower growth in waste electrical and electronic equipment recycling site (Zhang et al. 2018a). Inoculation of mycorrhizal fungi reduced the accumulation of heavy metal such as Cd, Cu, Pb, Cr, Zn and Ni than that of control plants, by phytostabilization mechanism. AMF have been demonstrated to alleviate heavy metal stress in plant by retaining metals in roots and reducing the translocation to shoot (Deng et al. 2004; Janouskova et al. 2005). Zhipeng et al. (2016) reported that Pb concentration in pakchoi shoots was significantly reduced by the addition of *Funneliformis mosseae*, *Glomus versiforme* and *Rhizophagus intraradices*. Similarly, inoculation of these three AMF species significantly reduced the Cd accumulation in shoot than root. It is worth to mention that AMF inoculum would be a better option to enhance the yield and also reduce the metal toxic issues in contaminated soils.

Accumulation of heavy metals in roots of sunflower was found higher than that of shoots that indicates the reduction in translocation of metal from root to shoot by AMF (Zhang et al. 2018a). Interestingly, AMF inoculation reduced the accumulation of As in root as well as in shoots of the plant grown in As contaminated sites, due to the dilution effect for improving P mobilization from soil (Chen et al. 2007). These dilution effect may be another reason for reduced metal accumulation in plants. AMF induces the production of antioxidant enzymes such as peroxidase and catalase. Peroxidase protect the plant against free radical formed during metal stress conditions (Latef 2013) and inoculation of AMF was found to increase the expression of genes related to synthesis of antioxidant enzymes (Ferrol et al. 2016). These studies proven that AMF enhances plant stress tolerance by reducing reactive oxygen species produced during stress condition. AMF inoculation was found to increase catalase activity in sunflower (Zhang et al. 2018a) grown in electronic equipment recycling site and pigeon pea grown in Cd and Pb contaminated soils (Garg and Aggarwal 2011).

4.3.1.1 Glomalin Related Soil Protein

Glomalin is a glycoprotein produced by AMF hypha, which is first identified in 1996 during a work to produce monoclonal antibodies for AMF. The antibody reacts with substances on AMF hypha and named as glomalin after it was first identified in the order Glomales (Wright et al. 1996). The name Glomalin related soil protein (GRSP) was used instead of Glomalin due to the soil extraction procedure and Bradford assay measures all proteinaceous materials in the extract (Rillig 2004). Three types of glomalin have been identified based on solubility characters (i) easily extractable glomalin related soil protein (EE GRSP), (ii) Total Glomalin related soil protein (T GRSP) (iii) A "scum" at the air water interface that occurs during harvesting of hypha from pot cultured AMF (Nichols 2003). The role of GRSP in the ecosystem is still not clear, however the hypothetical roles has evolved from that of an active secretion to enhance soil aggregation, to a hydrophobin that modifies water acquisition (Rillig 2005) and GRSP specifically related to fungal metabolism (Gillespie et al. 2011). It also correlated to soil aggregate stability, long term C and nitrogen storage and responds to land use changes (Rillig et al. 2001; Wilson et al. 2009).

4.3.1.2 GRSP in Heavy Metal Remediation

Earlier studies showed that the GRSP produced by AMF plays an important role in carbon sequestration (Zhang et al. 2015). In addition, GRSP has the ability to sequester the metals such as Pb, Cd, Zn, and Mn and reduce the availability of metals to the plants and reduce the metal toxicity in plants (Cornejo et al. 2008; Vodnik et al. 2008) (Fig. 4.6). Recently Wu et al. (2014) reported that GRSP produced by arbuscular mycorrhizal fungi sequestered 0.21–1.78% of lead and 0.38–0.98% of cadmium in a field experiment. One gram of glomalin was found to sequester 4.3 mg Cu, 0.08 mg Cd and 1.12 mg Pb in metal polluted soil (Gonzalez-Chavez et al. 2004). In addition, *Gigaspora rosea* was found to sequester 28 mg Cu per gram glomalin.

Heavy metal sequestration ability of GRSP is affected by the heavy metal concentration, amount of GRSP and soil properties. Further, high GRSP content could facilitate more binding site for metals. Experiment done by Jia et al. (2016) proved the positive correlation between total GRSP in soil with Cd and Pb sequestration.

4.3.2 *Phytoextraction*

Plants can be used to remove the metal pollutants from soil and concentrating them in the shoot and the metal accumulated biomass can be harvested using standard agricultural methods and smelted to recover the metal (Pawlowska et al. 2000).

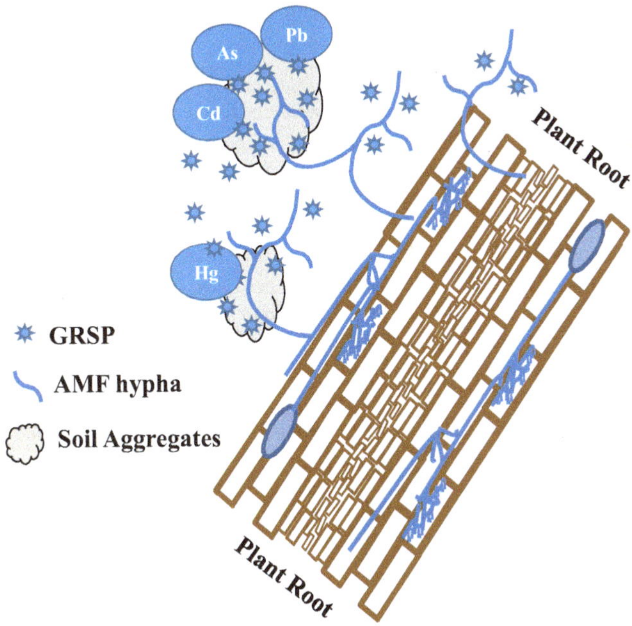

Fig. 4.6 Sequestration of heavy metal by glomalin related soil protein secreted by AMF

Phytoextraction is a most effective and attractive strategy to clean up the metal contaminated sites. This process highly depends on the plant which has high root to shoot metal transfer ability (hyperaccumulating plants). For achieving a desired reduction in soil metal concentration, it is required to select and cultivate hyperaccumulating plants for several cycle that includes harvest and removal of metal-enriched biomass. Hyperaccumulation of heavy metals in plants has been recorded in many plants (Table 4.2). This mechanism of metal accumulation is depends on the plant species, soil conditions and type and state of heavy metal present in the soil.

Phytoextraction of heavy metals from contaminated site is important in terms of protection of the environment and reduction of the heavy metal toxicity to plant and animals. These hyper accumulating plants naturally take up the heavy metals from soil and transport them to the shoot without exhibiting toxicity symptoms (Zhang et al. 2018b). Selection of hyper accumulating plant is very important in any successful phytoremediation process. Among the plants, grasses are the excellent candidate, because of their fibrous root system which can provide large surface area for root-metal contact (Kulakow et al. 2000). Use of indigenous plant species often favoured because it requires less management and adapt to the native climatic conditions (Sarma 2011). Hyperaccumulating plants should have high growth rate and biomass production, tolerance to the metals known to exist in the site and high translocation capacity for successful phytoextraction of metals from soil. The hyperaccumulating plants should have following criteria (i) The concentration of metals should be higher in shoot than roots; (ii) the concentration of

Table 4.2 List of heavy metal hyperaccumulating plants

Metal	Plants
Pb	*Brassica juncea*, water hyacinth (*Eichhornia crassipes*), Hydrilla (*Hydrilla verticillata*), sunflower (*Helianthus annuus*), *Lemna minor*, *Salvinia molesta*, *Spirodela polyrhiza*
Cd	Alpine pennycrest (*Thlaspi cacrulescens*), *Cardaminosis helleri*, Eel grass (*Vallisneria spiralis*), water hyssop (*Bacopa monnieri*), water hyacinth (*Eichhornia crassipes*), Hydrilla (*Hydrilla verticillata*), duck weed (*Lemna minor*), Giant duckweed (*Spirodela palyrhiza*)
As	Chinese brake fern (*Pteris vittata*), Fern (*Pteris cretica*)
Cr	Duck weed (*Lemna minor*), *Ceratophyllum demersum*, Giant reed (*Arundo donax*), cattail (*Typha angustifolia*), Alfo albo (*Medicago sativa*), water hyssop (*Bacopa monnieri*), *Pista stratiotes*, water fern (*Salvinia molesta*), *Spirodela polyrhiza*
Cu	*Aeolanthus bioformifollus*, *Lemna minor*, *Vigna radiate*, creosote bush (*Larrea tridentate*), water hyssop (*Bacopa monnieri*), Indian mustard (*Brassica juncea*)
Ni	*Phyllanthus serpentines*, *Lemna minor*, *Salvinia molesta*, *Brassica juncea*, *Spirodela polyrhiza*
Hg	*Lemna minor*, water lettuce (*Pistia stratiotes*), water hyacinth (*Eichhornia crassipes*), Hydrilla (*Hydrilla verticillata*)

Adopted from Malik and Biswas (2012)

metals in the shoot should be 50–100 times greater than the normal plants (iii) should have fast growth and high biomass (iv) should be easy to cultivate and fully harvestable (v) should have tolerance to the heavy metal present in the soil (Szczygłowska et al. 2011). Over 500 plant species belonging to 101 families have been reported with capabilities for phytoextraction which includes the members of the *Asteraceae*, *Brassicaceae*, *Caryophyllaceae*, *Cyperaceae*, *Cunouniaceae*, *Fabaceae*, *Flacourtiaceae*, *Lamiaceae*, *Poaceae*, *Violaceae* and *Euphobiaceae* family (Sarma 2011).

Earlier studies of phytoextraction have focused on the non mycorrhizal metallophytes families such as *Brassicaceae* or *Caryophyllaceae* (Szczygłowska et al. 2011). However, after considering the contribution of AMF in phytoextraction many studies have been done in mycorrhizal plants in recent years. Inoculation of AMF enhances the uptake and root to shoot translocation of As in *Pteris vittata* (Leung et al. 2006). Experimental evidence showed that mycorrhizal plants accumulated 88.1 g of As and non-mycorrhizal plant accumulated only 60.4 mg per kg of plant biomass grown in contaminated soils (Leung et al. 2006). Turnau and Mesjasz-Przybylowicz (2003) reported that *Berkheya coddii* plant accumulated 30% more Ni when establishing the symbiotic association with mycorrhiza.

Soil amendment of chelating agents enhances the bioavailability of heavy metals and improve the phytoextraction process. Use of ethylene diamine tetraacetic acid (EDTA) and ethylene diamine dissuccinate (EDDS) enhance the bioavailability of Pb, Zn, and Cd and increased the plant shoot accumulation of heavy metals (Grcman et al. 2003).

4.4 Conclusion

It is highly important to study the native mycorrhizal population to know the impact of metals on AMF before initiating the phytoremediation process. In addition, understanding of metal tolerance mechanism in AMF is essential, because native mycorrhizal fungi shows high metal tolerance and may exhibit better performance in metal remediation than non-native mycorrhizal fungi. Even though mycorrhizal fungi is reported to enhance the phytoremediation process, plant, AMF and metal interaction decide the phytostabilization and phytoextraction process. Studies on the tripartite interaction among plant species, AMF species and heavy metals may enhance our in depth knowledge on how plant and AMF symbiosis decide to remove the heavy metal (phytoextraction) and sequester the heavy metal in the soil (phytostabilization).

References

Allen, M. F., Swenson, W., Querejeta, J. I., Egerton-Warburton, L. M., & Treseder, K. K. (2003). Ecology of mycorrhizae: A conceptual framework for complex interactions among plants and fungi. *Annual Review of Phytopathology, 41*, 271–303.

Audet, P., & Charest, C. (2007). Dynamics of arbuscular mycorrhizal symbiosis in heavy metal phytoremediation: Meta-analytical and conceptual perspectives. *Environmental Pollution, 147*, 609–614.

Barceló, J., & Poschenrieder, C. (2003). *Phytoremediation: principles and perspectives.* Contributions to Science, 2. Institut d'Estudis Catalans, Barcelona, pp. 333–344.

Budel, B., Weber, B., Kuhl, M., Pfanz, H., Sultemeyer, D., & Wessels, D. (2004). Reshaping of sandstone surfaces by cryptoendolithic cyanobacteria: Bioalkalization causes chemical weathering in arid landscapes. *Geobiology, 2*, 261–268.

Chen, B., Xiao, X., Zhu, Y. G., Smith, F. A., Xie, Z. M., & Smith, S. E. (2007). The arbuscular mycorrhizal fungus *Glomus mosseae* gives contradictory effects on phosphorus and arsenic acquisition by *Medicago sativa* Linn. *Science of the Total Environment, 379*, 226–234.

Cornejo, P., Meier, S., Borie, G., Rilig, M. C., & Borie, F. (2008). Glomalin-related soil protein in a Mediterranean ecosystem affected by a copper smelter and its contribution to cu and Zn sequestration. *Science of the Total Environment, 406*, 154–160.

Deng, H., Ye, Z. H., & Wong, M. H. (2004). Accumulation of lead, zinc, copper and cadmium by 12 wetland plant species thriving in metal-contaminated sites in China. *Environmental Pollution, 132*, 29–40.

Ferrol, N., Tamayo, E., & Vargas, P. (2016). The heavy metal paradox in arbuscular mycorrhizas: From mechanisms to biotechnological applications. *Journal of Experimental Botany, 67*, 6253–6265.

Frank, A. B. (1885). Uber di auf werzelsymbiose beruhende Ernahrung gewisser Baume durch unterirdischeplize. *Berichte der Deutschen Botanischen Gesellschaft, 3*, 128–145.

Garg, N., & Aggarwal, N. (2011). Effect of mycorrhizal inoculations on heavy metal uptake and stress alleviation of *Cajanus cajan* (L.) Mill sp. genotypes grown in cadmium and lead contaminated soils. *Plant Growth Regulation, 66*, 9–26.

Gaur, A., & Adholeya, A. (2004). Prospects of arbuscular mycorrhizal fungi in phytoremediation of heavy metal contaminated soils. *Current Science, 86*, 528–534.

Gillespie, A. W., Farrell, R. E., Walley, F. L., Ross, A. R., Leinweber, P., Eckhardt, K.-U., Regierd, T. Z., & Blyth, R. R. (2011). Glomalin-related soil protein contains non-mycorrhizal-related heat-stable proteins, lipids and humic materials. *Soil Biology and Biochemistry, 43*, 766–777.
Gohre, V., & Paszkowski, U. (2006). Contribution of the arbuscular mycorrhizal symbiosis to heavy metal phytoremediation. *Planta, 223*, 1115–1122.
Gonzalez-Chavez, M. C., Carrillo-Gonzalez, R., Wright, S. F., & Nichols, K. A. (2004). The role of glomalin, a protein produced by arbuscular mycorrhizal fungi, in sequestering potentially toxic elements. *Environmental Pollution, 130*, 317–323.
Gonzalez-Chavez, M. C., Carrillo-Gonzalez, R., & Gutierrez- Castorena, M. C. (2009). Natural attenuation in a slag heap contaminated with cadmium: The role of plants and arbuscular mycorrhizal fungi. *Journal of Hazardous Materials, 161*, 1288–1298.
Gonzalez-Chávez, M. C., Ortega-Larrocea, M. P., Carrillo-González, R., López-Meyer, M., Xoconostle-Cázares, B., Gomez, S. K., Harrison, M. J., Figueroa-López, A. M., & Maldonado-Mendoza, I. E. (2011). Arsenate induces the expression of fungal genes involved in as transport in arbuscular mycorrhiza. *Fungal Biology, 115*, 1197–1209.
Grcman, H., Vodnik, D., Velikonja-Bolta, S., & Lestan, D. (2003). Ethylenediaminedissuccinate as a new chelate for environmentally safe enhanced lead phytoextraction. *Journal of Environmental Quality, 32*, 500–506.
Hassan, S. E. D., Boon, E., St-Arnaud, M., & Hijri, M. (2011). Molecular biodiversity of arbuscular mycorrhizal fungi in trace metal polluted soils. *Molecular Ecology, 20*, 3469–3483.
Helgason, T., & Fitter, A. H. (2009). Natural selection and the evolutionary ecology of the arbuscular mycorrhizal fungi (phylum Glomeromycota). *Journal of Experimental Botany, 60*, 2465–2480.
Hossain, M. A., Piyatida, P., Teixeira Da Silva, J. A., & Fujita, M. (2012). Molecular mechanism of heavy metal toxicity and tolerance in plants: Central role of glutathione in detoxification of reactive oxygen species and methylglyoxal and in heavy metal chelation. *Journal of Botany, 872875*, 1–37. https://doi.org/10.1155/2012/872875.
Janouskova, M., Pavlikova, D., Macek, T., & Vosatka, M. (2005). Arbuscular mycorrhiza decreases cadmium phytoextraction by transgenic tobacco with inserted metallothionein. *Plant and Soil, 272*, 29–40.
Jia, X., Zhao, Y., Liu, T., Huang, S., & Chang, Y. (2016). Elevated CO_2 increases glomalin-related soil protein (GRSP) in the rhizosphere of *Robinia pseudoacacia* L. seedlings in Pb- and Cd-contaminated soils. *Environmental Pollution, 218*, 349–357.
Joner, E. J., Briones, R., & Leyval, C. (2000). Metal-binding capacity of arbuscular mycorrhizal mycelium. *Plant and Soil, 226*, 227–234.
Kaldorf, M., Kuhn, A. J., Schroder, W. H., Hildebrandt, U., & Bothe, H. (1999). Selective element deposits in maize colonized by a heavy metal tolerance conferring arbuscular mycorrhizal fungus. *Journal of Plant Physiology, 154*, 718–728.
Krishnamoorthy, R. (2015). *Exploring the biodiversity of arbuscular mycorrhizal fungi and associated endobacteria to improve maize growth under salt stress condition*. Dissertation for the degree of doctor of philosophy, Chungbuk National University, South Korea.
Krishnamoorthy, R., Kim, C. G., Subramanian, P., Kim, K. Y., Selvakumar, G., & Sa, T. M. (2015). Arbuscular mycorrhizal fungi community structure, abundance and species richness changes in soil by different levels of heavy metal and metalloid concentration. *PLoS One*, 1–15. https://doi.org/10.1371/journal.pone.0128784.
Krishnamoorthy, R., Venkateswaran, V., Senthilkumar, M., Anandham, R., Selvakumar, G., Kim, K. Y., Kang, Y. Y., & Sa, T. M. (2017). Potential microbiological approaches for the remediation of heavy metal-contaminated soils. In D. Singh, H. Singh, & R. Prabha (Eds.), *Plant-microbe interactions in agro-ecological perspectives* (pp. 341–366). Singapore: Springer.
Kroopnick, P. M. (1994). Vapor abatement cost analysis methodology for calculating life cycle costs for hydrocarbon vapor extracted during soil venting. In D. L. Wise & D. J. Trantolo (Eds.), *Remediation of hazardous waste* (pp. 779–790). New York: Marcel Dekker.

Kulakow, P. A., Schwab, A. P., & Banks, M. K. (2000). Screening plant species for growth on weathered, petroleum hydrocarbon-contaminated sediments. *International Journal of Phytoremediation, 2*, 297–317.

Latef, A. A. A. (2013). Growth and some physiological activities of pepper (*Capsicum annuum* L.) in response to cadmium stress and mycorrhizal symbiosis. *Journal of Agricultural Science and Technology, 15*, 1437–1448.

Leung, H. M., Ye, Z. H., & Wong, M. H. (2006). Interactions of mycorrhizal fungi with *Pteris vittata* (as hyperaccumulator) in as-contaminated soils. *Environmental Pollution, 139*, 1–8.

Malik, N., & Biswas, A. K. (2012). Role of higher plants in remediation of metal contaminated sites. *Scientific Reviews and Chemical Communications, 2*, 141–146.

Martin, F., Perotto, S., & Bonfante, P. (2007). Mycorrhizal fungi: A fungal community at the interface between soil and roots. In R. Pinton, Z. Varanini, & P. Nannipieri (Eds.), *The rhizosphere: Biochemistry and organic substances at the soil-plant interface* (pp. 201–236). New York: Marcel Dekker.

Nichols, K. (2003). *Characterization of glomalin, a glycoprotein produced by arbuscular mycorrhizal fungi*. PhD Dissertation, University of Maryland, College Park, Maryland.

Nogueira, M. A., Nehls, U., Hampp, R., Poralla, K., & Cardoso, E. J. B. N. (2007). Mycorrhiza and soil bacteria influence extractable iron and manganese in soil and uptake by soybean. *Plant and Soil, 298*, 273–284.

Pawlowska, T. E., & Charvat, I. (2004). Heavy-metal stress and developmental patterns of arbuscular mycorrhizal fungi. *Applied and Environmental Microbiology, 70*, 6643–6649.

Pawlowska, T. E., Chaney, R. L., Chin, M., & Charvat, I. (2000). Effects of metal phytoextraction practices on the indigenous community of arbuscular mycorrhizal fungi at a metal-contaminated landfill. *Applied and Environmental Microbiology, 66*, 2526–2530.

Rajkumar, M., Sandhya, S., Prasad, M. N. V., & Freitas, H. (2012). Perspectives of plant associated microbes in heavy metal phytoremediation. *Biotechnology Advances, 30*, 1562–1574.

Rascio, N., & Navari-Izzo, F. (2011). Heavy metal hyperaccumulating plants: How and why do they do it? And what makes them so interesting? *Plant Science, 180*, 169–181.

Renker, C., Blanke, V., & Buscot, F. (2005). Diversity of arbuscular mycorrhizal fungi in grassland spontaneously developed on area polluted by a fertilizer plant. *Environmental Pollution, 135*, 255–266.

Rillig, M. C. (2004). Arbuscular mycorrhizae, glomalin and soil quality. *Canadian Journal of Soil Science, 84*, 355–363.

Rillig, M. C. (2005). A connection between fungal hydrophobins and soil water repellency? *Pedobiologia, 49*, 395–399.

Rillig, M. C., & Mummey, D. L. (2006). Mycorrhizas and soil structure. *New Phytologist, 171*, 41–53.

Rillig, M. C., Wright, S. F., Nichols, K. A., Schmidt, W. F., & Torn, M. S. (2001). Large contribution of arbuscular mycorrhizal fungi to soil carbon pools in tropical forest soils. *Plant and Soil, 233*, 167–177.

Sanders, I., & Croll, D. (2010). Arbuscular mycorrhiza: The challenge to understand the genetics of the fungal partner. *Annual Review of Genetics, 44*, 271–292.

Sarma, H. (2011). Metal Hyperaccumulation in plants: A review focusing on phytoremediation technology. *Journal of Environmental Science and Technology, 4*, 118–138.

Sieverding, E., & Oehl, F. (2006). Revision of Entrophospora and description of Kuklospora and Intraspora, two new genera in the arbuscular mycorrhizal Glomeromycetes. *Journal of Applied Botany and Food Quality, 80*, 69–81.

Singh, H. (2006). Mycorrhizal fungi in rhizosphere bioremediation. In H. Singh (Ed.), *Mycoremediation: Fungal bioremediation* (pp. 533–572). New York: Wiley.

Smith, S. E., & Read, D. J. (2008). *Mycorrhizal symbiosis*. Cambridge, London: Academic.

Srivastava, S., Mishra, S., Tripathi, R. D., Dwivedi, S., Trivedi, P. K., & Tandon, P. K. (2007). Phytochelatins and antioxidant systems respond differentially during arsenite and arsenate stress in *Hydrilla verticillata* (L.f.) Royle. *Environmental Science & Technology, 41*, 2930–2936.

Szczygłowska, M., Piekarska, A., Konieczka, P., & Namieśnik, J. (2011). Use of Brassica plants in the phytoremediation and biofumigation processes. *International Journal of Molecular Sciences, 12*, 7760–7771.

Tonin, C., Vandenkoornhuyse, P., Joner, E. J., Straczek, J., & Leyval, C. (2001). Assessment of arbuscular mycorrhizal fungi diversity in the rhizosphere of *Viola calaminaria* and effect of these fungi on heavy metal uptake by clover. *Mycorrhiza, 10*, 161–168.

Turnau, K., & Mesjasz-Przybylowicz, J. (2003). Arbuscular mycorrhiza of *Berkheya coddii* and other Ni-hyperaccumulating members of Asteraceae from ultramafic soils in South Africa. *Mycorrhiza, 13*, 185–190.

Vallino, M., Massa, N., Lumini, E., Bianciotto, V., Berta, G., & Bonfante, P. (2006). Assessment of arbuscular mycorrhizal fungal diversity in roots of *Solidago gigantea* growing in a polluted soil in northern Italy. *Environmental Microbiology, 8*, 971–983.

Villiers, F., Ducruix, C., Hugouvieux, V., Ezan, N. J. E., Garin, J., Junot, C., & Bourguignon, J. (2011). Investigating the plant response to cadmium exposure by proteomic and metabolomic approaches. *Proteomics, 11*, 1650–1663.

Vodnik, D., Grcman, H., Macek, I., Van, J., Elteren, J. T., & Kovacevic, M. (2008). The contribution of glomalin-related soil protein to Pb and Zn sequestration in polluted soil. *Science of the Total Environment, 392*, 130–136.

Weissenhorn, I., Leyval, C., & Berthelin, J. (1995). Bioavailability of heavy metals and abundance of arbuscular mycorrhiza in a soil polluted by atmospheric deposition from a smelter. *Biology and Fertility of Soils, 19*, 22–28.

Wilson, G. W. T., Rice, C. W., Rillig, M. C., Springer, A., & Hartnett, D. C. (2009). Soil aggregation and carbon sequestration are tightly correlated with the abundance of arbuscular mycorrhizal fungi: Results from long-term field experiments. *Ecology Letters, 12*, 452–461.

Wright, S. F., Franke-Snyder, M., Morton, J. B., & Upadhyaya, A. (1996). Time-course study and partial characterization of a protein on hyphae of arbuscular mycorrhizal fungi during active colonization of roots. *Plant and Soil, 181*, 193–203.

Wu, Z. P., Mcgrouther, K., Huang, J. D., Wu, P. B., Wu, W. D., & Wang, H. L. (2014). Decomposition and the contribution of glomalin-related soil protein (GRSP) in heavy metal sequestration: Field experiment. *Soil Biology and Biochemistry, 68*, 283–290.

Zarei, M., Konig, S., Hempel, S., Nekouei, M. K., Savaghebi, G., & Buscot, F. (2008). Community structure of arbuscular mycorrhizal fungi associated to *Veronica rechingeri* at the Anguran zinc and lead mining region. *Environmental Pollution, 156*, 1277–1283.

Zarei, M., Hempel, S., Wubet, T., Schäfer, T., Savaghebi, G., Jouzani, G. S., Nekouei, M. K., & Buscot, F. (2010). Molecular diversity of arbuscular mycorrhizal fungi in relation to soil chemical properties and heavy metal contamination. *Environmental Pollution, 158*, 2757–2765.

Zenk, M. H. (1996). Heavy metal detoxification in higher plants: A review. *Gene, 179*, 21–30.

Zhang, H., Xu, W., Guo, J., He, Z., & Ma, M. (2005). Coordinated responses of phytochelatins and metallothioneins to heavy metals in garlic seedlings. *Plant Science, 169*, 1059–1065.

Zhang, J., Tang, X. L., He, X. H., & Liu, J. X. (2015). Glomalin-related soil protein responses to elevated CO_2 and nitrogen addition in a subtropical forest: Potential consequences for soil carbon accumulation. *Soil Biology and Biochemistry, 83*, 142–149.

Zhang, Y., Hu, J., Bai, J., Wang, J., Yin, R., Wang, J., & Lin, X. (2018a). Arbuscular mycorrhizal fungi alleviate the heavy metal toxicity on sunflower (*Helianthus annuus* L.) plants cultivated on a heavily contaminated field soil at a WEEE-recycling site. *Science of the Total Environment, 628–629*, 282–290.

Zhang, J., Martinoia, E., & Lee, Y. (2018b). Vacuolar transporters for cadmium and arsenic in plants and their applications in phytoremediation and crop development. *Plant and Cell Physiology, pcy006*. https://doi.org/10.1093/pcp/pcy006.

Zhipeng, W., Weidong, W., Shenglu, Z., & Shaohua, W. (2016). Mycorrhizal inoculation affects Pb and Cd accumulation and translocation in Pakchoi (*Brassica chinensis* L.). *Pedosphere, 26*, 13–26.

Zhu, X. F., Zheng, C., Hu, Y. T., Jiang, A., Liu, Y., Dong, N. Y., Yang, J. L., & Zheng, S. J. (2011). Cadmium-induced oxalate secretion from root apex is associated with cadmium exclusion and resistance in *Lycopersicon esulentum*. *Plant, Cell & Environment, 34*, 1055–1064.

Chapter 5
Phytoremediation: An Alternative Tool Towards Clean and Green Environment

Sandhya Misra and Krishna G. Misra

Abstract Wetlands being the most productive and ecologically sensitive and adaptive ecosystems are constantly being challenged with anthropogenic pressures due to their wide variety of services they provide to mankind. The vast expansions of human population and associated activities have put a tremendous amount of pressure on these naturally occurring resources. Uncontrolled discharge of effluents in water from various sources resulted into altered nature of the associated ecosystems giving rise to several health issues and problems. Hence, realising the urgent need of protecting these ecologically fragile ecosystems several adaptive measures have been taken. In this connection, it is found that the available conventional methods are not feasible on various grounds like their cost, their by-products, time frame, etc. Therefore, the use of plants emerged as the alternative and promising tool for safe and sustainable ecosystem supporting life.

Keywords Phytoremediation · Wetlands · Hyperaccumulator plants · Phytostabilization · Rhizodegradation · Rhizofiltration · Phytodegradation · Phytovolatilization · Xenobiotics

5.1 Introduction

Wetlands being the most productive, ecologically sensitive and adaptive ecosystems provides ample amount of services to human society (Russi et al. 2013). Rapid rise of human population and related activities have put a tremendous pressure on these natural occurring ecosystems The controlled and uncontrolled activities like disposal of waste, accidental and process spillage, mining and smelting of metalliferous ores, sewage sludge application to agricultural soils, soils

S. Misra (✉)
Indian Institute of Technology Gandhinagar, Palaj, Gujarat, India

Birbal Sahni Institute of Palaeosciences, Lucknow, India

K. G. Misra
Birbal Sahni Institute of Palaeosciences, Lucknow, India

S. Shah et al. (eds.), *Sustainable Green Technologies for Environmental Management*, https://doi.org/10.1007/978-981-13-2772-8_5

contamination from industrial, military and agricultural activities either due to ignorance, lack of vision, or carelessness are responsible for the migration of contaminants into non-contaminated sites as dust or leachate and contribute towards contamination of our ecosystem (Kumar et al. 2013). This build-up of toxic pollutants (metals, radionuclides and organic contaminants in soil, surface water and ground water) not only resulted into altered nature but also caused a major strain on ecosystems. Hence, the deteriorating condition of soil and ground water has become a major concern across the globe.

Thus, there is an urgent need to look for the long term sustainable and remedial measures. In this connection some conventional practices, such as 'pump-and-treat' and 'dig-and-dump' techniques are available. Other methods for remediation include excavation, soil washing, thermal treatment, electro-reclamation, chemical, and other biological techniques. However, all of these have their own limitations in terms of their potentiality and coverage area and tend to be more expensive. Additionally, these approaches to remediation often make the soil infertile and unsuitable for agriculture and other uses by destroying the microenvironment. Hence, there is a need to find an alternative, taking into account the probable end use of the site once it has been remediated. In this regard, green plants seem to be the alternative tool, where the natural potential of plants to take up and absorb mineral and toxins from the environment can be used as remedial measure. Hence, the use of green plants to treat soil and ground water pollution popularly known as phyto-remediation emerged as potential and environmentally sound technology for pollution prevention, control and remediation (Rai 2008; Mander and Mitsch 2009; Vangronsveld et al. 2009).

The term phytoremediation ("phyto" means plants & the latin suffix "remedium" means to clean or restore) is a process in which, green plants are used to remove, transfer, stabilize, and/or destroy pollutants or contaminants like metals, pesticides, explosives, and oil, etc., taking advantage of their natural abilities to take up, accumulate, and/or degrade constituents of their soil and water environments (Pilon-Smits and Freeman 2006). However, good and effective results are found at places/environment where contaminant levels were low because high concentrations put limits on plant growth and take too long to clean up. Thus, the plants not only put a check on contaminants but also help to prevent wind, rain, and groundwater flow from carrying contaminants away from the site to surrounding areas or deeper underground. Choice of plants for this technology largely depends on the state of polluted/contaminated ecosystem, whether it is soil or water. Water is an important component of living systems and contamination of water especially due to anthropogenic activities (human induced) viz., discharge of municipal sewage and industrial activities, synthetic fertilizers, herbicides, insecticides and plant residues released from agricultural activities deteriorated water quality in both urban and rural areas respectively. This has raised a concern among society worldwide. Hence, in the present chapter, an attempt has been made to present brief review of various aspects related to phytoremedation in an aquatic ecosystem like wetlands.

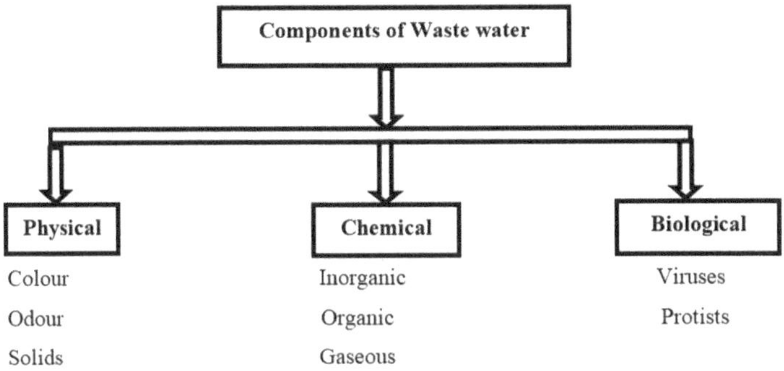

Fig. 5.1 Various components of waste water

5.2 Types of Contaminants in an Aquatic Ecosystems

Depending on the sources of pollutant coming from Industrial, municipal and domestic waste water, contaminants are categorised into organic and inorganic components (Fig. 5.1).

5.2.1 Physical Contaminants

Physical contaminants coming from different sources mainly add sludge and create anaerobic conditions, which result into variable colour, odour and suspended waste in a polluted aquatic ecosystem.

5.2.2 Chemical Contaminants

On the basis of their nature, components in this category can further be classified into organic and inorganic. Industrial effluents consist of inorganic contaminants such as heavy metals, ammonia, nitrate, nitrite, sulphate and cyanide, while oil, grease, refractory compounds, organochlorides and nitro compounds are the major organic contaminants (Nwoko 2010). Heavy metals, chlorides, sulphate and nitrates are inorganic contaminants commonly reported in groundwater and majority of wastewaters, while pesticides, pharmaceuticals, solvents, food additives, surfactants and petroleum products are the major organic contaminants (Kolpin et al. 2002; Lin et al. 2008; Stuart et al. 2011).

5.2.2.1 Organic Contaminants

Numerous measures like Total Organic Carbon (TOC), Biochemical Oxygen Demand (BOD) and Chemical oxygen demand (COD) are used to determine the organic content and hence determines the quality of water. Presence of persistent organic pollutants (POPs) popularly called *xenobiotics* are severely affecting the aforementioned parameters and hence are gaining concern from public health and safety point of view. Xenobiotics like pesticides, polychlorinated dibenzodioxins and dibenzofurans (often referred to as *dioxins*) as well as polychlorinated biphenyls (PCBs), aldrin, toxaphene, DDT, chlordane, dieldrin, endrin, HCB, heptachlor and mirex are extremely stable and persistent, toxic to humans and other living biota and have biomagnified along trophics webs and transported over long distance (Carvalho 2006). These xenobiotics are very resilient to biotic and abiotic degradation and cause detectable harmful effects even at relatively low concentrate ions (Larsen 2006). Apart from all these, other contaminants include pesticides, detergents, antimicrobials, food additives, solvents and sterols. Various such organic contaminants reported in waste water are well cited in the work carried out by Petrović et al. (2003), Bolong et al. (2009). In their work they have found that the there are many other contaminants like Endocrine-disrupting chemicals (EDS), pharmaceutical products, personal care products, surfactants, and various industrial additives whose presence is affecting public health and safety. Most of these organic contaminants (including chlorinated solvents, pesticides and hydrocarbons) are known carcinogens and neurotoxins. They cause damage to the central nervous system, irritation of respiratory and gastrointestinal systems and immunological, reproductive and endocrine disorders in children.

5.2.2.2 Inorganic Components

In polluted water, inorganic contaminants are found in the form of metals, ions/ nutrients and radionuclide. Heavy metals like cadmium (Cd), chromium (Cr), copper (Cu), mercury (Hg), lead (Pb), cobalt (Co), iron (Fe), nickel (Ni), manganese (Mn), zinc (Zn) and arsenic (As) are highly toxic and even their adverse effect on ecology and human life are well reported and it was found that accumulation of heavy metals in living organisms is not only toxic but also causes various diseases and disorders (Chaney 1988; Wani et al. 2012). Excessive nitrogen and phosphorus loading in wastewater is a major threat to water quality and leads to increased rates of eutrophication which has been identified as a major environmental threat in both freshwater and marine waters all over the globe. Hence the product of the decay of nitrogenous organic wastes in the form of ions /nutrients like ammonia, nitrite, nitrate, chloride, sulphate, phosphorus and cyanide (CN) are the major contaminants commonly found in groundwater and wastewater. Radioactive isotopes/ radionuclides like isotopes of radium (Ra), uranium (U) and radon (Rn) occurring naturally or produced due to anthropogenic activity like nuclear power generation (Manahan 1994) are reported in drinking water and found harmful to life.

5.2.3 Biological Contaminants

Among the biological contaminants, viruses, protists and other pathogens such as bacteria are frequently encountered in wastewater. Their excessive presence causes not only contamination in water bodies but also severely affect humans and aquatic biota.

5.3 Remedial Measures to Treat Polluted Aquatic System

5.3.1 Conventional Methods

For the treatment of organic and inorganic contaminants present in polluted waters, conventional practices, such as 'pump-and-treat' and 'dig-and-dump' techniques have been used since long. It was found that these conventional methods of treatment are often expensive and have their own limitations in terms of their potentiality and coverage area. Additionally, these conventional approaches to remediation often make the soil infertile and unsuitable for agriculture and other uses by destroying the microenvironment. Most of these technologies are based on physical and chemical methods that require input of chemicals, which makes the technology expensive. Moreover, they produce adverse impacts on aquatic ecosystems and human health. Over the time, all over the world considerable attention has been paid to select alternate methods/materials particularly biological methods for wastewater treatment, taking into account the probable end use of the site once it has been remediated.

In this regard, interest has been generated in the use of bio-sorbents for treating industrial, municipal and domestic wastewaters (Salt et al. 1995). Various biological agents including microbes (bacteria), algae, fungi, plants and agricultural residues possess potential for removing various contaminants from environment and treating wastewater generated from dairies, tanneries, sugar factories, pulp and paper industries, palm oil mills, distilleries, etc. Hence, phytoremediation is considered environmentally sound technology for pollution prevention, control and remediation.

5.3.2 Use of Green Plants (Local Flora)

In the era of bioremediation, vegetation/plants as a biological resource with immense capacity for removing variable contaminants from various components of ecosystem have been studied by several workers in diversified geographical parts of country and across the globe. Terrestrial as well as aquatic plant species show ability to remove/transform/degrade contaminants (Fig. 5.2). Crop plants, tree species, weeds and other wild plants with their natural ability to remove or degrade selected

Terrestrial plants (**a**) *Thlaspi*, (**b**) Pteris, (**c**) *Helianthus* and (**d**) *Populus*

Free-floating aquatic plant species: (**a**) *Lemna*, (**b**) *Eichhornia*, (**c**) *Azolla* and (**d**) *Salvinia*

Submerged aquatic plant (**a**) *Ceratophyllum*, (**b**) *Potamogeton*, (**c**) *Myriophyllum* & (**d**) *Hydrilla*

Emergent aquatic plant species: (**a**) *Elodea*, (**b**) *Scirpus*, (**c**) *Typha* and (**d**) *Phragmites*

Fig. 5.2 Various plant species suitable for phytoremediation

contaminants present in soil, sludge, sediment, groundwater, surface water and wastewater by utilizing their metabolic and hydraulic processes have been demonstrated and thereby found improving the environment. Thus, the plants not only put a check on contaminants but also help to prevent wind, rain, and groundwater flow from carrying contaminants away from the site to surrounding areas or deeper underground.

Hence the plant root systems together with the translocation, bioaccumulation and contaminant degradation abilities aid the technique. However, good and effective results depend on several factors like:

(a) ***Plant–***The physical characteristics of plant and its genetic adaptations and biological processes including metabolic and absorption capabilities should be taken into consideration while choosing a plant for phytoremediation. Higher biomass production and species with higher adaptability to climatic and soil conditions are a necessary requirement for good phytoremediation capacity. The seeds or plants should be from, or adapted to, the climate of the phytoremediation site. Therefore, the selection of plant variety is critical to ensure superior and effective remediation. For metal remediation, identification and selection of suitable hyper-accumulator plant species is required (Schnoor et al. 1995).

(b) ***Physical properties***–Physicochemical parameters are yet another important factor taken into consideration as pH, organic matter, redox potential, contaminant concentration and the mineral content of the soil and water affect the removal/degradation of the contaminant.

(c) ***Root zone***–Degradation of contaminants in the soil is facilitated by plant enzymes and root exudates. Hence, root length and root diameter plays an important role in determining the contaminant uptake and degradation capacities of a plant.

(d) ***Chelating agent***–To fasten up the heavy metal-uptake capacity of the plant, chelating agents such as EDTA and micronutrients increase the bio-availability of contaminants especially heavy metals and thus speed up the remediation process.

(e) ***Plant biomass***–Plant species with higher biomass production have potential to remove contaminant and thus it is a necessary requirement for successful phytoremediation.

5.3.3 Methods Involved in Phytoremedation

Through phytoremediation technology both organic and inorganic contaminants can be treated to successful range depending on the uptake mechanisms of plant and the contaminant targeted (Barceló and Poschenrieder 2003). Treatment of organic contaminants mainly involves the process of phytostabilization, rhizodegradation, rhizofiltration, phytodegradation and phytovolatilization mechanisms, while in case of inorganic contaminants processes like phytostabilization, rhizofiltration, phytoaccumulation and phytovolatilization give good results (Table 5.1 and Fig. 5.3).

Table 5.1 List of plants species and the processes they use for the contaminants removal

Processes	Contaminants	Media	Plants used	Remarks
Phytoextraction	Inorganic components like radionuclides and metals (Cr, Cu, Ni, Zn, Cd, Ag)	Soil, sediment, sludge	*Alyssum*, *Brassica*, *Thapsi*, sunflower	Process of uptake of contaminants into plant roots by accumulation or harvestable shoots
Rhizofiltration	Inorganic pollutants like metals and radionuclides(^{137}Cs, ^{230}Pb, ^{238}U)	Ground water, surface water, waste water	*Eichhornia*, *Lemna*	Process involves removal of contaminants through plant roots
Phytovolatilization	Organic/inorganic contaminants (Se, Hg, As)	Soil and sediments	*Poplar*, *Phragmites*, *scirpus*	Done by volatilization of contaminants by leaves
Phytostabilization	Metals (As, Cd, Cr, Cu, Pb, Zn)	Soil, sediment, sludge	*Sunflower*, *Chenopodium*	Process involves stabilization of contaminants by binding/ complexation
Rhizo/ Phytodegradation	Organic contaminants like aromatic, aliphatic and petroleum hydrocarbons, chlorinated solvents, TNT pesticides	Soil sediment sludge	*Grasses*, *alfaalfa*, hybrid poplar, *Brassica*, *Typha*, *Jatropha*, *Cassia*	Microbial degradation in the rhizosphere stimulated by plants

5.3.3.1 Phytoextraction

It is also referred to as phytoaccumulation, phytoabsorption, phytosequestration, phytomining or biomining. It involves uptake/absorption and translocation of contaminants/ hyper-accumulated metal contaminants and/or excess nutrients in harvestable root and shoot tissue. This process is applicable for metals (Ag, Cd, Co, Cr, Cu, Hg, Mn, Mo, Ni, Pb, Zn), metalloids (As, Se), radionuclides (90 Sr, 137 Cs, 234 U, 238 U), non-metals (B, Mg) and organic contaminants present in soils, sediments and sludge (Brooks 1998). There exist a variety of plants called hyper accumulator that can accumulate high quantities of metals than required for plant growth ranging from 10 to 100-fold (Erakhrumen 2007; Wani et al. 2012). The minimum concentration of metal a plant needs to contain to be termed a hyper accumulator varies for each metal (Reeves and Brooks 1983; Baker and Brooks 1989). Maximum number of hyper accumulators (Table 5.2) has been reported from families Brassicaceae, Lamiaceae, Scrophulariaceae, Cyperaceae, Poaceae, Apocynaceae, Euphorbiaceae, Flacourtiaceae, Fabaceae and Violaceae.

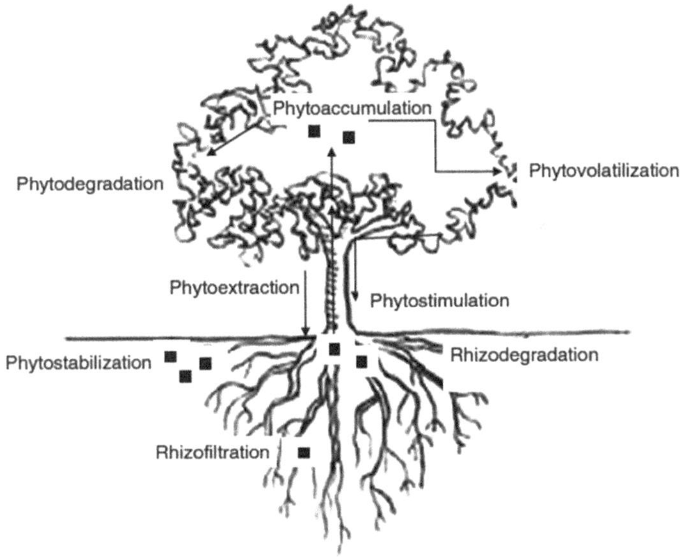

Fig. 5.3 Phytoremediation process. (adopted from Dhir 2013)

Table 5.2 List of hyperaccumulator plant species and the element they accumulate

Hyperaccumulator species	Element
Pteris vittata	As
Eichhornia crassipes, *Thlaspi caerulescens*	Cd
Alyssum sp.	Co
Elodea nuttallii	Cu
Alyssum sp., *Phytolacca acinosa*	Mn
Berkheya coddii, *Alyssum bertolonii*	Ni
Spirodela polyrhiza, *Dicoma niccolifera*, *Sutera fodina*	Cr
Thlaspi rotundifolium, *Minuartia verna*, *Brassica* sp., *Sesbania drummondii*	Pb
Astragalus bisulcatus, *Stanleya pinnata*	Se
Thlaspi caerulescens, *Eichhornia crassipes*	Zn

5.3.3.2 Phytostabilization

Plants are used to immobilize the contaminants in the soil and groundwater through absorption and accumulation in plant tissues, adsorption onto roots or precipitation within the root zone preventing their migration in soil. The plant root exudates stabilize, demobilize and bind the contaminants in the soil matrix, thereby reducing their bioavailability. This process is suitable for organic contaminants and metals present in soils, sediments and sludges.

5.3.3.3 Rhizofiltration

This process involves the removal of contaminants in surface water by plant roots by adsorption or precipitation onto plant roots or absorption followed by sequestration in the roots. Metals like Pb, Cd, Cu, Fe, Ni, Mn, Zn, Cr, excess nutrients and radionuclide (90 Sr, 137 Cs, 238 U, 236 U) present in groundwater, surface water and wastewater can easily be removed through this process (Dushenkov et al. 1995, 1997a, b). It is generally applicable for treating large volumes of water with low contaminant concentrations (ppb). It can be conducted *in situ* or *ex situ* to remediate contaminated surface water bodies. Efficiency of this process largely depends on the parameters such as pH, flow rate and contaminant concentration. In this method, proper disposal of the contaminated plant biomass could be a limitation.

5.3.3.4 Phytovolatilization

In this technique, growing trees and other plants take up water along with the contaminants, pass them through their leaves and volatilize into the atmosphere at comparatively low concentrations. This process is successfully used for removing metal contaminants present in groundwater, soils, sediments and sludge medium.

5.3.3.5 Rhizodegradation

This process involves the breakdown of contaminants in the soil through microbial activity localized in the root zone. Here microorganisms consume and digest organic substances for nutrition and energy. This method gives promising results for the treatment of a wide range of organic contaminants such as petroleum hydrocarbons, PAHs, pesticides, chlorinated solvents, PCP, poly chlorinated biphenyls (PCBs), benzene, toluene, o-xylene and surfactants can be removed by this technique (Donnelly et al. 1994; Gilbert and Crowley 1997).

5.3.3.6 Phytodegradation

In this process, enzymes produced and released by the plant are used to metabolize and degrade contaminants in the soil, sediments, sludges, groundwater or surface water. Nitroreductase enzyme present in *Myriophyllum aquaticum* degrades TNT concentrations (Schnoor et al. 1995). Hybrid poplar trees metabolized TNT to 4-amino-2, 6-dinitrotoluene (4-ADNT), 2-amino-4,6-dinitrotoluene (2-ADNT) and other unidentified compounds in laboratory hydroponic and soil experiments (Thompson et al. 1998). Thus, it is a successful method to remove organic compounds such as munitions (trinitrotoluene), chlorinated solvents, herbicides, insecticides and inorganic nutrients (Burken and Schnoor 1997; Thompson et al. 1998; Campos et al. 2008).

5.4 Current State of Phytoremediation Technology

A review of available literature shows that there has been a great progress in using this green technology for restoring and removing contaminants from several sites in diversified geographical parts of world. In this regard, a list of biological entities and terrestrial and aquatic plant species that have been exploited for phytoremediation/bioremedation is compiled (Table 5.3).

A review of work done in this field of phytoremediation by workers across the world document the advantage of this technology over the available conventional ones. But there are studies (Henry 2000; Prasad 2004) that also documented the disadvantage associated with various processes involved in phytoremediation (Table 5.4).

These phytoremediation methods proved to be very promising; from cost point of view they are 4–1000 times cheaper (Sadowsky 1999) and even from ecological point of view they are not that hazardous. Apart from having advantages over other technologies phytoremediation has its own limitations which cannot be neglected and need attention. These technologies have their limitation to a particular site, and the processes involved in degradation of chemical alters the nature of soil that could affect food chain. Further, there are chances of air pollution while burning the leaves and twigs of plants containing dangerous chemicals. Phytoremediation has been studied extensively in research and small-scale demonstrations, but full-scale applications are currently limited to a small number of projects (Cunningham and Ow 1996).

5.5 Conclusion

Wetlands are ecologically fragile ecosystems providing innumerable ecosystem services. However, anthropogenic pressure limit the functioning of these ecologically sensitive and adaptive ecosystems. Rampant discharge of effluents into water bodies from various sources resulted in imbalances in functioning of ecosytems giving rise to several health issues and ecological degradation. In these pressing circumstances, phytoremediation has emerged as an alternative and promising technology for safe and sustainable ecosystem supporting life. Phytoremediation is a process in which, green plants are used to remove, transfer, stabilize, and/or destroy pollutants or contaminants like pesticides, metals, etc., by tapping their natural abilities to take up, accumulate, and/or degrade constituents of their soil and water environments. Through processes like phytostabilization, phytodegradation, phyto-accumulation, phytovolatilization, rhizodegradation, rhizofiltration, organic and inorganic contaminants can be successfully treated. However, phytoremediation has its own set of drawbacks which warrants attention, and extensive study. In near future, the focus of the research should be in the direction of designing a safe and clean method to dispose the waste from this technology. Further, for wider use, we need to look into

Table 5.3 List of plants/biological entities used for bioremediation/phytoremediation

Category	Organism/Taxa	Targeted pollutant/application	References
Bacteria	*Flavobacterium* spp.	Organophosphates	Dhir (2013)
	Cunniughamella elegans and *Candida tropicalis*	PCBs (polychlorinated biphenyls) and PAHs (polycyclic aromatic hydrocarbons)	Dhir (2013)
	Alcaligenes spp. and *Pseudomonas* spp.	PCBs, halogenated hydrocarbons, alkylbenzene sulphonates, organophosphates, benzene, anthracene, phenolic Compounds, 2,4-D, DDT and 2,4,5-trichlorophenoxyacetic Acid etc.	Dhir (2013)
	Actinomycetes	Raw rubber	Dhir (2013)
	Nocardia tartaricans	Chemical detergents (ethylbenzene)	Dhir (2013)
	Arthrobacter and *Bacillus*	Endrin	Dhir (2013)
	Closteridium	Lindane	Dhir (2013)
	Trichoderma and *Pseudomonas*	Malathion	Dhir (2013)
Fungi	*Phanerochaete chrysosporium*	Halocarbons such as lindane and pentachlorophenol	Dhir (2013)
	P. sordida and *Trametes hirsuta*	DDT, DDE, PCBs, 4,5,6-trichlorophenol, 2,4,6-trichlorophenol, dichlorphenol and chlordane	Dhir (2013)
	Zylerion xylestrix	Pesticides/herbicides (aldrin, dieldrin, parathion and malathion)	Dhir (2013)
	Yeast (*Saccharomyces*)	DDT	Dhir (2013)
	Mucor	Dieldrin	Dhir (2013)
Algae	*Selanastrum capricornatum* Cyanobacteria(blue-green algae) *Microcystis aeruginosa*	Benzene, toluene, naphthalene, pyrene, Acrylonitrile, Dibenzanthraceae, Fluoroanthene, petroleum hydrocarbons	Dhir (2013)
	Chlamydomona sp. *Chlorella* sp. *Chlorococcum* sp. *Cylindrotheca* sp. *Dunaliella* sp. *Euglena gracilis* *Scenedesmus obliquus* *Selenastrum capricornutum*	DDT, parathion, phenol, benzene, Toluene, chlorobenzene, 1,2-dichlorobenzene, nitrobenzene Naphthalene, 2,6-dinitrotoluene, Phenanthrene, di-nbutylphthalate, Toxaphene, methoxychlor	Dhir (2013)

(continued)

Table 5.3 (continued)

Category	Organism/Taxa	Targeted pollutant/application	References
Terrestrial plant	Alfalfa, *Populus deltoids, Brassica juncea, Helianthus annuus, Thlaspi* sp., *T. caerulescens* & *T. rotundifolium*	Water clean up	Dhir (2013)
	Lolium perenne, Festuca	Rhizodegradation and phytostabilization	Dhir (2013)
	Teris vittata, Sutera fodina, Alyssum and *Thlaspi rotundifolium*	Remove heavy metals such as as, Cr, Ni, Zn and cd	Dhir (2013)
	Zea mays, Setaria faberi, Solanum melongena, Spinacia oleracea, Raphanus sativus, Ocimum basilicum and *Oryza sativa*	Transform and bioaccumulate Herbicides and pesticides like DDT and endosulfan	Dhir (2013)
	Hordeum vulgare and *Conyza canadensis*	Sequestration of herbicide metolachlor and glyphosate in vacuoles	Dhir (2013)
	Vetiveria zizanioides	Soils contaminated with polyaromatic hydrocarbons	Dhir (2013)
	Pteris vittata	Extract arsenic (As) in the above-ground part.	Dhir (2013)
	Populus × *Canadensis (Hybrid)*	Soil and groundwater contamination with petroleum-related organics,	Dhir (2013)
	Brassica juncea	Treat the radionuclide strontium (Sr89/90)-contaminated soil	Dhir (2013)
	Helianthus annuus	Treat lead-contaminated soil	Dhir (2013)
Submerged aquatic macrophytes (with heavy metal accumulation potential)	*Ceratophyllum submersum*	Ni	Kara (2010)
	Ceratophyllum demersum L.	Arsenate, Pb Cr	Xue et al. (2012), Saygideger et al. (2004), Osmolovskaya and Kurilenko (2005)

(continued)

Table 5.3 (continued)

Category	Organism/Taxa	Targeted pollutant/application	References
	Myriophyllum spicatum	Co, Ni, Cu, Zn	Wang et al. (1996), Sivaci et al. (2004), Lesage et al. (2008)
	Potamogeton pectinatus	Cd, Pb, Cu, Zn, Mn	Rai et al. (2003), Tripathi et al. (2003), Singh et al. (2005)
	Hydrilla verticillata	Cu, As	Srivastava et al. (2010, 2011)
	Eichornia crassipes, Ipomea aquatica, Wolffia arrhiza, Trapa bispinosa, Pistia stratiotes, Colocasia esculenta, Cyperus rotundus, Scirpus sp. *Sagittaria montevidensis, Cynodon dactylon*	Ca Cr, Cu, Pb, Zn, Mn and Fe	Mazumdar and Das (2014), Chatterjee et al. (2011)
Free-floating macrophytes	*Azolla pinnata*	Hg, Cd	Rai (2008)
Emergent species (with heavy metal accumulation potential)	*Typha latifolia*	As, Zn, Cu, Si	Ye et al. (1997), Qian et al. (1999), Nguyen et al. (2009), Manios et al. (2003), Afrous et al. (2011)
	Typha angustifolia	Pb	Panich-pat et al. (2005)
	Elodea densa	Hg	Molisani et al. (2006)
	Phragmites australis	As, Hg	Windham et al. (2001, 2003), Afrous et al. (2011)
	Scirpus maritimus	As, Hg	Afrous et al. (2011)
	Spartina alterniflora	As, Hg, Cu, Pb Al, Fe, Zn, Cr, Se	Carbonell et al. (1998), Ansede et al. (1999), Windham et al. (2001, 2003)
	Spartina patens	Cd, As	Zayed et al. (2000), Carbonell et al. (1998)

(continued)

Table 5.3 (continued)

Category	Organism/Taxa	Targeted pollutant/application	References
Aquatic plant species (with potential for removing explosives)	*Myriophyllum aquaticum*	TNT, RDX, HMX	Best et al. (1997, 1999a, b), Hughes et al. (1997), Rivera et al. (1998), Pavlostathis et al. (1998), Bhadra et al. (1999, 2001)
	Myriophyllum spicatum	TNT	Hughes et al. (1997)
	Potamogeton nodosus	TNT, RDX	Best et al. (1997, 1999b), Bhadra et al. (1999)
	Ceratophyllum demersum	TNT, RDX	Best et al. (1997, 1999b), Bhadra et al. (2001)
	Elodea canadensis	RDX, HMX	Rivera et al. (1998), Best et al. (1999a, b)
	Phalaris arundinacea	TNT, RDX	Best et al. (1999a, b)
	Typha angustifolia	TNT, RDX	Best et al. (1999a, b)
	Sagittaria latifolia	TNT, RDX	Best et al. (1997), Bhadra et al. (2001)
	Scirpus cyperinus	TNT, RDX	Best et al. (1997,1999a,b)
Aquatic plant species with the potential for accumulating radionuclides	*Lemna minor*	140 La, 99 Tc, 60 Co	Hattink et al. (2000), Hattink and Wolterbeek (2001), Weltje et al. (2002), Popa et al. (2006)
	Lemna gibba	60 Co, 32 P, 134 Cs	El-Shinawy and Abdel-Malik (1980)
	Azolla caroliniana	137 Cs, 60 Co	Popa et al. (2004)
	Ceratophyllum demersum	137 Cs, 60 Co, 32 P, 60 Co, 134 Cs, 89 Sr	El-Shinawy and Abdel-Malik (1980), Abdelmalik et al. (1980), Shokod'Ko et al. (1992), Bolsunovskiˇ et al. (2002)

(continued)

Table 5.3 (continued)

Category	Organism/Taxa	Targeted pollutant/application	References
	Potamogeton pectinatus	238 U, 137 Cs, 90 Sr	Kondo et al. (2003)
	Potamogeton lucens	90 Sr	Bolsunovskiǐ et al. (2002)
	Elodea canadensis	137 Cs, 90 Sr, 241 Am	Shokod'Ko et al. (1992), Bolsunovskiǐ et al. (2002), Bolsunovsky et al. (2005)
Aquatic plant species (with the potential removing/ accumulating various organic contaminants)	*Eichhornia crassipes*	Ethion, dicofol, cyhalothrin, pentachlorophenol	Roy and Hanninen (1994), Xia et al. (2002a, b)
	Lemna gibba	Phenol, 2,4,5-trichlorophenol (TCP)	Ensley et al. (1994), Sharma et al. (1997), Hafez et al. (1998), Tront and Saunders (2006)
	Lemna minor	2,4,5-trichlorophenol (TCP), halogenated phenols	Tront et al. (2001), Day and Saunders (2004), Tront and Saunders (2006), Tront et al. (2007)
	Spirodela oligorrhiza	Organophosphorus and organochlorine compounds (o,p′-DDT, p,p′-DDT), chlorobenzenes	Gobas et al. (1991), Rice et al. (1997), Gao et al. (2000a, b)
	Myriophyllum aquaticum	Simazine, o,p-2 DDT, p,p-2 DDT, hexachloroethane (HCA), perchlorate	Knuteson et al. (2002), Nzengung et al. (1999), Gao et al. (2000a)
	Potamogeton crispus	Phenol	Barber et al. (1995)
	Ceratophyllum demersum	Organophosphorus and organochlorine compounds, chlorobenzenes	Gobas et al. (1991), Wolf et al. (1991), Rice et al. (1997), Gao et al. (2000a, b)

(continued)

Table 5.3 (continued)

Category	Organism/Taxa	Targeted pollutant/application	References
	Elodea canadensis	Phenanthracene, organophosphorus and organochlorine compounds, chlorobenzenes Hexachloroethane (HCA), DDT, carbon tetrachloride	Gobas et al. (1991), Wolf et al. (1991), Rice et al. (1997), Machate et al. (1997), Gao et al. (2000a, b), Nzengung et al. (1999), Gao et al. (2000a), Garrison et al. (2000)
	Pontederia cordata	Oryzalin (herbicide)	Fernandez et al. (1999)
	Scirpus lacustris	Phenanthracene	Machate et al. (1997)

Table 5.4 Advantages and disadvantages of various methods used in phytoremediation techniques. (From Singh et al. 2011)

Method	Advantages	Disadvantages
Phytoextraction	It is a cost effective method	Time taking process as the metal hyper-accumulators are usually slow and secondly they grow with small biomass and shallow root systems
	Permanents removal of the contaminants	This process needs the harvestation and removal of plant biomass followed by metal reclamation or proper disposal of the biomass
Rhizofiltration	Both terrestrial and aquatic plants can be used onsite and offsite	There is a constant need to maintain pH
	Do not require translocation of contaminants to the shoots	Requires efforts and time in maintaining the initial stage of plant in a nursery or green house
Phytosatbilization	There is no requirement of disposal of hazardous biomass	Contaminant/pollutant stay back in the soil
	Useful to control soil erosion and reduce water availability in the system	Continuous monitoring is required
Phytovolatilization	It can transform the contaminants to less toxic forms	There is a risk of accumulation of hazardous elements in vegetation
	Effective and rapid degradation of the released contaminants	Plant tissues reported to have low level of metabolites

much detail right from the stage of plant selection to remediation of sites and their sustainability.

Acknowledgements Authors are grateful to the directors of respective institutes for extending the facilities and encouragement for the compilation of this paper. Google website is duly acknowledged for the pictures and other information. Misra S. acknowledge SERB- DST sponsored DST Fast Track Project YSS/2015/001193 for the necessary logistic supports available under this project.

References

Abdelmalik, W. E. Y., El-Shinawy, R. M. K., Ishak, M. M., & Mahmoud, K. A. (1980). Uptake of radionuclides by some aquatic macrophytes of Ismailia Canal, Egypt. *Hydrobiology, 69*, 3.

Afrous, A., Manshouri, M., Liaghat, A., Pazira, E., & Sedghi, H. (2011). Mercury and arsenic accumulation by three species of aquatic plants in Dezful, Iran. *African Journal of Agricultural Research, 6*(24), 5391–5397.

Ansede, J., Pellechia, P., & Yoch, D. (1999). Selenium biotransformation by the salt marsh cordgrass *Spartina alterniflora*: Evidence for dimethylselenoniopropionate formation. *Environmental Science & Technology, 33*, 2064–2069. https://doi.org/10.1021/es9812296.

Baker, A. J. M., & Brooks, R. R. (1989). Terrestrial higher plants which hyperaccumulate heavy elements: A review of their distribution, ecology and phytochemistry. *Biorecovery, 1*, 81–126.

Barber, J. T., Sharma, H. A., & Ensley, H. E. (1995). Detoxifi cation of phenol by the aquatic angiosperm, *Lemna gibba*. *Chemosphere, 31*, 3567.

Barceló, J., & Poschenrieder, C. (2003). Phytoremediation: Principles and perspectives. *Contributions to Science, 2*(3), 333–344.

Best, E. P. H., Zappi, M. E., Fredrickson, H. L., Sprecher, S. L., Larson, S. L., & Ochman, M. (1997). Screening of aquatic and wetland plant species for phytoremediation of explosives-contaminated groundwater from the Iowa army ammunition plant. *Annals of the New York Academy of Sciences, 829*, 179.

Best, E. P., Sprecher, S. L., Larson, S. L., Fredrickson, H. L., & Bader, D. F. (1999a). Environmental behavior of explosives in groundwater from the Milan army ammunition plant in aquatic and wetland plant treatments. Removal, mass balances and fate in groundwater of TNT and RDX. *Chemosphere, 38*(14), 3383–3396.

Best, E. P. H., Sprecher, S. L., Larson, S. L., Fredrickson, H. L., & Bader, D. F. (1999b). Environmental behavior of explosives in groundwater from the Milan army ammunition plant in aquatic and wetland plant treatments. Uptake and fate of TNT and RDX in plants. *Chemosphere, 39*, 2057.

Bhadra, R., Spanggord, R. J., Wayment, D. G., Hughes, J. B., & Shanks, J. V. (1999). Characterization of oxidation products of TNT metabolism in aquatic phytoremediation systems of *Myriophyllum aquaticum*. *Environmental Science & Technology, 33*, 3354.

Bhadra, R., Wayment, D. G., Williams, R. K., Barman, S. N., Stone, M. B., Hughes, J. B., & Shanks, J. V. (2001). Studies on plant-mediated fate of the explosives RDX and HMX. *Chemosphere, 44*, 1259.

Bolong, N., Ismail, A., Salim, M., & Matsuura, T. (2009). A review of the effects of emerging contaminants in wastewater and options for their removal. *Desalination, 239*, 229–246. https://doi.org/10.1016/j.desal.2008.03.020.

Bolsunovskiˇ, A. I., Ermakov, A. I., Burger, M., Degermendzhi, A. G., & Sobolev, A. I. (2002). Accumulation of industrial radionuclides by the Yenisei River aquatic plants in the area affected by the activity of the mining and chemical plant. *Radiatsionnaia Biologiia Radioecologiia, 42*, 194.

Bolsunovsky, A., Zotina, T., & Bondareva, L. (2005). Accumulation and release of 241Am by a macrophyte of the Yenisei River (Elodea canadensis). *Journal of Environmental Radioactivity, 81*, 33.

Brooks, R. R. (1998). Geobotany and hyperaccumulators. In R. R. Brooks (Ed.), *Plants that hyperaccumulate heavy metals* (pp. 55–94). Wallingford: CAB International.

Burken, J. G., & Schnoor, J. L. (1997). Uptake and metabolism of atrazine by poplar trees. *Environmental Science & Technology, 31*, 1399–1406.

Campos, M., Merino, I., Casado, R., Pacios, L. F., & Gómez, L. (2008). Review. Phytoremediation of organic pollutants. *Spanish Journal of Agricultural Research, 6*(Special issue), 38–47.

Carbonell, A. A., Aarabi, M. A., Delaune, R. D., Gambrell, R. P., & Patrick, W. H., Jr. (1998). Arsenic in wetland vegetation: Availability, phytotoxicity, uptake and effects on plant growth and nutrition. *Science of the Total Environment, 217*, 189.

Carvalho, F. P. (2006). Agriculture, pesticides, food security and food safety. *Environmental Science & Policy, 9*, 685–692.

Chaney, R. L. (1988). Metal speciation and interactions among elements affect trace element transfer in agricultural and environmental food-chains. In J. R. Kramer & H. E. Allen (Eds.), *Metal speciation: Theory, analysis and applications* (pp. 219–260). Chelsea: Lewis Publishers.

Chatterjee, S., Chetia, M., Singh, L., Chattopadhyay, B., Datta, S., & Mukhopadhyay, S. K. (2011). A study on the phytoaccumulation of waste elements in wetland plants of a Ramsar site in India. *Environmental Monitoring and Assessment, 178*, 361–371.

Cunningham, S. D., & Ow, D. W. (1996). Promises and prospects of phytoremediation. *Plant Physiology, 110*, 715–719.

Day, J. A., & Saunders, F. M. (2004). Glycoside formation from chlorophenols in *Lemna minor*. *Environmental Toxicology and Chemistry, 25*, 613.

Dhir, B. (2013). *Phytoremediation: Role of aquatic plants in environmental clean-up*. New Delhi: Springer.

Donnelly, P. K., Hedge, R. S., & Fletcher, J. S. (1994). Growth of PCB-degrading bacteria on compounds from photosynthetic plants. *Chemosphere, 128*, 984–988.

Dushenkov, V., Kumar, P. B. A. N., Motto, H., & Raskin, I. (1995). Rhizofiltration: The use of plants to remove heavy metals from aqueous streams. *Environmental Science & Technology, 29*, 1239–1245.

Dushenkov, S., Vasudev, D., Kapulnik, Y., Gleba, D., Fleisher, D., Ting, K. C., & Ensley, B. (1997a). Removal of uranium from water using terrestrial plants. *Environmental Science & Technology, 31*, 3468–3474.

Dushenkov, S., Vasudev, D., Kapulnik, Y., Gleba, D., Fleisher, D., Ting, K. C., & Ensley, B. (1997b). Phytoremediation: A novel approach to an old problem. In D. L. Wise (Ed.), *Global environmental biotechnology* (pp. 563–572). Amsterdam: Elsevier Science B.V.

El-Shinawy, R. M. K., & Abdel-Malik, W. E. Y. (1980). Retention of radionuclides by some aquatic fresh water plants. *Hydrobiology, 69*, 125.

Ensley, H. E., Barber, J. T., Polita, M. A., & Oliver, A. I. (1994). Toxicity and metabolism of 2, 4-dichlorophenol by aquatic angiosperm *Lemna gibba*. *Environmental Toxicology and Chemistry, 13*, 325.

Erakhrumen, A. (2007). Phytoremediation: An environmentally sound technology for pollution prevention, control and remediation in developing countries. *Educational Research Review, 2*(7), 151–156.

Fernandez, R. T., Whitwell, T., Riley, M. B., & Bernard, C. R. (1999). Evaluating semiaquatic herbaceous perennials for use in herbicide phytoremediation. *Journal of the American Society for Horticultural Science, 124*, 539.

Gao, J., Garrison, A. W., Mazur, C. S., Wolfe, N. L., & Hoehamer, C. F. (2000a). Uptake and phytotransformation of o, p′-DDT and p, p′-DDT by axenically cultivated aquatic plants. *Journal of Agricultural and Food Chemistry, 48*(12), 6121–6127.

Gao, J., Garrison, A. W., Hoehamen, C., Mazur, C. S., & Wolfe, N. L. (2000b). Uptake and phytotransformation of organophosphorus pesticide by axenically cultivated aquatic plants. *Journal of Agricultural and Food Chemistry, 48*, 6114.

Garrison, A. W., Nzengung, V. A., Avants, J. K., Ellington, J. J., Jones, W. J., Rennels, D., & Wolfe, N. L. (2000). Phytodegradation of p, p′-DDT and the enantiomers of o,p′-DDT. *Environmental Science & Technology, 34*, 1663.

Gilbert, E. S., & Crowley, D. E. (1997). Plant compounds that induce polychlorinated biphenyl biodegradation by Arthrobacter sp. strain B1B. *Applied and Environmental Microbiology, 63*, 1933–1938.

Gobas, E. A. P. C., McNeil, E. J., Lovett-Doust, L., & Haffner, G. D. (1991). Bioconcentration of chlorinated aromatic hydrocarbons in aquatic macrophytes. *Environmental Science & Technology, 25*, 924.

Hafez, N., Abdalla, S., & Ramadan, Y. (1998). Accumulation of phenol by *Potamogeton crispus* from aqueous industrial waste. *Bulletin of Environmental Contamination and Toxicology, 60*, 944–948. https://doi.org/10.1007/s001289900719.

Hattink, J., & Wolterbeek, H. T. (2001). Accumulation of 99 Tc in duckweed *Lemna minor* L. as a function of growth rate and 99 Tc concentration. *Journal of Environmental Radioactivity, 57*, 117–138.

Hattink, J., De Goeij, J. J. M., & Wolterbeek, H. T. (2000). Uptake kinetics of 99 Tc in common duckweed. *Environmental and Experimental Botany, 44*, 9–13.

Henry, J. R. (2000). *An overview of phytoremediation of lead and mercury*. NNEMS Report, Washington D.C, pp. 3–9. http://www.bvsde.paho.org/bvsarp/i/fulltext/over/over.pdf

Hughes, J. B., Shanks, J. E., Vanderford, M. Y., Lauritzen, J., & Bhadra, R. (1997). Transformation of TNT by aquatic plants and plant tissue cultures. *Environmental Science & Technology, 31*, 266–271.

Kara, Y. (2010). Bioaccumulation of nickel by aquatic macrophytes. *Desalination and Water Treatment, 19*, 325–328.

Knuteson, S. L., Whitwell, T., & Klaine, S. J. (2002). Influence of plant age and size on simazine uptake and toxicity. *Journal of Environmental Quality, 31*, 2090.

Kolpin, D., Furlong, E., Meyer, M. T., Thurman, E. M., Zaugg, S. D., Barber, L. B., & Buxton, H. T. (2002). Pharmaceuticals, hormones, and other organic wastewater contaminants in U.S. streams, 1999–2000: A national reconnaissance. *Environmental Science & Technology, 36*, 1202–1211.

Kondo, K., Kawabata, H., Ueda, S., Hasegawa, H., Inaba, J., Mitamura, O., Seike, Y., & Ohmomo, Y. (2003). Distribution of aquatic plants and absorption of radionuclides by plants through the leaf surface in brackish Lake Obuchi, Japan, bordered by nuclear fuel cycle facilities. *Journal of Radioanalytical and Nuclear Chemistry, 257*, 305.

Kumar, M. D., Panda, R., Niranjan, V., & Bassi, N. (2013). Technology choices and institutions for improving economic and livelihood benefits from multiple uses tanks in western Orissa. In M. D. Kumar, M. V. K. Sivamohan, & N. Bassi (Eds.), *Water management, food security and sustainable agriculture in developing economies*. Oxford: Routledge.

Larsen, J. C. (2006). Risk assessments of polychlorinated dibenzo-p-dioxins, polychlorinated dibenzofurans, and dioxin-like polychlorinated biphenyls in food. *Molecular Nutrition & Food Research, 50*, 885–896.

Lesage, E., Mundia, C., Rousseau, D. P. L., Van de Moortel, A. M. K., Laing, G. D., Tack, F. M. G., De Pauw, N., & Verloo, M. G. (2008). Removal of heavy metals from industrial effluents by the submerged aquatic plant *Myriophyllum spicatum* L. In J. Vyamazal (Ed.), *Wastewater treatment, plant dynamics and management in constructed and natural wetlands* (pp. 211–221). Dordrecht: Springer.

Lin, A. Y., Yu, T., & Lin, C. (2008). Pharmaceutical contamination in residential, industrial, and agricultural waste streams: Risk to aqueous environments in Taiwan. *Chemosphere, 74*, 131–141.

Machate, T., Noll, H., Behrens, H., & Kettrup, A. (1997). Degradation of phenanthracene and hydraulic characteristics in constructed wetland. *Water Research, 31*, 554.
Manahan, S. (1994). *Environmental chemistry* (6th ed.p. 811). New York: Lewis Publishers.
Mander, Ü., & Mitsch, W. (2009). Pollution control by wetlands. *Ecological Engineering, 35*, 153–158. https://doi.org/10.1016/j.ecoleng.2008.10.005.
Manios, T., Stentiford, E. I., & Millner, P. (2003). Removal of heavy metals from a metaliferous water solution by *Typha latifolia* plants and sewage sludge compost. *Chemosphere, 53*, 487–494.
Mazumdar, K., & Das, S. (2014). Phytoremediation of Pb, Zn, Fe, and Mg with 25 wetland plant species from a paper mill contaminated site in north East India. *Environmental Science and Pollution Research, 22*, 701–710. https://doi.org/10.1007/s11356-014-3377-7.
Molisani, M. M., Rocha, R., Machado, W., Barreto, R. C., & Lacerda, I. D. (2006). Mercury contents in aquatic macrophytes from two reservoirs in the para'ıba do sul: Guandu river system, Se, Brazil. *Brazilian Journal of Biology, 66*, 101.
Nguyen, T. T. T., Davy, F. B., Rimmer, M., & De Silva, S. (2009). *Use and exchange of genetic resources of emerging species for aquaculture and other purposes*. FAO/ NACA expert meeting on the use and exchange of aquatic genetic resources relevant for food and agriculture, 31 March–02 April 2009, Chonburi, Thailand.
Nwoko, C. O. (2010). Trends in phytoremediation of toxic elemental and organic pollutants. *African Journal of Biotechnology, 9*(37), 6010–6016.
Nzengung, V. A., Lee, N. W., Rennels, D. E., McCutcheon, S. C., & Wang, C. (1999). Use of aquatic plants and algae for decontamination of waters polluted with chlorinated alkanes. *International Journal of Phytoremediation, 1*, 203.
Osmolovskaya, N., & Kurilenko, V. (2005). Macrophytes in phytoremediation of heavy metal contaminated water and sediments in urban inland ponds. *Geophysical Research Abstracts, 7*, 10510.
Panich-Pat, T., Srinives, P., Kruatrachue, M., Pokethitiyook, P., Upathamd, S., & Lanzae, G. R. (2005). Electron microscopic studies on localization of lead in organs of *Typha angustifolia* grown on contaminated soil. *Science Asia, 31*, 49–53.
Pavlostathis, S. G., Comstock, K. K., Jacobson, M. E., & Saunders, F. M. (1998). Transformation of 2,4,6-trinitrotoluene by the aquatic plant *Myriophyllum aquaticum. Environmental Toxicology and Chemistry, 17*, 2266.
Petrović, M., Gonzalez, S., & Barceló, D. (2003). Analysis and removal of emerging contaminants in wastewater and drinking water. *Trends in Analytical Chemistry, 22*, 685–696.
Pilon-Smits, E., & Freeman, J. (2006). Environmental cleanup using plants: Biotechnological advances and ecological considerations. *Frontiers in Ecology and the Environment, 4*, 203–210. https://doi.org/10.1890/1540-9295(2006)004[0203:ecupba]2.0.co;2.
Popa, K., Cecal, A., Humelnicu, D., Caraus, I., & Draghici, C. L. (2004). Removal of 60 Co^{2+} and 137 Cs^{+} ions from low radioactive solutions using *Azolla caroliniana* wild. Water fern. *Central European Journal of Chemistry, 2*, 434.
Popa, K., Palamaru, M. N., Iordan, A. R., Humelnicu, D., Drochioiu, G., & Cecal, A. (2006). Laboratory analyses of $60Co^{2+}$, $65Zn^{2+}$ and $(55+59)Fe^{3+}$ radioactions uptake by *Lemna minor. Isotopes in Environmental and Health Studies, 42*, 87.
Prasad, M. N. V. (2004). Phytoremediation of metals in the environment for sustainable development. *Proceedings of the Indian National Science Academy, 70*, 71–98.
Qian, J. H., Zayed, A., Zhu, M. L., Yu, M., & Terry, N. (1999). Phytoaccumulation of trace elements by wetland plants, III: Uptake and accumulation of ten trace elements by twelve plant species. *Journal of Environmental Quality, 28*, 1448.
Rai, P. (2008). Technical note: Phytoremediation of hg and cd from industrial effluents using an aquatic free floating Macrophyte Azolla Pinnata. *International Journal of Phytoremediation, 10*, 430–439. https://doi.org/10.1080/15226510802100606.

Rai, U. N., Tripathi, R. D., Vajpayee, P., Pandey, N., Ali, M. B., & Gupta, D. K. (2003). Cadmium accumulation and its phytotoxicity in *Potamogeton pectinatus* (Potamogetonaceae). *Bulletin of Environmental Contamination and Toxicology, 70*, 566.

Reeves, R. D., & Brooks, R. R. (1983). Hyperaccumulation of lead and zinc by two metallophytes from a mining area of Central Europe. *Environmental Pollution Series A, 31*, 277–287.

Rice, P. J., Anderson, T. A., & Coats, J. R. (1997). Phytoremediation of herbicide-contaminated surface water with aquatic plants. In E. L. Kruger, T. A. Anderson, & J. R. Coats (Eds.), *Phytoremediation of soil and water contaminants*. Washington, DC: American Chemical Society.

Rivera, R., Medina, V. F., Larson, S. L., & McCutcheon, S. C. (1998). Phytotreatment of TNT-contaminated groundwater. *Journal of Soil Contamination, 7*, 511.

Roy, S., & Hanninen, O. (1994). Pentachlorophenol: Uptake/ elimination, kinetics and metabolism in an aquatic plant, *Eichhornia crassipes*. *Environmental Toxicology and Chemistry, 13*, 763.

Russi, D., ten Brink, P., Farmer, A., Badura, T., Coates, D., Förster, J., Kumar, R., & Davidson, N. (2013). *The economics of ecosystems and biodiversity for water and wetlands*. London, Brussels/Gland: IEEP/Ramsar Secretariat.

Sadowsky, M. J. (1999). *Phytoremediation: past promises and future practices*. In Proceedings of the 8th international symposium on microbial ecology, Halifax, pp. 1–7.

Salt, D. E., Blaylock, M., Kumar, P. B. A. N., Dushenkov, V., Ensley, B. D., Chet, I., & Raskin, I. (1995). Phytoremediation: A novel strategy for the removal of toxic metals from the environment using plants. *Biotechnology, 13*, 468–475.

Saygideger, S., Dogan, M., & Keser, G. (2004). Effect of lead and pH on lead uptake, chlorophyll and nitrogen content of Typha latifolia L. and Ceratophyllum demersum L. *International Journal of Agriculture and Biology, 6*, 168–172.

Schnoor, J. L., Licht, L. A., McCutcheon, S. C., Wolfe, N. L., & Carreira, L. H. (1995). Phytoremediation of organic and nutrient contaminants. *Journal of Environmental Science and Technology, 29*, 318A–323A.

Sharma, H. A., Barber, J. T., Ensley, H. E., & Polito, M. A. (1997). Chlorinated phenols and phenols by Lemna gibba. *Environmental Toxicology and Chemistry, 16*, 346.

Shokod'Ko, T. I., Drobot, P. I., Kuzmenko, M. I., & Shklyar, A. Y. (1992). Peculiarities of radionuclides accumulation by higher aquatic plants. *Hydrobiological Journal, 28*, 92.

Singh, N. K., Pandey, G. C., Rai, U. N., Tripathi, R. D., Singh, H. B., & Gupta, D. K. (2005). Metal accumulation and ecophysiological effects of distillery effluent on *Potamogeton pectinatus* L. *Bulletin of Environmental Contamination and Toxicology, 74*, 857.

Singh, D., Gupta, R., & Tiwari, A. (2011). Phytoremediation of lead from wastewater using aquatic plants. *International Journal of Biomedical Research*. https://doi.org/10.7439/ijbr.v2i7.124.

Sivaci, E. K., Sivaci, A., & Sokman, M. (2004). Biosorption of cadmium by Myriophyllum spicatum and *Myriophyllum triphyllum* orchard. *Chemosphere, 56*, 1043.

Srivastava, S., Mishra, S., Dwivedi, S., & Tripathi, R. (2010). Role of thiol metabolism in arsenic detoxification in *Hydrilla verticillata*(L.f.) Royle. *Water, Air, and Soil Pollution, 212*, 155–165.

Srivastava, S., Srivastava, M., Suprasanna, S., & D'Souza, F. (2011). Phytofiltration of arsenic from simulated contaminated water using *Hydrilla verticillata* in field conditions. *Ecological Engineering, 37*, 1937–1941.

Stuart, M. E., Manamsa, K., Talbot, J. C., & Crane, E. J. (2011). *Emerging contaminants in g groundwater*. Groundwater science programme open report OR/11/013. British Geological Survey.

Thompson, P. L., Ramer, L. A., & Schnoor, J. L. (1998). Uptake and transformation of TNT by hybrid poplar trees. *Environmental Science & Technology, 32*, 975–980.

Tripathi, R. D., Rai, U. N., Vajpayee, M. B., Ali, M. B., Khan, E., Gupta, D. K., Mishra, S., Shukla, M. K., & Singh, S. N. (2003). Biochemical responses of *Potamogeton pectinatus* L. exposed to higher concentration of zinc. *Bulletin of Environmental Contamination and Toxicology, 71*, 255.

Tront, A. M., & Saunders, F. M. (2006). Role of plant activity and contaminant speciation in aquatic plant assimilation of 2,4,5-trichlorophenol. *Chemosphere, 64*(3), 400–407.
Tront, J. M., Day, J. A., & Saunders, M. F. (2001). *Trichlorophenol removal with Lemna minor*. In: Proceedings of the water environment federation (Vol. 40, p. 929). San Diego: WEFTEC.
Tront, J. M., Reinhold, D. M., Bragg, A. W., & Saunders, F. M. (2007). Uptake of halogenated phenols by aquatic plants. *Journal of Environmental Engineering, 133*, 955.
Vangronsveld, J., Herzig, R., Weyens, N., et al. (2009). Phytoremediation of contaminated soils and groundwater: Lessons from the field. *Environmental Science and Pollution Research, 16*, 765–794. https://doi.org/10.1007/s11356-009-0213-6.
Wang, T. C., Weissman, J. C., Ramesh, G., Varadarajan, R., & Benemann, J. R. (1996). Parameters for removal of toxic heavy metals by water Milfoil (*Myriophyllum spicatum*). *Bulletin of Environmental Contamination and Toxicology, 57*, 779–786.
Wani, S. H., Sanghera, G. S., Athokpam, H., Nongmaithem, J., Nongthongbam, R., Naorem, B. S., & Athokpam, H. S. (2012). Phytoremediation: Curing soil problems with crops. *African Journal of Agricultural Research, 7*(28), 3991–4002.
Weltje, L., Brouwer, A. H., Verburg, T. G., Wolterbeek, H. T., & de Goeij, J. J. M. (2002). Accumulation and elimination of lanthanum by duckweed (*Lemna minor* L.) as influenced by organism growth and lanthanum sorption to glass. *Environmental Toxicology and Chemistry, 21*, 1483–1489.
Windham, L., Weis, J. S., & Weis, P. (2001). Lead uptake, distribution and effects in two dominant salt marsh macrophytes *Spartina alterniflora* (cordgrass) and *Phragmites australis* (common reed). *Marine Pollution Bulletin, 42*, 811.
Windham, L., Weis, J. S., & Weis, P. (2003). Uptake and distribution of metals in two dominant salt marsh macrophytes, *Spartina alterniflora* (cordgrass) and *Phragmites australis* (common reed). *Estuarine, Coastal and Shelf Science, 56*, 63.
Wolf, S. D., Lassiter, R. R., & Wooten, S. E. (1991). Predicting chemical accumulation in shoots of aquatic plants. *Environmental Toxicology and Chemistry, 10*, 655.
Xia, J., Wu, L., & Tao, Q. (2002a). Phytoremediation of methyl parathion by water hyacinth(*Eichhornia crassipes* Solm.). *Chemical Abstracts, 137*, 155879.
Xia, J., Wu, L., & Tao, Q. (2002b). Phytoremediation of some pesticides by water hyacinth (*Eichhornia crassipes* Solm.). *Chemical Abstracts, 138*, 390447.
Xue, P., Yan, C., Sun, G., & Luo, Z. (2012). Arsenic accumulation and speciation in the submerged macrophyte *Ceratophyllum demersum* L. *Environmental Science and Pollution Research International, 19*, 3969–3976.
Ye, Z. H., Baker, A. J. M., Wong, M. H., & Willis, A. J. (1997). Zinc, lead and cadmium tolerance, uptake and accumulation by *Typha latifolia*. *The New Phytologist, 136*, 469.
Zayed, A., Pilon-Smits, E., de Souza, M., Lin, Z. Q., & Terry, N. (2000). Remediation of selenium polluted soils and waters by phytovolatilization. In N. Terry & G. Barnuelos (Eds.), *Phytoremediation of contaminated soil and water* (p. 61). Boca Raton: Lewis.

Chapter 6
Biodegradation of Polycyclic Aromatic Hydrocarbons (PAHs): A Sustainable Approach

Shaili Srivastava and Madan Kumar

Abstract Polycyclic aromatic hydrocarbons (PAHs) are aromatic hydrocarbons having two or more fused benzene rings. PAH are found in environment from natural as well as anthropogenic sources. They are widely distributed environmental contaminants that have detrimental biological effects, including toxicity, mutagenicity, and carcinogenicity. PAHs are thermodynamically more stable and resistant to microbial degradation due to their hydrophobic nature and their stabilization due to presence of multiple benzene rings and low aqueous solubility. Despite these properties, a variety of bacterial, fungal and algal species are reported for biodegradation. Most of studies involved in PAH microbial degradation is based on enzymes involved in PAH metabolism and their mineralization. Several bacteria have been found to degrade PAH such as *Sphingomonas* sp., *Psedomonas* sp., *Alcaligens eutrophus, Burkhelderia* sp. *Mycobacterium, Rhodococcus, Nocardioides, Mycobacterium, Rhodococcus, Nocardioides* and *Novosphingobium,* etc. There are several biochemical pathways and gene reported which are responsible for bacterial degradation of PAHs. Many fungi metabolize polycyclic aromatic hydrocarbons with enzymes that include lignin peroxidase, manganese peroxidase, laccase, cytochrome P450, and epoxide hydrolase. The products include *trans*-dihydrodiols, phenols, quinones, dihydrodiol epoxides, and tetraols, which may be conjugated to form glucuronides, glucosides, xylosides, and sulfates. The fungal and bacterial metabolites generally are less toxic than the parent hydrocarbons. Cultures of fungi that degrade polycyclic aromatic hydrocarbons may be useful for bioremediation of contaminated soils, sediments, and waters. Microalgae and eukaryotic algae sp. have been also reported for their bioaccumulation, biotransformation and degradation capability of PAH. While mechanism of biodegradation pathways from algae are not very specific and vary from species to species. In case of algal biodegradation of PAH it works more precisely in combination with bacterial co-culture.

S. Srivastava (✉)
Amity School of Earth and Environmental Sciences, Amity University, Gurugram, Haryana, India

M. Kumar
School of Environmental Sciences, Jawaharlal Nehru University, New Delhi, India

S. Shah et al. (eds.), *Sustainable Green Technologies for Environmental Management*, https://doi.org/10.1007/978-981-13-2772-8_6

Keywords Biodegradation · Polycyclic aromatic hydrocarbons (PAHs) · Microbial consortium

6.1 Introduction

Aromatic hydrocarbons are most abundant and diverse organic pollutants present in the environment. Aromatic compounds are defined as the organic molecules with at least one benzene ring. The major sources of aromatic compounds are living organisms (aromatic amino acids), plants (secondary metabolites), lignin and fossil reserves (Fuchs et al. 2011). The aromatic compounds exist as complex mixtures in the environment. Aromatic compounds can be grouped into three major classes: polycyclic aromatic hydrocarbons (PAHs), heterocyclic compounds and substituted aromatics (Seo et al. 2009). These compounds are of great environmental concern and their removal from the polluted system is essential due to their recalcitrance, bioaccumulation potential and toxicity to ecosystems including humans. Their removal from the contaminated system has led to development of various physical, chemical, thermal and biological processes. Among these, biological processes are the most common, cost effective and sustainable approach for degradation (Haritash and Kaushik 2009; Vila et al. 2015; Kuppusamy et al. 2017). In this chapter, we have focused on sources, properties, and biodegradtion of PAHs.

6.2 Sources of PAHs

PAHs are ubiquitous recalcitrant environmental pollutants with known mutagenic, teratogenic and carcinogenic properties. PAHs comprise of fused benzene rings having only carbon and hydrogen in the structure. On the basis of benzene rings present in the structure: PAHs can be classified as low molecular weight (LMW) and high molecular weight (HMW). The LMW consists of 2–3 rings (naphthalene, phenanthrene anthracene fluorine) and HMW contains 4–7 rings such as pyrene, chrysene, coronenes (Kuppusamy et al. 2017). These benzene rings are arranged in linear, angular and cluster form in the structure. PAHs are mainly produced as complex mixture and there are more than 100 types of PAHs present in the environment. The United States Environmental Protection Agency (US EPA) report 2008 labeled 28 compounds as hazardous (Gan et al. 2009). The 16 PAH are categorized as pollutants of high concern by US EPA. The World Health Organization also documented 8 PAHs (benzo[a]anthracene, chrysene, benzo[b]fluoranthene, benzo[k] fluoranthene, benzo[a]pyrene, indeno[1,2,3-cd], pyrene, Dibenz(a,h)Anthracene (DBahA) and benzo[ghi]perylene) has potential expected to be carcinogenic pollutants (Bansal and Kim 2015).

PAHs are group of organic compounds that are produced naturally by volcanic eruptions, forest fires, oil seeps, exudates from trees and through anthropogenic sources such as incomplete burning of crude oil, coal, gas, wood, garbage, cooking at high temperatures, exhaust from vehicles, discharge from industry and treatment plants (DHHS 1995; Gehle 2009). On the basis of their source of origin, PAHs can be classified as petrogenic, pyrogenic and biogenic PAHs. The pyrogenic PAHs are produced from the incomplete combustion of organic compounds at high temperature or prolonged heating at low temperature either at low or absence of oxygen. Incomplete combustion taking place either by natural or anthropogenic means are the largest contributor of PAHs to the environment. The example of pyrogenic processes is thermal cracking of petroleum, distillation of coal for coke and coal tar production, forest fire, vehicular emission, etc. Petrogenic PAHs are generated from crude oil, crude oil products such as petroleum product and oil spills. The biogenic sources include plants and microbes and their degradation products, industries, incinerators, wastewater treatment plants, emission from burning of motor fuels, grilled and smoked food, cigarette, tobacco, wood burning (Haritash and Kaushik 2009; Abdel-Shafy and Mansour 2016).

6.3 Properties, Fate and Exposure of PAHs

Physical and chemical property of PAH depends on the number of rings or molecular weight. These compounds are solids having relatively high molecular weight, low volatility at room temperature, less or insoluble in water, soluble in organic solvents and prone to photo-oxidative degradation (DHHS 1995). Their aqueous solubility, volatility and reactivity decreases with increase in molecular weight (Seo et al. 2009). In pure form these chemicals are colorless, white or pale yellow-green with faint or pleasant odour. These are used in synthesis of medicine, dyes, pesticides and plastics (DHHS 1995). The physical and chemical properties of 16 most priority PAHs are presented in Table 6.1 as listed by US EPA. These are present ubiquitously in interstellar space and the environment i.e., air, water, and soil or sediment. The soil is the reservoir of PAHs with almost 90% of total PAH burden in the environment (Kuppusamy et al. 2017). The global annual emission of 16 PAHs was 504 Gg in 2007 and the major contributors were biomass (residential/ commercial) burning, open field biomass burning, deforestation, wildfire and petroleum combustion by motor vehicles. The HMW carcinogenic PAH emission was 6.19% of total PAH emission. The carcinogenic PAH emission was higher in developing countries (6.22%) compared to developed countries (5.735) due to difference in advancement in technology and energy structure (Shen et al. 2013). The exposure of PAHs takes place through air, water, soil and food sources and the route of exposure is inhalation, ingestion and dermal contact in both occupational and non-occupational conditions (Abdel-Shafy and Mansour 2016).

Table 6.1 Physical and chemical property of 16 high priority pollutants documented by US EPA

PAHs Name	Formula	Cas registry	B. P. (°C)	M.P. (°C)	Solubility (mg/L)	V.P. (mmHg at 25 °C)	Toxic equivalent factor (TEF)	Estimated half-lives (days)
Naphthalene	C10H8	91-20-3	218	80.2	31	8.52×10^{-2}	n.d.	5.66
Acenaphthene	C12H10	83-32-9	279	93.4	3.93	2.5×10^{-3}	0.001	18.77
Acenaphthylene	C12H8	208-96-8	280	91.8	1.93	6.68×10^{-3}	0.001	30.7
Anthracene	C14H10	120-12-7	342	216.4	0.076	6.53×10^{-6}	0.01	123
Phenanthrene	C14H10	85-01-8	340	100.5	1.2	1.2×10^{-4}	0.001	14.97
Fluorene	C13H10	86-73-7	295	216–7	1.68–1.98	6×10^{-4}	0.001	15.14
Fluoranthene	C16H10	206-44-0	375	108.8	0.20–0.26	9.22×10^{-6}	0.001	191.4
Benzo[a]anthracene	C18H12	56-55-3	438	158	0.01	4.11×10^{-3}	0.1	343.8
Chrysene	C18H12	218-01-9	448	254	1.5×10^{-3}	6.23×10^{-9}	0.01	343.8
Pyrene	C16H10	129-00-0	150.4	393	0.132	4.5×10^{-6}	0.001	283.41
Benzo[a]pyrene	C20H12	50-32-8	495	179	3.8×10^{-3}	5.49×10^{-9}	1	421.6
Benzo[b]fluoranthene	C20H12	205-99-2	481	168.3	0.0012	5×10^{-7}	n.d.	284.7
Benzo[k]fluoranthene	C20H12	207-08-9	480	215.7	7.6×10^{-4}	9.7×10^{-10}	0.1	284.7
Dibenzo[a,h] anthracene	C22H14	53-70-3	524	262	5×10^{-4}	9.55×10^{-10}	n.d.	511.4
Benzo[g,h,i]perylene	C22H12	191-24-2	500	277	2.6×10^{-5}	1×10^{-10}	n.d.	517.1
Indenol[1,2,3-cd] pyrene	C22H12	193-39-5	536	161–3	0.062	1.25×10^{-3}	n.d.	349.2

6.4 Biodegradation of PAHs by Bacteria

The remediation of PAHs in the environment takes place by various established physical, chemical, thermal and biological processes and their integration such as chemical oxidation, incineration, solvent extraction, thermal conduction, phytoremediation, bioaugmentation, biostimulation, bioreactors and composting (Haritash and Kaushik 2009; Kuppusamy et al. 2017). Compared to various physic-chemical processes, biological processes is cost effective, sustainable and green approach to environment. Physico-chemical processes only cause transformation from one compound to another less toxic form but in case of biological processes, complete mineralization of contaminants in the form of CO_2 and water can be observed. In this chapter, we are discussing biological processes for pollutants remediation. The contaminants are utilized by microorganisms as source of energy and carbon in the environment. The bioremediation of PAHs pollutants in the environment can be done by *ex situ* and *in situ* treatment methods. The bioremediation of pollutants by microorganisms is very promising technology and this research field is very active as new contaminants and microbes are being identified and reported on regular basis. The versatility and adaptability of microbes to utilize diverse range of contaminants present in diverse environment for energy and carbon source is the key to bioremediation. The biological degradation of contaminants can be aerobic and anaerobic process.

6.4.1 Aerobic Degradation of PAH by Bacteria

The microbial degradation of recalcitrant environment pollutants such as PAHs takes place favorably under aerobic conditions and are extensively documented. In aerobic degradation process, oxygen molecule is the final electron acceptor and also acts as co-substrate for hydroxylation and oxygenolytic cleavage of aromatic rings (Ghosal et al. 2016). There are various microorganisms identified as PAH degraders and these microbes belongs to *Pseudomonas, Alcaligenes*, *Sphingomonas*, *Mycobacterium*, *Streptomyces*, *Burkholderia*, *Polaromonas*, *Ralstonia*, *Rhodococcus*, *Stenotrophomonas Flavobacterium, Bacillus, Rhodotorula, Acenitobacte*, *Klebsiella* and *Arthrobacter* (Haritash and Kaushik 2009; Seo et al. 2009). The degradation of four or more ring HMW PAHs containing four or more ring was commonly observed in genus *Sphingomonads, Actinobacteria* and *Mycobacterium* as sole energy and carbon source (Vila et al. 2015).

6.4.1.1 Pathways and Enzymes Involved in Aerobic Degradation of PAHs

Microorganisms can degrade PAHs through both aerobic and anaerobic pathways but the aerobic pathway is studied extensively. In the aerobic degradation pathway, the PAHs is activated by incorporating molecular oxygen directly with the help of mono or di oxygenases. The multicomponent dioxygenase usually consists of reductase, ferrodoxin and terminal iron sulfur containing terminal oxygenase (Haritash and Kaushik 2009; Ghosal et al. 2016). These ring hydroxylating oxygenases (RHD) belongs to rieske-type non-heme iron oxygenase family. PAH specific RHD can be distinguished phylogenetically in Gram positive and Gram negative bacterial strains. The dioxygenase hydroxylates the aromatic ring of PAH and results in formation of *cis*-dihydrodiol and gets rearomatized to diol intermediate by dehydrogenases. These diol are cleaved by intadiol or extradiol ring cleaving dioxygenase by ortho or meta cleavage pathway and forms common intermediate catechol followed by their conversion in TCA intermediates (Seo et al. 2009; Fuchs et al. 2011). Bacterial monooxygenase or cytochrome P450 activates catabolic process by introduction of one oxygen molecule in the aromatic ring and convert into arene epoxide intermediate and its subsequent conversion into*trans*-dihydrodiols (Moody et al. 2004).

Naphthalene is the simplest PAH and there are several microorganism discovered and reported with detailed mechanism of degradation, metabolic intermediates formed, enzymes involved and regulation of genes. The proposed pathway for degradation of naphthalene has been represented in Fig. 6.1.

The gene for naphthalene degradation can be divided into *nah*-like in *Pseudomonas* and non *nah*-like gene in other genus. In *Pesudomonas putida* strain G7, the catabolic gene is present in three operons encoding enzymes for upper pathway (naphthalene to salicylate), lower pathway (salicylate to TCA by meta-cleavage) and third *nah*R (a LysR trans-acting positive transcriptional regulator). In this operon, a terminal dioxygenase is reported with highly conserved structure having alpha and beta chains. The *nah*Ac gene (naphthalene 1,2 dioxygenase) was used as biomarker for PAH degradation as well as to establish correlation between PAH degradation and its levels present in the environment (Cébron et al. 2008; Lu et al. 2011). *Ralstonia* sp. U2 contains *nag* gene cluster similar in order to that of *nah* from *Pseudomonas with additional nagG* and *nagH* gene for salicylate to gentisate conversion (Zhou et al. 2001). The PAH catabolic gene reported from other bacteria are *phn* gene from *Burkholderia* sp. strain RP007, *phd* genes of *Nocardioides* sp. KP7, *nar* genes from *Rhodococcus* sp. NCIMB12038, *phd* genes from *Comamonas teststeroni* strain GZ39. These genes were detected while degradation of PAH (naphthalene, phenanthrene and anthracene). The involvement of *nid* genes in PAH such as fluoranthene, phenanthrene, pyrene and benzo(a) pyrene degradation has been reported from *Mycobacterium* sp. strain PYR-1 (Moody et al. 2001).

Phenanthrene is a LMW, three ring containing PAH pollutant ubiquitously present in an environment. It is a model compound for the degradation study of carcinogenic compounds (PAH) as its structure contains reactive 'K-' region and '-bay' regions commonly observed in other carcinogenic compounds (Seo et al. 2009; Ghosal et al. 2016). The catabolic pathway and enzymes for phenanthrene degradation has been extensively studied in *Mycobacterium* sp. strain PYR-1 (Fig. 6.2). The degradation of phenanthrene has been widely studied and the catabolizing microbes isolated are *Burkholderia, Arthrobacter, Mycobacterium*, *Pseudomonas*, *Sphingomonas, Acidovorax* and *Pseudomonas* (Tian et al. 2002). The utilization of phenanthrene has

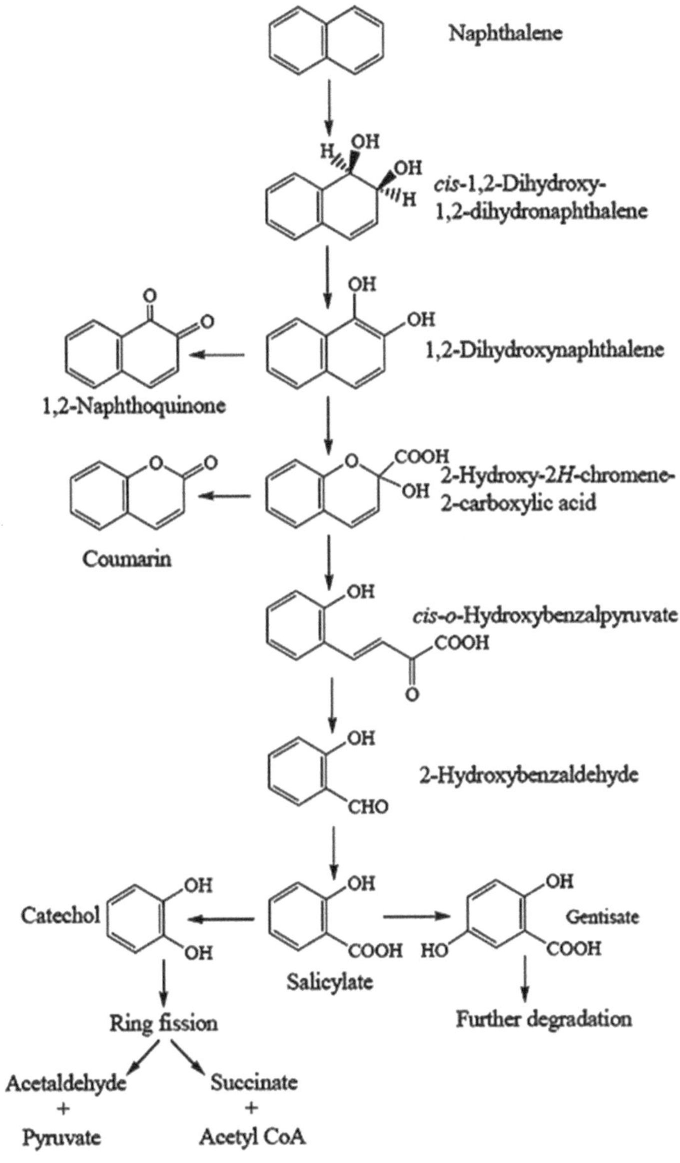

Fig. 6.1 Proposed pathway for aerobic degradation of Naphthalene. (Adapted from Seo et al. 2009)

been studied in detail from *Mycobacterium* sp. strain PYR-1. The catabolic activation of ring occurs at K- region by mono and dioxygenases, followed by ortho and meta ring cleavage (Moody et al. 2001; Haritash and Kaushik 2009; Seo et al. 2009).

Pyrene is the model compound used for the degradation study of HMW PAH. This compound contains four benzene rings in the structure and is highly

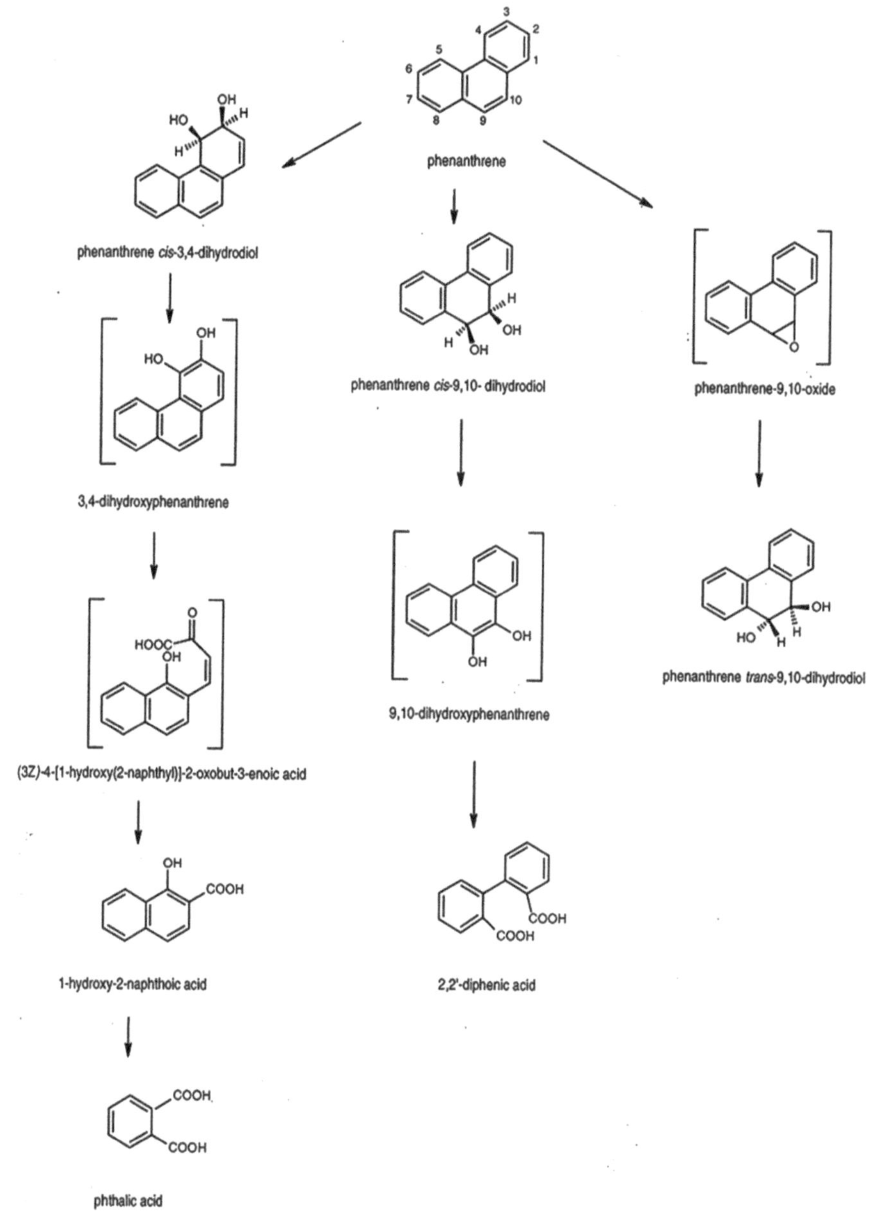

Fig. 6.2 Proposed pathway for aerobic degradation of phenanthrene by *Mycobacterium* sp. strain PYR-1. (Adapted from Moody et al. 2001)

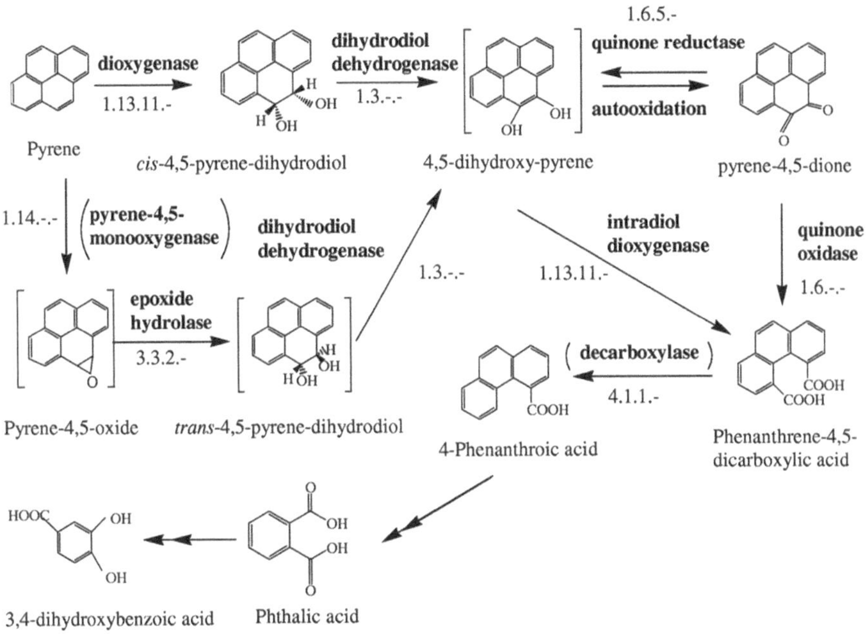

Fig. 6.3 Proposed pathway for aerobic degradation of pyrene by *Mycobacterium* sp. strain KMS. (Adapted from Liang et al. 2006)

hydrophobic in nature. The catabolism of pyrene as sole carbon and energy source has been studied in *Mycobacterium, Pseudomonas*, *Burkholderia, Bacillus, Cycloclastics* and *Sphingomonas* (Seo et al. 2009). The enzymes and pathways involved in pyrene degradation from *Mycobacterium* sp. strain KMS has been comprehensively studied and the degradation pathway was proposed (Fig. 6.3). The enzymes such as aromatic-ring-hydroxylating dioxygenase, dihydrodiol dehydrogenase, oxidoreductase, and epoxide hydrolase was found to be induced in presence of pyrene and the catabolic intermediates have also been detected (Liang et al. 2006; Seo et al. 2009).

Benzo[*a*]pyrene (BAP) is a highly recalcitrant five membered aromatic ring PAH and one of the most potent carcinogen. The genera reported for degraded of BAP are *Mycobacterium* sp., *Stenotrophomonas* and *Sphingomonas*. The metabolism of BAP has been extensively studied in *Mycobacterium vanbaalenii* PYR-1 and the enzymes and pathways involved in degradation has also been elucidated (Fig. 6.4). The catabolic activation of BAP ring (C-4,5, C-9,10, and C-11,12) takes place by stereo- and regioselective dioxygenases and monooxygenases and the degradation intermediates and pathways also proposed (Moody et al. 2004; Seo et al. 2009).

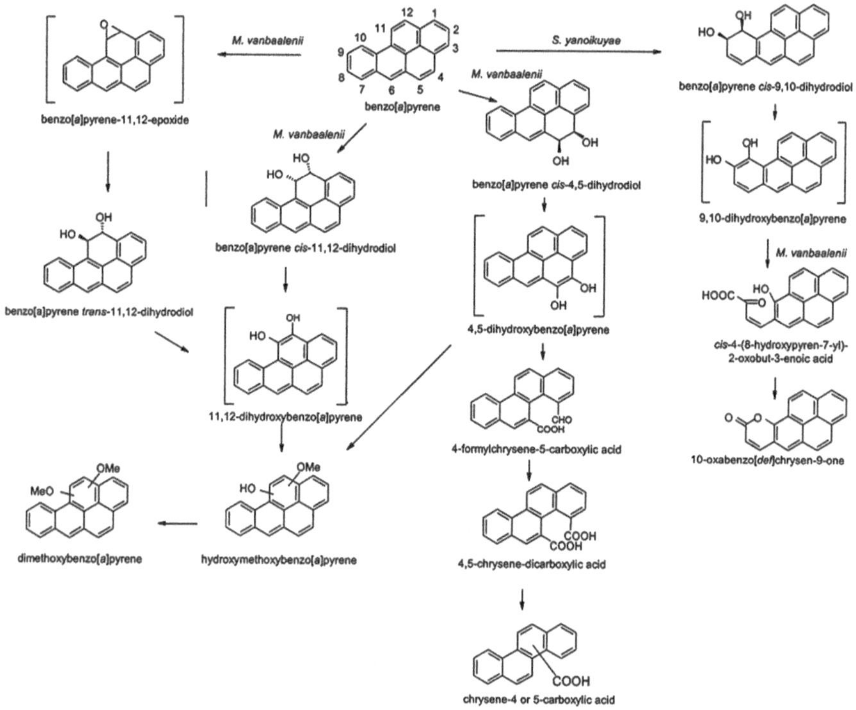

Fig. 6.4 Proposed pathway for the aerobic degradation of benzo[a]pyrene by *Mycobacterium vanbaalenii* PYR-1. (Adapted from Moody et al. 2004)

6.4.2 *Anaerobic Degradation of PAHs by Bacteria*

Aerobic degradation of PAH is most extensively studied from various environments and the pathway of degradation has also been elucidated. In aerobic process, the degradation takes place in presence of oxygen where oxygen helps in ring opening and acts as terminal electron acceptor (Meckenstock and Mouttaki 2011). The degradation of contaminants in places where there is less or lack of oxygen such as ocean sediment, aquifers, lakes and lower soil layer takes place by anaerobic mechanism. In anaerobic degradation process, nitrogen, sulphur, metal ions (ferric and manganese), CO_2, chlorate, perchlorate, trimethylamine oxide and fumarate can act as terminal electron acceptor (TEA) in place of oxygen (Meckenstock and Mouttaki 2011; Nzila 2018). The TEAs capture electron released during the degradation of organic compounds and helps in synthesis of ATP from ADP (Fig. 6.5). The ananerobic degradation of polyaromatic compounds is an important bioremediation process in the environment. In 1988, first report of anaerobic degradation can be of naphthalene was reported by Mihelcic and Luthy (1988) and later on described with sulphate, ferric ion and manganese as electron acceptor (Mihelcic and Luthy 1988; Meckenstock and Mouttaki 2011).

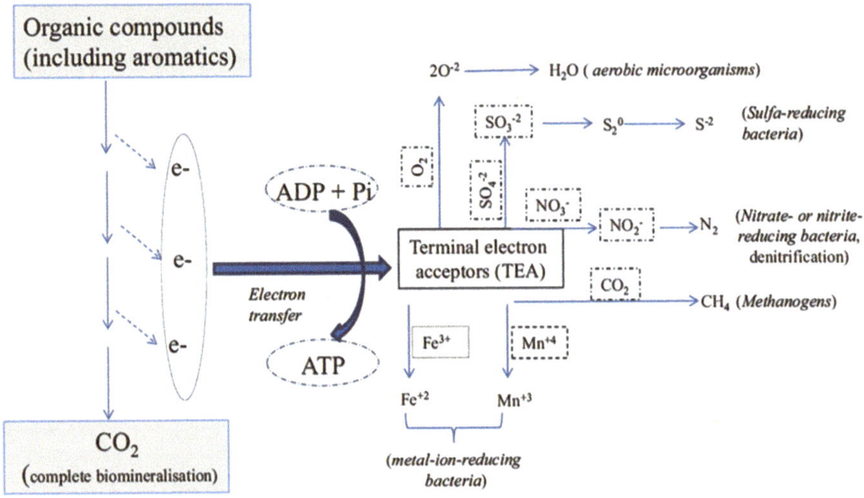

Fig. 6.5 Representation of anaerobic biodegradation of organic pollutants including PAHs. (Adapted from Nzila 2018)

But since then the degradation of aromatic pollutants under anaerobic conditions is limited to LMW PAH and very few studies have reported HMW PAH degradation. Limited biotransformation of naphthalene, phenanthrene, anthracene, pyrene, acenaphthylene and acenaphthene by methanogenic consortia was observed (Christensen et al. 2004; Maillacheruvu and Pathan 2009). Biodegradation of ^{14}C-labelled pyrene was studied under nitrate reducing condition and the emission of ^{14}C CO_2 was observed (Nieman et al. 2001). Gas chromatography (GC) or high-pressure liquid chromatography (HPLC) was used to study the degradation of anthracene using sediment from aquifer under nitrogen reducing conditions (Wang et al. 2012). The PAH degradation by single strain *Microbacterium* sp., *Cellulosimicrobium cellulans* CWS2, *Pseudomonas* sp. and *Pseudomonas* sp. JP1 under nitrate and sulphate reducing conditions were observed (Nzila 2018). The mixture of PAHs and their degradation by consortia from sludge sample of a municipal sewage treatment plant, petrochemical sludge, sediments from a rice field, marine sediments, mangrove sediment, municipal solid waste compost with different TEAs and their combinations were studied (Nzila 2018).

6.4.2.1 Pathways and Enzymes for Anaerobic PAHs Degradation

The degradation of PAHs by anaerobic process is challenging due to stability of C-C and C-H bonds in benzene ring. The metabolic activation of PAH takes place by carboxylation, methylation and fumarate addition in absence of oxygen by anaerobic microorganisms (Zhang and Young 1997; Meckenstock and Mouttaki 2011). ^{13}C-labelled buffer study under sulphate reducing conditions identified major

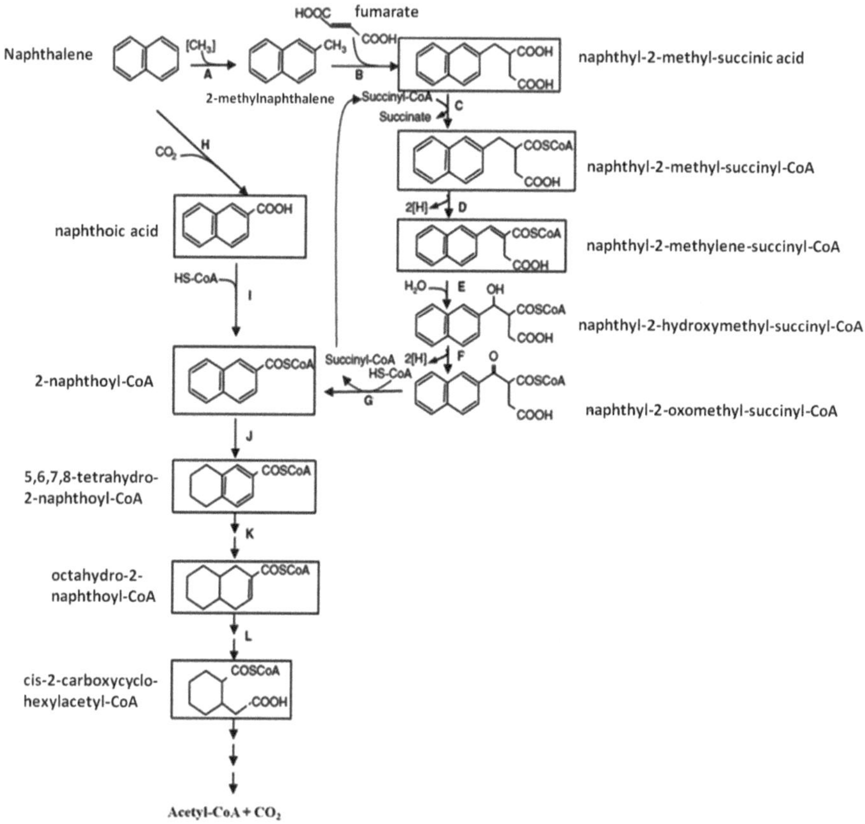

Fig. 6.6 Pathways and enzymes involved in anaerobic degradation of proposed anaerobic naphthalene and 2-methylnaphthalene degradation pathways
Enzymes represented as G, naphthalene carboxylase; H, naphthoyl-CoA ligase; (*A*) naphthalene methyl-transferase; (*B*) napthyl-2-methylsuccinyl synthase; (*C*) napthyl-2-methyl-succinyl-CoA transferase; (*D*) naphthyl-2-methyl-succinyl-CoA dehydrogenase; (*E*) naphthyl-2-methylene succinyl-CoA hydratase; (*F*) naphthyl-2-hydroxymethyl-succinyl-CoA dehydrogenase; (*G*) napthyl-2-oxomethyl-succinyl-CoA thiolase; (*H*) naphthoate carboxylase; (*I*) naphthoyl-CoA ligase; (*J* and *K*) 2-napthtoyl-CoA reductase; (*L*) enoyl-CoA hydratase. Cis-2-carboxycyclo-hexylacetic acid is then further degraded to form acetyl-CoA and CO2. (Adapted from Meckenstock and Mouttaki 2011)

metabolite as 2-napthonic acid and phenanthrene-2-carboxylic acid from naphthalene and phenanthrene respectively with labelled carboxylic acid in the structure (Zhang and Young, 1997).Methylation of napthalene and other hetercyclic compounds by methyltransferases followed by fumarate addition to 2-naphthoyl-CoA production was also proposed pathway for degradation of PAH (Annweiler et al. 2001; Safinowski and Meckenstock 2006). The carboxylases perform initial carboxylation of naphthalene to 2-naphthoic acid followed by 2-naphthoyl CoA, cyclohexane ring formation, ring cleavage and its subsequent multistep conversion to CO_2 as shown in Fig. 6.6 (Meckenstock and Mouttaki 2011).

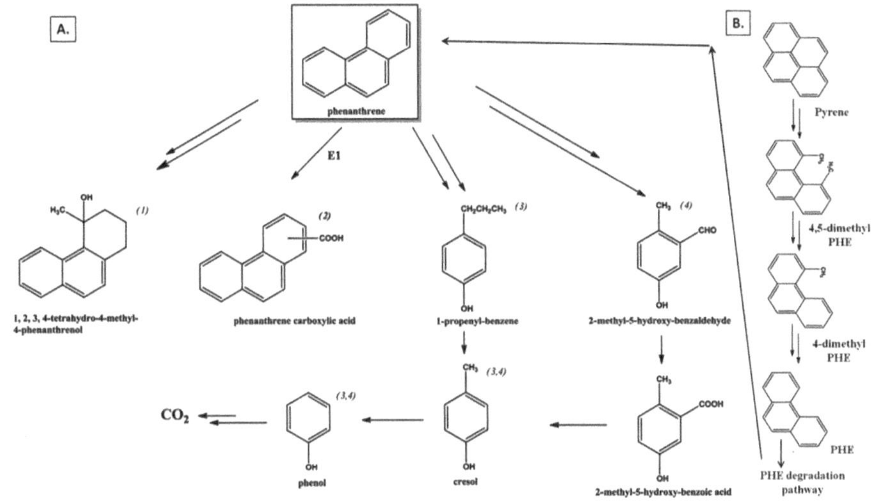

Fig. 6.7 Pathways for degradation of phenanthrene (**a**) and pyrene (**b**) and their metabolites detected during degradation.PHE (Phenanthrene). (Modified from Nzila 2018)

The mechanism of phenanthrene degradation takes place first by hydroxylation, methylation and carboxylation of first ring while other rings remains intact and then same process goes for other rings till all are cleaved. The metabolites detected from phenanthrene and pyrene degradation are 1,2,3,4-tetrahydro-4-methyl-4-phenanthrenol, p-cresol, phenol, 2-methyl-5-hydroxy-benzaldehyde 1-propenyl-benzene, anthraquinone and 1-anthraquinone-carboxylic acid (Tsai et al. 2009; Liang et al. 2014). Phenanthrene and pyrene degradation products are shown in Fig. 6.7.

Anthracene degradation detected anthraquinone and 1-anthraquinone-carboxylic acid as catabolic intermediates during degradation. The degradation of benzopyrene resulted in generation of ethyl chrysene, 1,12-dimethylbenz[a]anthracene, 7,8,9,10-tetrahydrobenzo[a]pyrene pyrene, phenanthrene, four naphthalene derivatives [1-(2-hydroxypropyl)-naphthalene, 1,7-dimethyl-naphthalene, 1-methyl-naphthalene, 2-hydroxy-3-(3-methyl-2-butenyl)-1,4-naphthalenedione] and two benzoic acid derivatives i.e., diethyl phthalate, 2-acetyl-3-methoxybenzoicacid by *Pseudomonas* sp. JP1 and *Microbacterium* sp. CSW (Liang et al. 2014; Qin et al. 2017). The anaerobic degradation of benzopyrene by *Microbacterium* sp. CSW is represented in Fig. 6.8.

The anaerobic process is advantageous compared to aerobic process i.e., removal or degradation of pollutants, nitrate and sulphate removal from site and opportunity for production of biomethane as alternate energy source while degradation. But the major disadvantage of anaerobic process is extremely slow growth rate and very less efficiency. The limitations of anaerobic degradation can be overcome by bio-augmentation, surfactant or biosurfactant addition, pH, introduction of consortia of microorganism, co-metabolism and improving biomethane production.

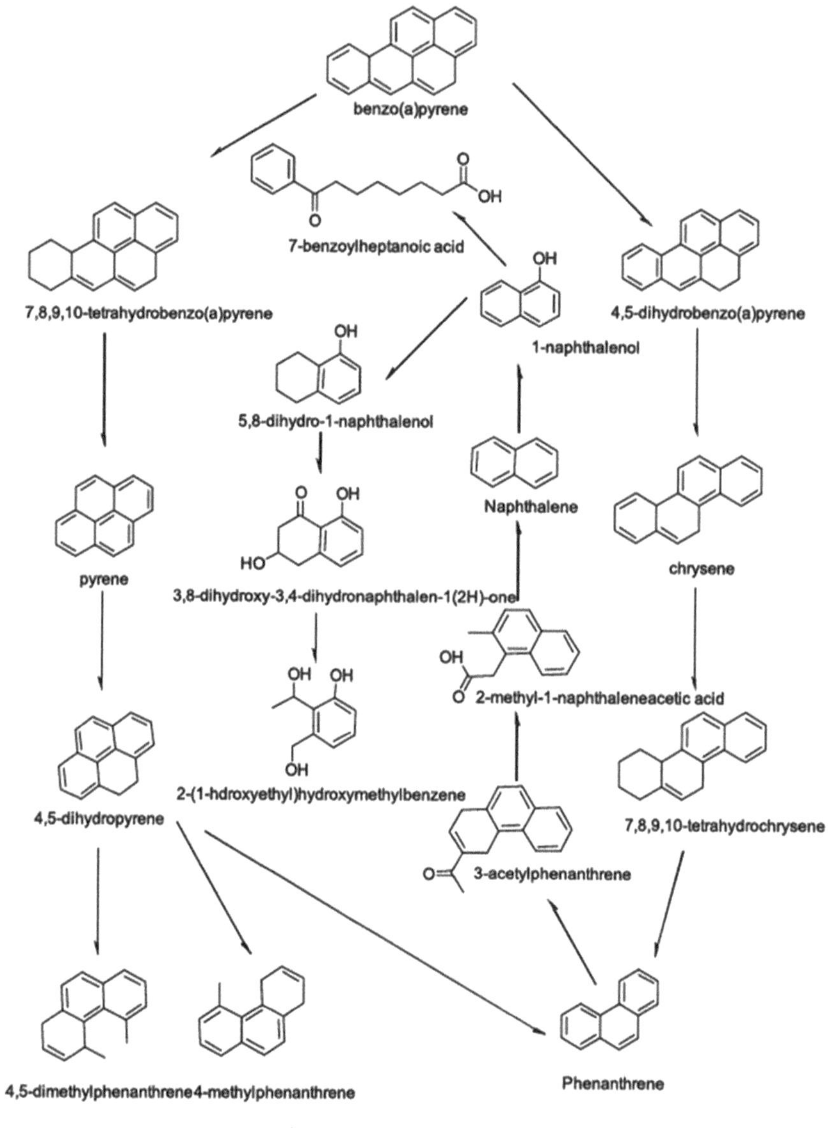

Fig. 6.8 Proposed pathway of anaerobic benzopyrene degradation by M.CSW3 CSW3 under nitrate-reducing conditions. (Qin et al. 2017)

6.5 PAHs Degradation by Complex Microbial Communities

The contaminants emitted and present in the environment are in mixture form and there are several microbes present in the environment to degrade them. Pure culture based methods ignore the interaction going on between the diverse microbial communities in the environment so culture independent approach is applied to study the microbial community. These approaches identify the microbial communities and their abundance on the polluted site. The microbes in communities are interconnected through metabolic pathways. With the advancement in 'omics' approach such as genomics, proteomics and metabolomics and their application in environment sample will help to understand the functionally relevant communities, their interaction and construction of their metabolic network in the environment.

The techniques commonly used for identification of microbial communities are denaturing gradient gel electrophoresis (DGGE), pyrosequencing, Stable Isotope Probing (SIP), Microarrays, functional gene (RHD) as molecular marker and their combination RHD-SIP, BACTRAPS (in situ protein-SIP and 16S rRNA gene pyrosequencing), metagenomics, mettranscriptomics, metaproteomics in combination with metabolomics (Vila et al. 2015).

Massive marine oil spills are usually followed by intensification in hydrocarbon degradation research. Analyses of sand from a beach affected by the Prestige oil spill had shown high relative abundances of Sphingomonadaceae and Mycobacterium that could be associated to PAH degradation (Vila et al. 2015). The link of community dynamics to depletion of specific fuel components and to single PAH exposure revealed a role in PAH utilization for the gammaproteobacteria *Methylophaga* and *Marinobacter*, and members of Actinobacteria (Vila et al. 2015). The detection of a NidA dioxygenase gene in subcultures with pyrene identified an uncultured Gordonia as a key HMW PAH degrader. A similar community composition was found in a beach affected by the surface oil slick from Deepwater Horizon oil spill. The reduced complexity of microbial consortia enriched from natural communities facilitates the correlation between specific populations and functions. Jones et al. (2014) used substrate enrichment, co-incubation experiments and pyrosequencing to identify the genera *Cupriavidus* (Betaproteobacteria) and Luteimonas (Gammaproteobacteria) as the most likely related to benzo(a)pyrene cometabolism in a PAH-polluted soils. However, the most relevant studies on the identification of soil and groundwater PAH-degrading phylotypes are based on DNA-SIP.

6.6 Biodegradation by Fungi

Polycyclic aromatic hydrocarbons (PAHs) are ubiquitous contaminants in the environment with potential mutagenicity and carcinogenicity, which are generated from natural combustion processes as well as from human activities (Cerniglia 1992). Some PAHs are acutely toxic, mutagenic and carcinogenic (Boonchan et al. 2000). The deleterious properties of PAHs have made their remediation a critical need. Bioremediation is one of the promising technologies to reclaim PAH-contaminated sites due to its relatively low cost and limited impact on the environment (Liebeg and Cutright 1999). Diverse fungi capable of utilizing PAHs have been investigated as well. Some filamentous fungi, basidiomycetes, white-rot fungi and deuteromycetes have been shown to remove PAHs more competently than bacteria. PAHs susceptible to fungal biodegradation include naphthalene, phenanthrene, anthracene, pyrene, benzo[a]pyrene, fluorene, dibenzothiophene, catechol, benzo[α]anthracene, chrysene, benzo[β]fluoranthene and benzo[k]-fluoranthene (Zheng and Obbard 2002). At least two mechanisms are involved in PAH biodegradation: one utilizes the cytochrome P-450 system (Yadav et al. 2006) and the other uses the soluble extracellular enzymes of lignin catabolism, including lignin peroxidase, manganese peroxidase (Steffen et al. 2003) and laccase (Gianfreda et al. 1999).

6.6.1 Fungal Species

In contrast to bacteria, fungi do not utilize PAHs as their sole sources of carbon and energy but transform PAHs co-metabolically to detoxified chemical products. Recent studies have reported several fungal species with the capacity to degrade a series of PAHs, such as naphthalene, phenanthrene, fluoranthene, chrysene, pyrene and benzo[a]pyrene (Kiehlmann et al. 1996; Saraswathy and Hallberg 2002; Mollea et al. 2005; Mineki et al. 2015). Compared to bacteria, there are advantages in using fungi for bioremediation as they possess extracellular enzymes and their mycelia provide deeper penetration and larger surface area for absorption in soil. PAHs biodegradation has been reported by using different species of White rot fungi (WRF) such as *Phanerochaete chrysosporium, Pleurotus ostreatus, Bjerkandera adusta, Irpex lacteus, Trametes versicolor*, etc. (Wang et al. 2008; Mir-Tutusaus et al. 2014). A diverse group of ligninolytic and non-ligninolytic fungi (Marco-urrea et al. 2015) are able to oxidize PAHs (Table 6.2).

Many of bacterial strains are also able to degrade five-benzene-ring PAHs partially, forming oxidized products. In contrast to bacteria, fungi generally do not utilize PAHs as their sole carbon and energy source but transform these compounds cometabolically to detoxified metabolites (Sutherland 1992). The most extensive studies have focused on white rot fungi such as *Phanerochaete chrysosporium, Pleurotus ostreatus and Trametes versicolor*. These fungi are able to degrade some five-benzene-ring PAHs and detoxify PAH-polluted soils and sediments due to the production of extracellular lignin-degrading enzymes. Nonlignolytic fungi, such as *Cunninghamella elegans, Penicillium janthinellum,* and *Syncephalastrum* sp., can transform a variety of PAHs, including pyrene, chrysene, and benzo[a]pyrene, to polar metabolites (Pothuluri et al. 1994). Biodegradation of PAHs by the white rot fungus *Phanerochaete chrysosponum* has clearly been verified in several studies that report mineralization of ^{4}C-labeled PAH (Sanglard et al. 1986).

Extensive biodegradation of PAH, e.g., fluorene and benzo[a]pyrene (B[a]P) was reported from *P. chrysosponum, Trametes versicolor TV1* and *Chrysosponum lignorum* CL1 (Morgan et al. 1991). Field et al. (1992) has been studied on different isolates of fungal strains such as *Bjerkandera adusta* CBS 595.78, *Polyporus, pinsitus* CBS 678.70, *Trametes* sp. strain Naald 11; *Trametes* sp. strain Eik 39; *Trametes*

Table 6.2 Polycyclic aromatic hydrocarbon utilized by different fungal species

S. no.	Compound	Fungal sp./Strain
1	Acenaphtene	*Cunninghamella elegans*
2	Anthracene	*Bjerkandera* sp., *Cunninghamella elegans, Naematoloma frowardii, Phanerochaete chrysosporium, Phanerochaete laevis, Pleurotus ostreatus, Pleurotus sajor-caju*
3	Phenanthrene	*Aspergillus niger, Cunninghamella elegans, Naematoloma frowardii, Phanerochaete chrysosporium, Phanerochaete laevis, Pleurotus ostreatus, Syncephalastrum racemosum, Trametes versicolor*
4	Fluorene	*Cunninghamella elegans, Laetiporus sulphureus, Phanerochaete chrysosporium, Pleurotus ostreatus, Trametes versicolor*
5	Fluoranthene	*Cunninghamella elegans, Naematoloma frowardii, Laetiporus sulphureus, Penicillium* sp., *Pleurotus ostreatus*
6	Pyrene	*Aspergillus niger, Agrocybe aegerita, Candida parapsilopsis, Crinipellis maxima, Crinipellis perniciosa, Crinipellis stipitaria, Crinipellis zonata, Cunninghamella elegans, Fusarium oxysporum, Kuehneromyces mutablis, Marasmiellus ramealis, Marasmius rotula, Mucor* sp., *Naematoloma frowardii, Penicillium janczewskii, Penicillium janthinellum, Phanerochaete chrysosporium, Pleurotus ostreatus, Syncephalastrum racemosum, Trichoderma harzianum*
7	Benzo[a] anthracene	*Candida krusei, Cunninghamella elegans, Phanerochaete chrysosporium Phanerochaete laevis, Pleurotus ostreatus, Rhodotorula minuta, Syncephalastrum racemosum, Trametes versicolor*
8	Benzo[a] pyrene	*Aspergillus ochraceus, Bjerkandera adusta, Bjerkandera* sp., *Candida maltosa, Candida tropicalis, Chrysosporium pannorum, Cunninghamella elegans, Mortierella verrucosa, Naematoloma frowardii, Neurospora crassa, Penicillium janczewskii, Penicillium janthinellum,*
9	Chrysene	*Cunninghamella elegans, Penicillum janthinellum, Syncephalastrum racemosum*
10	Benzo[a] pyrene	*Cunninghamella elegans*

Modified from Kadri et al. (2017)

sp. strain Berk 41; an unidentified strain of the order *Aganicales*, strain Beuk 47; *Bjerkandera* sp. strain Bos 55; *Daedaleopsis confragosa* GM 2; and *Stereum* sp. strain Schim 22. These strains were originally isolated from rotting pine needles, rotting oak wood, rotting birch wood, rotting beech wood, forest litter, forest soil, and forest litter, respectively. He found that PAH biodegradation is a ubiquitous phenomenon among white rot fungi. All strains tested degraded anthracene, and some of the strains degraded Benzo alpha pyrene, significantly beyond the level in the poisoned mycelium controls. Although PAH biodegradation was found to be a universal characteristic of the white rot fungi tested, two distinct patterns of PAH biodegradation could be distinguished.

6.6.2 Fungal Enzymes and Degradation Pathways

Fungi is found to possess inherent potential and efficiency to degrade PAHs. Ligninolytic fungi was of much interest to the scientists as they produce extracellular enzymes with reduced substrate specificity. This feature of ligninolytic fungi provided a necessary fillip for the organisms to degrade various organopollutants (Hammel 1995). Extracellular peroxidases produced by these fungi causes the initial oxidation of PAHs (Zhang et al. 2015). While fungal lignin peroxidases oxidize quite a number of PAHs directly, fungal manganese peroxidases co-oxidize them indirectly through enzyme-mediated lignin peroxidation. Further, it was reported that white rot fungi oxidize anthracene to anthraquinone (Vyas et al. 1994). Basically, the ligninolytic system contains three principal enzyme groups namely lignin peroxidase (LiP), Mn-dependent peroxidase (MnP), phenol oxidase (laccase, tyrosinase), and H_2O_2 producing enzymes (Novotný et al. 2004). Ligninolysis is an oxidative process and it is found to operate under nutrient limiting conditions for instance nitrogen limiting conditions (Novotný et al. 2004). In effect, the fungal mediated degradation of PAHs is slow and it was found that fungi have potential to degrade diverse group of xenobiotics. Ability of many soil fungi to live, grow, propagate (Lee et al. 2015a, b) and capability to degrade PAHs through fungal enzymes like cytochrome P450 monooxygenase, epoxide hydrolases, lipases, proteases and dioxygenases have been studied in depth (Balaji et al. 2014).

The two main enzyme groups involved in the initial attack on PAHs by fungi are the cytochrome P-450 monoxygenases and the lignin peroxidases (Fig. 6.9). Both

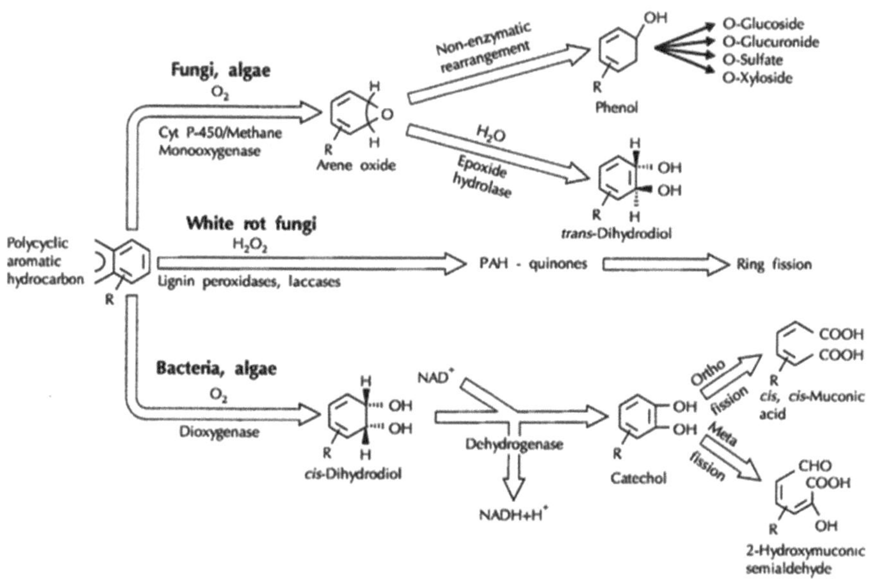

Fig. 6.9 Initial steps in the microbial pathways for oxidation of polycyclic aromatic hydrocarbons. (Adopted from Cerniglia 1993)

enzymes are relatively non-specific for the PAHs that they metabolize. Cytochromes P-450 incorporate one atom of molecular oxygen into the PAH molecule to form an arene oxide, which then undergoes either spontaneous isomerization to form a phenol, with subsequent conjugation with sulfate, glucuronic acid, glucose or xylose, or enzymatic hydration to form a trans-dihydrodiol. A non-ligninolytic fungus, *Cunninghamella elegans*, has been shown to metabolize PAHs that range in size from naphthalene to benzo[a]pyrene (Sutherland 1992).

Lignin peroxidases and laccases are extracellular enzymes produced by the white rot fungi (Field et al. 1992). The lignin peroxidases have been shown to oxidize PAHs that have ionization potentials of less than about 7.6 eV. The extracellular lignin peroxidases initiate a free radical attack on PAHs, by a single electron transfer, to form quinones. *Phanerochaete chrysosporium* was reported by Hammel et al. (1991) to oxidize anthracene via 9,10-anthraquinone to phthalate. In a similar study, they showed that *P. chrysosporium* oxidizes phenanthrene at its C9 and C10 positions to give 2,2′-diphenic acid as a ring cleavage product (Hammel et al. 1991). These results suggest that both lignin peroxidase and other enzymes may be involved in the degradation of PAHs. Sutherland (1992) reported that *P. chrysosporium* under non-lignolytic conditions metabolizes phenanthrene to phenols and trans-dihydrodiols, which suggests a cytochrome P-450 mediated reaction. Thus, *P. chrysosporium* contains several enzymes that may be involved in the degradation of PAHs.

Ligninolytic enzymes enable a one electron radical oxidation of contaminant resulting in the formation of aryl cation radicals and subsequently quinones (Vyas et al. 1994). It was reported that pure culture of *P. chrysosporium* degrade anthracene to anthraquinone (Hammel et al. 1991) which was broken down into phthalic acid and carbon dioxide. Further, it was found that purified forms of lignin peroxidase and manganese peroxidase were found to oxidize anthracene, pyrene, fluorene and benzo[a]pyrene to quinones (Hammel et al. 1991, 1992; Hammel, 1995; Bogan and Lamar 1996) (Fig. 6.10).

Hence, several systems are involved in the degradation of PAHs with fungal enzymes including intracellular cytochrome P450 and extracellular lignin peroxidase, manganese peroxidase and laccase. The pathways of degradation of PAHs vary for each enzymes and dependent on nutritional and environmental conditions of fungal strains (Fig. 6.11).

PAHs biodegradation occurs both under aerobic and anaerobic conditions. Clemente et al. (2001) reported the highest degradation of naphthalene (69%) and phenanthrene (12%) by a fungal strain that had manganese peroxidase enzyme system. Soil fungi like *Aspergillus* sp., *Trichocladium canadense,* and *Fusarium oxysporum* coulddegrade polycyclic aromatic hydrocarbons of low-molecular-weight PAHs (2–3 rings) and produce ligninolytic enzymes and activities (LiP, MnP, laccase). Under anaerobic conditions, ligninolytic enzyme activity was not normally observed (Silva et al. 2009).

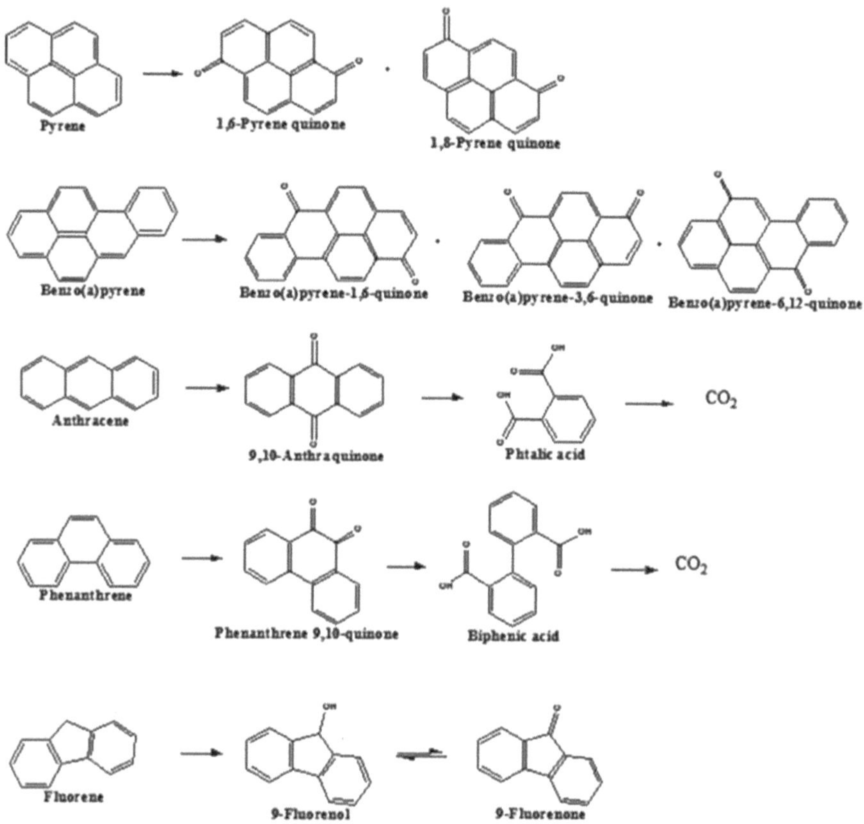

Fig. 6.10 Oxidation of PAHs by lignolytic fungi. (Kadri et al. 2017)

6.7 Algal Degradation of PAHs

Currently, the use of microalgae in bioremediation of colored wastewater has attracted great interest due to their central role in carbon dioxide fixation. In addition, the algal biomass generated has great potential as feedstock for biofuel production. These bioremediation capabilities of microalgae are useful for environmental sustainability. Compared to bacteria and fungi, relatively little attention has been paid to the biodegradation of PAHs by microalgae (cyanobacteria, diatoms etc.). Microalgae are one of the major primary producers in aquatic ecosystems, and play vital roles in the fate of PAHs in those environments. Several strains of microalgae are known to metabolize naphthalene, phenanthrene, anthracene, BaP and other PAHs (El-sheekh et al. 2012) (Table 6.3).

The biotransformation pathway of naphthalene by microalgae *Oscillatoria* sp., strain JCM are shown in the Fig. 6.12 (Cerniglia et al. 1980). Under photoautotrophic growth conditions, strain JCM has been reported to oxidize naphthaleneto1-

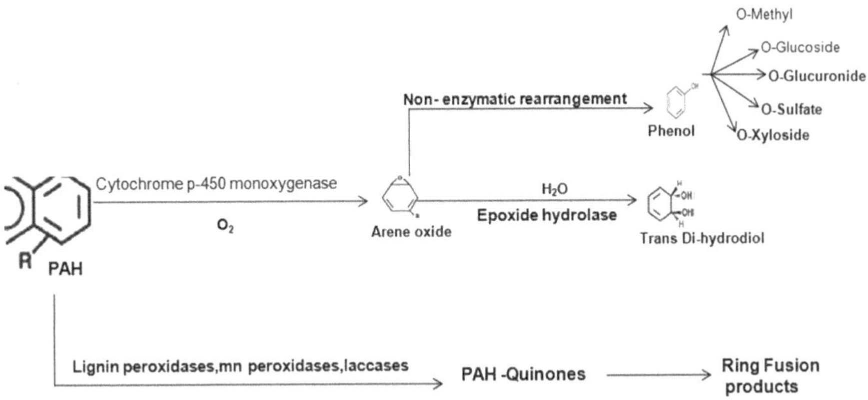

Fig. 6.11 Different pathways for the fungal metabolism of polycyclic aromatic hydrocarbons. (Modified from Kadri et al. 2017)

Table 6.3 The bioaccumulation and biotransformation of selected pesticides by eukaryotic algae

Algae	Bioaccumulation	Biotransformation
Chlamydomonas	Mirex	Lindane, naphthalene, phenol
Chlorella sp	Toxaphene, methoxychlor	Lindane, chlordimeform
Chlorococcum sp.	Mirex	
Cylindrotheca	DDT	
Dunaliella	Mirex	DDT, naphthalene
Euglena gracilis	DDT, parathion	Phenol
Scenedesmus obliquus	DDT, parathion	Naphthalene sulfonic acids
Selenastrum capricornutum	Benzene, toluene, chlorobenzene, 1,2-dichlorobenzene, nitrobenzene, naphthalene, 2,6-dinitrotoluene, phenanthrene, di-n-butylphthalate, pyrene	Benzo[a]pyrene

Adapted from Kobayashi and Rittman (1982)

naphthol, whereas marine cyanobacterium *Agmenellum quadruplicatum* strain PR-6 can convert phenanthrene to phenanthrene trans-9,10-dihydrodiol and 1-methoxyphenanthrene (Narro et al. 1992a, b).

The microalgae *Scenedesmus obliquus* GH2 is used to construct an artificial microalgal-bacterial consortium for crude-oil degradation (Tang et al. 2010). Addition of the bacterial consortium in different amendments significantly enhanced degradation efficiency of both aliphatic and aromatic hydrocarbons of crude oil. Another consortium of pre-isolated oil-degrading bacteria in association with three species of plants effectively remediated contaminated silt-loam soil more than silt, loam and sandy loam with an average 80% reduction of total petroleum hydrocarbon (Ghosh and Syed 2001).

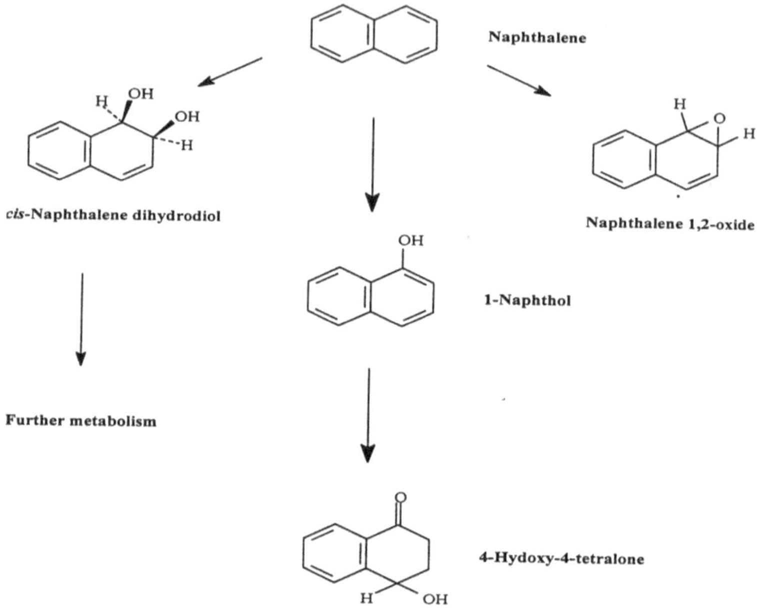

Fig. 6.12 Proposed naphthalene biotransformation pathway by the cyanobacteria, *Oscillatoria* sp. strain JCM. (Adapted from Cerniglia et al. 1980)

The phytoremediation of PAHs by algae have limited success due to the high toxicity (Dhankher et al. 2012). The accumulation and biodegradation of two typical polycyclic aromatic hydrocarbons (PAHs), phenanthrene (PHE) and fluoranthene *(FLA),* by the diatoms was studied by Hong et al. (2008), using two algal species *Skeletonema costatum* and *Nitzschia* sp. It was found that the accumulation and degradation abilities of *Nitzschia* sp. were higher than those of *S. costatum.* Degradation of FLA by the two algal species was slower, indicating that FLA was a more recalcitrant PAH compound. The microalgal species also showed comparable higher efficiency in the removal of the PHE and FLA mixture compared with PHE or FLA alone, suggesting that the presence of one PAH stimulated the degradation of the other.

Muñoz et al. (2003) suggested that it is possible to use microalgae to produce the O_2 required by acclimatized bacteria to biodegrade hazardous pollutants such as polycyclic aromatic hydrocarbons, phenolics, and organic solvents. When PAHs are taken up by microorganisms, they activated in aerobic metabolism by insertion of two oxygen atoms by bacteria and green algae to produce either cis-dihydrodiols or phenols.

Jinqi and Houtian (1992) investigated the degradation of azo dyes by *Chlorella vulgaris* and *C. pyrenoidosa* and found that certain dyes, such as Eriochrome blueSE and blackT, could be decolorized and actually used as carbon and nitrogen sources, but this was dependent on the chemical structure of the dyes (Fig. 6.13). The degradation was found to be an inducible catabolic process. They also found that the

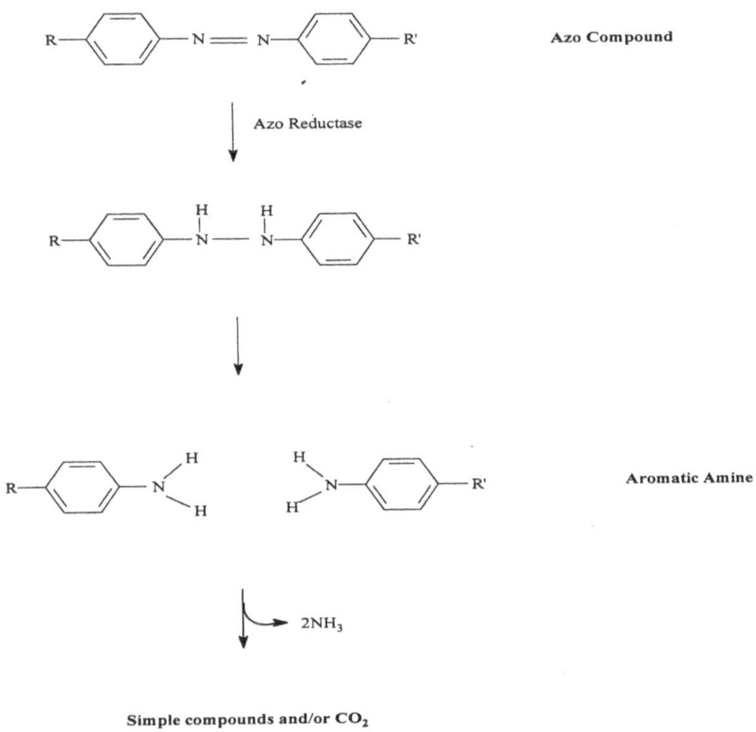

Fig. 6.13 Proposed degradation of azo dyes by eukaryotic algae. (Adapted from Jinqi and Houtian 1992)

algae degraded aniline, a potential degradation product of the azo dye breakdown. In another study, *Ochromonas danica*, a nutritionally versatile chrysophyte, grew heterotrophically on phenol or p-cresol as the sole source of carbon up to concentrations of 4 mM.

There are few examples of algae degrading aromatic compounds. (1980) and Jinqi and Houtian, (1992) have shown the removal of pollutants, accumulation of catabolic intermediates and the involved processes (Figs. 6.12 and 6.13, respectively). More detailed studies were carried out by some researchers (Lindquist and Warshawsky 1985a, b; Warshawsky et al.1995; Schoeny et al. 1988) who examined the effects of the chlorophyte alga, *Selenastrum capricornutum*, on benzo[a]-pyrene. They found that the alga used a dioxygenase system to oxidize the compound to cis-dihydrodiols which were then converted to sulfate ester and α and β-glucoside conjugates. The presence of this ring hydroxylating dioxygenase system is of particular importance as this mechanism is typically found only in bacteria and not in eukaryotes, where trans-dihydrodiols typically originate from epoxidation by the action of cytochrome P-450 monooxygenases and epoxide hydrolases on the PAH molecule.

6.8 Conclusion

During the past decade, a variety of microorganisms have been isolated and characterized for their ability to degrade different PAHs. Furthermore, many metabolic enzymes for the degradation of different PAHs have been isolated from microorganisms and several novel pathways have been elucidated based on the identification of initial ring oxidation and ring cleavage products. The genes responsible for PAHs catabolic pathways are always localized as gene clusters, and some gene clusters have been cloned and sequenced. The advancement in genetic, genomic, proteomic and metabolomic approaches, which are employed to study catabolism of organic pollutants have contributed remarkably in our understanding on the physiology, ecology, biochemistry of PAHs degrading microorganisms. However, detailed research is a prerequisite to determine exactly what is going on in PAH-contaminated environment. In addition, there are still various aspects of bioremediation of PAHs that remain unknown or otherwise have insufficient information, which requires further study. Enzymatic bioremediation is the tool to convert PAHs to less harmful/non-harmful forms with less chemicals, energy, and time. It is a solution to degrade/remove contaminants in an eco-friendly way. Microbial degradation represents the major mechanism responsible for the ecological recovery of PAH-contaminated sites. Some microorganisms are known to excrete biosurfactants which enhance the bioavailability of organic pollutants. Many microorganisms exhibit chemotaxis toward pollutants. These strategies lead to enhanced degradation of organic pollutants. The addition of small amount of biosurfactant, which increases the bioavailability of PAHs, or some merely toxic chemicals, like salicylicacid, which induce PAHs catabolic operons may enhance biodegradation of PAHs in the environment. It has been seen that organic amendments influence the indigenous microbial community as well as efficiency of bioremediation of PAHs in contaminated soil. Research has brought to the light the ability of diverse group of fungi in the bioremediation of PAHs contaminated sites through enzyme systems like MnP, LiP, laccase and other fungal enzymes, such as cytochrome P450 monooxygenase, epoxide hydrolases, lipases, protease and dioxygenase. Eukaryotic algae are capable of biotransforming and biodegrading aromatic pollutants commonly found in natural and wastewaters. However there are still some persistent organic pollutants that are difficult to degrade by the microalgae. The genetic engineering can solve this problem and offers a promising tool to improve the absorption and bioremediation of many organic pollutants and increase microalgal tolerance to these pollutants. From this chapter, we can conclude that the organisms like bacteria, fungi, and microalgae has immense potential in the biodegradation of many organic pollutants.

References

Abdel-Shafy, H., & Mansour, M. (2016). A review on polycyclic aromatic hydrocarbons: Source, environmental impact, effect on human health and remediation. *Egyptian Journal of Petroleum, 25*, 107–123. https://doi.org/10.1016/j.ejpe.2015.03.011.

Annweiler, E., Michaelis, W., & Meckenstock, R. (2001). Anaerobic cometabolic conversion of Benzothiophene by a sulfate-reducing enrichment culture and in a tar-oil-contaminated aquifer. *Applied and Environmental Microbiology, 67*, 5077–5083. https://doi.org/10.1128/aem.67.11.5077-5083.2001.

Balaji, V., Arulazhagan, P., & Ebenezer, P. (2014). Enzymatic bioremediation of polyaromatic hydrocarbons by fungal consortia enriched from petroleum contaminated soil and oil seeds. *Journal of Environmental Biology, 35*(3), 521–529.

Bansal, V., & Kim, K. (2015). Review of PAH contamination in food products and their health hazards. *Environment International, 84*, 26–38. https://doi.org/10.1016/j.envint.2015.06.016.

Bogan, B. W., & Lamar, R. T. (1996). Polycyclic aromatic hydrocarbon-degrading capabilities of *Phanerochaete laevis* HHB-1625 and its extracellular ligninolytic enzymes. *Applied and Environmental Microbiology, 62*, 1597–1603.

Boonchan, S., Britz, M., & Stanley, G. (2000). Degradation and mineralization of high-molecular-weight polycyclic aromatic hydrocarbons by defined fungal-bacterial cocultures. *Applied and Environmental Microbiology, 66*, 1007–1019. https://doi.org/10.1128/aem.66.3.1007-1019.2000.

Cébron, A., Norini, M., Beguiristain, T., & Leyval, C. (2008). Real-time PCR quantification of PAH-ring hydroxylating dioxygenase (PAH-RHDα) genes from gram positive and gram negative bacteria in soil and sediment samples. *Journal of Microbiological Methods, 73*, 148–159. https://doi.org/10.1016/j.mimet.2008.01.009.

Cerniglia, C. (1992). Biodegradation of polycyclic aromatic hydrocarbons. *Biodegradation, 3*, 351–368. https://doi.org/10.1007/bf00129093.

Cerniglia, C. (1993). Biodegradation of polycyclic aromatic hydrocarbons. *Current Opinion in Biotechnology, 4*, 331–338. https://doi.org/10.1016/0958-1669(93)90104-5.

Cerniglia, C., Van Baalen, C., & Gibson, D. (1980). Metabolism of naphthalene by the cyanobacterium Oscillatoria sp., strain JCM. *Microbiology, 116*, 485–494. https://doi.org/10.1099/00221287-116-2-485.

Christensen, N., Batstone, D., He, Z., et al. (2004). Removal of polycyclic aromatic hydrocarbons (PAHs) from sewage sludge by anaerobic degradation. *Water Science and Technology, 50*, 237–244. https://doi.org/10.2166/wst.2004.0580.

Clemente, A., Anazawa, T., & Durrant, L. (2001). Biodegradation of polycyclic aromatic hydrocarbons by soil fungi. *Brazilian Journal of Microbiology, 32*, 255–261. https://doi.org/10.1590/s1517-83822001000400001.

Dhankher, O. P., Pilon-Smits, E. A. H., Meagher, R. B., & Doty, S. (2012). Biotechnological approaches for phytoremediation. In A. Atman & P. M. Hasegawa (Eds.), *Plant biotechnology and agriculture, prospects for the 21st century* (pp. 309–328). Amsterdam: Academic.

DHHS, U. (1995). *Toxicological profile for polycyclic aromatic hydrocarbons*. Atlanta: Agency for Toxic Substances and Disease Registry.

El-Sheekh, M., Ghareib, M., & El-Souod, G. (2012). Biodegradation of phenolic and polycyclic aromatic compounds by some algae and cyanobacteria. *Journal of Bioremediation & Biodegradation, 03*. https://doi.org/10.4172/2155-6199.1000133.

Field, J. A., DeJong, E., Costa, G. F., & DeBont, J. A. M. (1992). Biodegradation of polycyclic aromatic hydrocarbons by new isolates of white rot fungi. *Applied and Environmental Microbiology, 58*, 2219–2226.

Fuchs, G., Boll, M., & Heider, J. (2011). Microbial degradation of aromatic compounds- from one strategy to four. *Nature Reviews. Microbiology, 9*, 803–816. https://doi.org/10.1038/nrmicro2652.

Gan, S., Lau, E., & Ng, H. (2009). Remediation of soils contaminated with polycyclic aromatic hydrocarbons (PAHs). *Journal of Hazardous Materials, 172*, 532–549. https://doi.org/10.1016/j.jhazmat.2009.07.118.

Gehle, K. (2009). Case studies in environmental medicine toxicity of polycyclic aromatic hydrocarbons (PAHs). *Agency for Toxic Substances and Disease Registry (ATSDR), 72*, 1355–1358.

Ghosal, D., Ghosh, S., Dutta, T., & Ahn, Y. (2016). Current state of knowledge in microbial degradation of polycyclic aromatic hydrocarbons (PAHs): A review. *Frontiers in Microbiology, 7*. https://doi.org/10.3389/fmicb.2016.01369.

Ghosh, S., & Syed, H. (2001, November 5–8). *Influence of soil characteristics on bioremediation of petroleum-contaminated soil*. Geological Society of America Annual Meeting. Boston, MA, USA.

Gianfreda, L., Xu, F., & Bollag, J. (1999). Laccases: A useful group of oxidoreductive enzymes. *Bioremediation Journal, 3*, 1–26. https://doi.org/10.1080/10889869991219163.

Hammel, K. (1995). Mechanisms for polycyclic aromatic hydrocarbon degradation by ligninolytic fungi. *Environmental Health Perspectives, 103*, 41–43. https://doi.org/10.1289/ehp.95103s441.

Hammel, K. E., Green, B., & Gai, W. Z. (1991). Ring fission of anthracene by a eukaryote. *Proceedings of the National Academy of Sciences, 88*, 10605–10608.

Hammel, K. E., Gai, Z. G., Green, B., & Moen, M. A. (1992). Oxidative degradation of phenanthrene by the ligninolytic fungus *Phanerochaete chrysosporium*. *Applied and Environmental Microbiology, 58*, 1831–1838.

Haritash, A., & Kaushik, C. (2009). Biodegradation aspects of polycyclic aromatic hydrocarbons (PAHs): A review. *Journal of Hazardous Materials, 169*, 1–15. https://doi.org/10.1016/j.jhazmat.2009.03.137.

Hong, Y., Yuan, D., Lin, Q., & Yang, T. (2008). Accumulation and biodegradation of phenanthrene and fluoranthene by the algae enriched from a mangrove aquatic ecosystem. *Marine Pollution Bulletin, 56*, 1400–1405. https://doi.org/10.1016/j.marpolbul.2008.05.003.

Jinqi, L., & Houtian, L. (1992). Degradation of azo dyes by algae. *Environmental Pollution, 75*, 273–278. https://doi.org/10.1016/0269-7491(92)90127-v.

Jones, M., Rodgers-Vieira, E., Hu, J., & Aitken, M. (2014). Association of growth substrates and bacterial genera with benzo[a]pyrene mineralization in contaminated soil. *Environmental Engineering Science, 31*, 689–697. https://doi.org/10.1089/ees.2014.0275.

Kadri, T., Rouissi, T., Kaur Brar, S., et al. (2017). Biodegradation of polycyclic aromatic hydrocarbons (PAHs) by fungal enzymes: A review. *Journal of Environmental Sciences, 51*, 52–74. https://doi.org/10.1016/j.jes.2016.08.023.

Kiehlmann, E., Pinto, L., & Moore, M. (1996). The biotransformation of chrysene to trans-1,2-dihydroxy-1,2-dihydrochrysene by filamentous fungi. *Canadian Journal of Microbiology, 42*, 604–608. https://doi.org/10.1139/m96-081.

Kobayashi, H., & Rittmann, B. (1982). Microbial removal of hazardous organic compounds. *Environmental Science & Technology, 16*, 170A–183A. https://doi.org/10.1021/es00097a002.

Kuppusamy, S., Thavamani, P., Venkateswarlu, K., et al. (2017). Remediation approaches for polycyclic aromatic hydrocarbons (PAHs) contaminated soils: Technological constraints, emerging trends and future directions. *Chemosphere, 168*, 944–968. https://doi.org/10.1016/j.chemosphere.2016.10.115.

Lee, H., Jang, Y., Lee, Y., et al. (2015a). Enhanced removal of PAHs by Peniophora incarnata and ascertainment of its novel ligninolytic enzyme genes. *Journal of Environmental Management, 164*, 10–18. https://doi.org/10.1016/j.jenvman.2015.08.036.

Lee, H., Yun, S., Jang, S., et al. (2015b). Bioremediation of polycyclic aromatic hydrocarbons in creosote-contaminated soil by Peniophora incarnata KUC8836. *Bioremediation Journal, 19*, 1–8. https://doi.org/10.1080/10889868.2014.939136.

Liang, Y., Gardner, D., Miller, C., et al. (2006). Study of biochemical pathways and enzymes involved in pyrene degradation by Mycobacterium sp. strain KMS. *Applied and Environmental Microbiology, 72*, 7821–7828. https://doi.org/10.1128/aem.01274-06.

Liang, L., Song, X., Kong, J., et al. (2014). Anaerobic biodegradation of high-molecular-weight polycyclic aromatic hydrocarbons by a facultative anaerobe Pseudomonas sp. JP1. *Biodegradation, 25*, 825–833. https://doi.org/10.1007/s10532-014-9702-5.

Liebeg, E., & Cutright, T. (1999). The investigation of enhanced bioremediation through the addition of macro and micro nutrients in a PAH contaminated soil. *International Biodeterioration & Biodegradation, 44*, 55–64. https://doi.org/10.1016/s0964-8305(99)00060-8.

Lindquist, B., & Warshawsky, D. (1985a). Identification of the 11,12-dihydro-11,12-dihydroxybenzo(a)pyrene as a major metabolite produced by the green alga, Selenastrumcapricornutum. *Biochemical and Biophysical Research Communications, 130*, 71–75. https://doi.org/10.1016/0006-291x(85)90383-3.

Lindquist, B., & Warshawsky, D. (1985b). Stereospecificity in algal oxidation of the carcinogen benzo(a)pyrene. *Experientia, 41*, 767–769. https://doi.org/10.1007/bf02012587.

Lu, X., Zhang, T., Han-Ping Fang, H., et al. (2011). Biodegradation of naphthalene by enriched marine denitrifying bacteria. *International Biodeterioration and Biodegradation, 65*, 204–211. https://doi.org/10.1016/j.ibiod.2010.11.004.

Maillacheruvu, K., & Pathan, I. (2009). Biodegradation of naphthalene, phenanthrene, and pyrene under anaerobic conditions. *Journal of Environmental Science and Health, Part A Environmental Science, 44*, 1315–1326. https://doi.org/10.1080/10934520903212956.

Marco-Urrea, E., García-Romera, I., & Aranda, E. (2015). Potential of non-ligninolytic fungi in bioremediation of chlorinated and polycyclic aromatic hydrocarbons. *New Biotechnology, 32*, 620–628. https://doi.org/10.1016/j.nbt.2015.01.005.

Meckenstock, R., & Mouttaki, H. (2011). Anaerobic degradation of non-substituted aromatic hydrocarbons. *Current Opinion in Biotechnology, 22*, 406–414. https://doi.org/10.1016/j.copbio.2011.02.009.

Mihelcic, J. R., & Luthy, R. G. (1988). Degradation of polycyclic aromatic hydrocarbon compounds under various redox conditions in soil-water systems. *Applied and Environmental Microbiology, 54*(5), 1182–1187.

Mineki, S., Suzuki, K., Iwata, K., et al. (2015). Degradation of polyaromatic hydrocarbons by fungi isolated from soil in Japan. *Polycyclic Aromatic Compounds, 35*, 120–128. https://doi.org/10.1080/10406638.2014.937007.

Mir-Tutusaus, J., Masís-Mora, M., Corcellas, C., et al. (2014). Degradation of selected agrochemicals by the white rot fungus Trametes versicolor. *Science of the Total Environment, 500–501*, 235–242. https://doi.org/10.1016/j.scitotenv.2014.08.116.

Mollea, C., Bosco, F., & Ruggeri, B. (2005). Fungal biodegradation of naphthalene: Microcosms studies. *Chemosphere, 60*, 636–643. https://doi.org/10.1016/j.chemosphere.2005.01.034.

Moody, J., Freeman, J., Doerge, D., & Cerniglia, C. (2001). Degradation of phenanthrene and anthracene by cell suspensions of Mycobacterium sp. strain PYR-1. *Applied and Environmental Microbiology, 67*, 1476–1483. https://doi.org/10.1128/aem.67.4.1476-1483.2001.

Moody, J., Freeman, J., Fu, P., & Cerniglia, C. (2004). Degradation of benzo[a]pyrene by Mycobacterium vanbaalenii PYR-1. *Applied and Environmental Microbiology, 70*, 340–345. https://doi.org/10.1128/aem.70.1.340-345.2004.

Morgan, P., Lewis, S., & Watkinson, R. (1991). Comparison of abilities of white-rot fungi to mineralize selected xenobiotic compounds. *Applied Microbiology and Biotechnology, 34*, 693–696. https://doi.org/10.1007/bf00167925.

Muñoz, R., Guieysse, B., & Mattiasson, B. (2003). Phenanthrene biodegradation by an algal-bacterial consortium in two-phase partitioning bioreactors. *Applied Microbiology and Biotechnology, 61*, 261–267. https://doi.org/10.1007/s00253-003-1231-9.

Narro, M. L., Cerniglia, C. E., VanBaalen, C., & Gibson, D. T. (1992a). Evidence for an NIH shift in oxidation of naphthalene by the marine cyanobacterium *Oscillatoria* sp. strain JCM. *Applied and Environmental Microbiology, 58*(4), 1360–1363.

Narro, M. L., Cerniglia, C. E., VanBaalen, C., & Gibson, D. T. (1992b). Metabolism of phenanthrene by themarine cyanobacterium Agmenellum QuadruplicatumPR-6. *Applied and Environmental Microbiology, 58*, 1351–1359.

Nieman, J. K., Sims, R. C., McLean, J. E., Sims, J. L., & Sorensen, D. L. (2001). Fate of pyrene in contaminated soil amended with alternate electron acceptors. *Chemosphere, 44*(5), 1265–1271.
Novotný, Č., Svobodová, K., Erbanová, P., et al. (2004). Ligninolytic fungi in bioremediation: Extracellular enzyme production and degradation rate. *Soil Biology and Biochemistry, 36*, 1545–1551. https://doi.org/10.1016/j.soilbio.2004.07.019.
Nzila, A. (2018). Biodegradation of high-molecular-weight polycyclic aromatic hydrocarbons under anaerobic conditions: Overview of studies, proposed pathways and future perspectives. *Environmental Pollution, 239*, 788–802.
Pothuluri, J. V., Selby, A., Evans, F. E., Freeman, J. P., & Cerniglia, C. E. (1994). Transformation of chrysene and other polycyclic aromatic hydrocarbon mixtures by the fungus *Cunninghamellaelegans*. *Canadian Journal of Botany, 73*, 1025–1033.
Qin, W., Zhu, Y., Fan, F., et al. (2017). Biodegradation of benzo(a)pyrene by Microbacterium sp. strain under denitrification: Degradation pathway and effects of limiting electron acceptors or carbon source. *Biochemical Engineering Journal, 121*, 131–138. https://doi.org/10.1016/j.bej.2017.02.001.
Safinowski, M., & Meckenstock, R. U. (2006). Methylation is the initial reaction in anaerobic naphthalene degradation by a sulfate reducing enrichment culture. *Environmental Microbiology, 8*(2), 347–352.
Sanglard, D. M., Leisola, S. A., & Fiechter, A. (1986). Role ofextracellular ligninases in biodegradation of benzo(a)pyrene byPhanerochaetechrysosponum. *Enzyme and Microbial Technology, 8*, 209–212.
Saraswathy, A., & Hallberg, R. (2002). Degradation of pyrene by indigenous fungi from a former gasworks site. *FEMS Microbiology Letters, 210*, 227–232.
Schoeny, R., Cody, T., Warshawsky, D., & Radike, M. (1988). Metabolism of mutagenic polycyclic aromatic hydrocarbons by photosynthetic algal species. *Mutation Research, 197*(2), 289–302.
Seo, J. S., Keum, Y. S., & Li, Q. X. (2009). Bacterial degradation of aromatic compounds. *Int J Env Res Pub He, 6*(1), 278–309.
Shen, H., Huang, Y., Wang, R., Zhu, D., Li, W., Shen, G., Wang, B., Zhang, Y., Chen, Y., Lu, Y., & Chen, H. (2013). Global atmospheric emissions of polycyclic aromatic hydrocarbons from 1960 to 2008 and future predictions. *Environmental Science & Technology, 47*(12), 6415–6424.
Silva, I. S., Grossman, M., & Durrant, L. R. (2009). Degradation of polycyclic aromatic hydrocarbons (2–7 rings) under microaerobic and very-low-oxygen conditions by soil fungi. *International Biodeterioration and Biodegradation, 63*, 224–229.
Steffen, K. T., Hatakka, A., & Hofrichter, M. (2003). Degradation ofbenzo[a]pyrene by the litter-decomposing basidiomyceteStrophariacoronilla: Role of manganese peroxidase. *Applied and Environmental Microbiology, 69*, 3957–3964.
Sutherland, J. B. (1992). Detoxification of polycyclic aromatic hydrocarbons by fungi. *Journal of Industrial Microbiology, 9*, 53–62.
Tang, X., He, L. Y., Tao, X. Q., Dang, Z., Guo, C. L., Lu, G. N., & Yi, X. Y. (2010). Construction of an artificial microalgal-bacterial consortium that efficiently degrades crude oil. *Journal of Hazardous Materials, 18*, 1158–1162.
Tian, L., Ma, P., & Zhong, J. J. (2002). Kinetics and key enzyme activities of phenanthrene degradation by Pseudomonas mendocina. *Process Biochemistry, 37*, 1431–1437.
Tsai, J. C., Kumar, M., Chang, S. M., & Lin, J. G. (2009). Determination of optimal phenanthrene, sulfate and biomass concentrations for anaerobic biodegradation of phenanthrene by sulfate-reducing bacteria and elucidation of metabolic pathway. *Journal of Hazardous Materials, 171*(1–3), 1112–1119.
Vila, J., Tauler, M., & Grifoll, M. (2015). Bacterial PAH degradation in marine and terrestrial habitats. *Current Opinion in Biotechnology, 33*, 95–102.
Vyas, B. R. M., Bakowski, S., Šašek, V., & Matucha, M. (1994). Degradation of anthracene by selected white rot fungi. *FEMS Microbiology Ecology, 14*, 65–70.

Wang, X., Gong, Z., Li, P., et al. (2008). Degradation of pyrene and benzo(a)pyrene in contaminated soil by immobilized Fungi. *Environmental Engineering Science, 25*, 677–684. https://doi.org/10.1089/ees.2007.0075.

Wang, Y., Wan, R., Zhang, S., & Xie, S. (2012). Anthracene biodegradation under nitrate-reducing condition and associated microbial community changes. *Biotechnology and Bioprocess Engineering, 17*(2), 371–376.

Warshawsky, D., Cody, T., Radike, M., Reilman, R., Schumann, B., LaDow, K., & Schneider, J. (1995). Biotransformation of benzo[a]pyrene and other polycyclic aromatic hydrocarbons and heterocyclic analogs by several green algae and other algal species under gold and white light. *Chemico-Biological Interactions, 97*(2), 131–148.

Yadav, J. S., Doddapaneni, H., & Subramanian, V. (2006). P450ome ofthe white rot fungus *Phanerochaetechrysosporium*: Structure, evolution and regulation of expression of genomic P450 clusters. *Biochemical Society Transactions, 34*, 1165–1169.

Zhang, X., & Young, L. Y. (1997). Carboxylation as an initial reaction in the anaerobic metabolism of naphthalene and phenanthrene by sulfidogenic consortia. *Applied and Environmental Microbiology, 63*(12), 4759–4764.

Zhang, S., Ning, Y., Zhang, X., et al. (2015). Contrasting characteristics of anthracene and pyrene degradation by wood rot fungus Pycnoporus sanguineus H1. *International Biodeterioration and Biodegradation, 105*, 228–232. https://doi.org/10.1016/j.ibiod.2015.09.012.

Zheng, Z., & Obbard, J. P. (2002). Polycyclic aromatic hydrocarbonremoval from soil by surfactant solubilization and*Phanerochaetechrysosporium* oxidation. *Journal of Environmental Quality, 31*(6), 1842–1847.

Zhou, N., Fuenmayor, S., & Williams, P. (2001). Nag genes of Ralstonia (formerly Pseudomonas) sp. strain U2 encoding enzymes for Gentisate catabolism. *Journal of Bacteriology, 183*, 700–708. https://doi.org/10.1128/jb.183.2.700-708.2001.

Chapter 7
Rain Gardens as Stormwater Management Tool

Piyush Malaviya, Rozi Sharma, and Pradeep Kumar Sharma

Abstract Stormwater runoff contributes significantly to urban flooding, groundwater pollution, reduction in water table, surface water quality impairment, etc. as it contains various pollutants that pose risks to life forms. Therefore, management practices must be implemented for mitigating stormwater pollution. Out of the several best management practices (BMPs), rain gardens (also known as bioretention systems (green infrastructures)) is one such practice that is being widely used these days to reduce non-point source pollution arising from urban areas. Physico-chemical and biological features of rain gardens positively helps in remediating contaminants, storing runoff water, reducing peak-flow, nutrient cycling, sequestring heavy metals and also provides supplementary benefits such as recreational facilities. In this chapter, information has been provided on stormwater pollution and use of rain gardens for stormwater treatment. The potential of rain gardens for stormwater treatment has also been critically examined by looking at the present research initiatives taken towards effective implementation of this Green Infrastructure (GI) technology.

Keywords Rain gardens · Green infrastructure · Stormwater management

7.1 Introduction

Proliferation of urban infrastructure has resulted in the qualitative degradation of ecosystem services due to the replacement of urban green surfaces/ permeable land surfaces with impermeable surfaces like paved roads, driveways, or roofs which are

P. Malaviya (✉) · R. Sharma
Department of Environmental Science, University of Jammu,
Jammu, Jammu and Kashmir, India

P. K. Sharma
Department of Environmental Science, Graphic Era University, Dehradun, Uttarakhand, India

S. Shah et al. (eds.), *Sustainable Green Technologies for Environmental Management*, https://doi.org/10.1007/978-981-13-2772-8_7

made up of concrete and asphalt. These impervious surfaces completely seal soil layers and block the infiltration of surface runoff. Improperly managed stormwater runoff, thereby, results in increased peak flows and runoff velocities and may cause floods and erosion, particularly in lower altitude areas and underground constructions. Impermeable surfaces also block the groundwater recharge pathway and hence reduce aquifer replenishment. Increased runoff from roads that are mostly travelled accelerates the mobilization and transport of pollutants resulting in degradation of the quality of water bodies that receive such pollutant loaded surface runoff. Hence, runoff from impermeable surfaces is the major contributor to the collapse of healthy freshwater ecosystems in urban areas. These problems eventually lead to increased pollutant load in the receiving water bodies, impairment of the hydrological characteristics of urban watersheds, ecosystem damage and threatening of public health. All these problems are not only because of the dramatic increase in impermeable ground surfaces, but also due to the encroachment of stream areas, changed rainfall patterns and insufficient sewer network capacity (Malaviya and Singh 2012).

Besides these consequences, stormwater has the potential to provide a non-potable water supply. If stormwater is treated properly, we can exploit this wastewater for numerous non-potable uses. It requires less treatment as compared to municipal wastewater treatment. However, the perceived risks, particularly those associated with public health, are needed to be properly addressed before utilising the stormwater (Lundy et al. 2017). Outdated techniques like end-of-pipe control technologies and automated effluent monitors do not work efficiently for the treatment of episodic and variable loading of pollutants in stormwater. Thus, structural best management practices (BMPs) such as filter strips and swales, infiltration systems, storage facilities and alternative road structures (Eriksson et al. 2007) are preferred over these traditional technologies for stormwater treatment. Out of the various BMPs, phytopurification or green infrastructure technologies act as most critical tools due to their cost effectiveness. Green infrastructures (GIs) like permeable pavement, rain gardens, and green roofs dovetail landscaping with stormwater management. Further, GIs provide opportunities for permeation of stormwater, evaporation and uptake by plants, targeting to improve water quality and the reduction of sewer overflows.

The most recommended green infrastructure technology involves rain gardens (bioretention systems) for successful use in the treatment of stormwater runoff (Alyaseri et al. 2017). Rain gardens serve as small sponges that soak stormwater into the ground through a soil-based medium, removes pathogens, reduces nutrients, organic substances and various heavy metals present in stormwater runoff. Viewing the importance of rain gardens (bioretention systems) for stormwater treatment, in the present chapter we endeavour to discuss about stormwater and its characteristics, rain gardens, their types, pros and cons, construction and also the important studies relating to stormwater and its treatment by rain gardens.

7.2 Urban Stormwater Pollution: Present Scenario

The term urban stormwater is defined as runoff concomitant with rain or snow storm and that can be measured in a downstream river, stream, trench, sewer or pipe shortly after the precipitation has reached the ground (NRC 2008). Water that has been permeated into the ground but reaches a stream channel within a day or so of the rainfall is also included in stormwater. Stormwater pollutant load is comprised of litter, debris and sediments which are visually apparent components and nutrients, heavy metals, coliforms and toxic chemicals (e.g., polycyclic aromatic hydrocarbons (PAHs), polychlorinated biphenyls, organochlorines) which forms hidden component (Taebi and Droste 2004).

Urban set-up has significantly affected the quality of ecosystem services. The quality of ecosystem services has been worsening day by day due to the replacement of urban green spaces with water-resistant surfaces that are made up of materials like concrete and asphalt, which completely seal soil layers (Ishimatsu et al. 2017). Subsequently, increased impermeable surfaces alter the hydrological regimes and local stream discharge resulting in higher flood peaks in urban streams and rivers (Gallagher et al. 2011). Due to the biodiversity loss and changed rainfall patterns, urban flooding caused by stromwater is threatening the life of urban dwellers.

Accumulation of surface water not only damages human and animal lives and property in nearby areas by flooding but also carries nutrients and chemicals to nearby wetlands and water bodies and pollute them (Fu et al. 2005). Oils and other chemicals deposited on the surfaces of roads are washed off by rain, and thus initial flush of water runoff containing a number of anthropogenic pollutants enters nearby streams, ponds, and wetlands (Hostetler 2009). Thus, urbanization induces an increase in pollutant transportation rates over impermeable surfaces in addition to increase in pollution sources (Fu et al. 2005). Runoff, therefore, needs to be slowed down and retained so as to prevent volumes of water polluting freshwater ecosystems or flooding of low land areas (Hostetler 2009; Van-meter et al. 2011).

The types of surfaces encountered determine the quality of the stormwater and the characteristics of the pollutants (e.g., roads, parking lots, roofing materials, recreational areas, etc) (Steuer et al. 1997). Stormwater runoff from urbanized watersheds is also perceived to be a source of pollutants to water bodies (U.S. EPA 1995a, b). NPS pollution mainly contributes to the impairment of receiving waters through the Unites States (U.S. EPA 2002). The list of 25 Selected Stormwater Priority Pollutants (SSPP) derived from DayWater project along with the bacterial indicators is presented in Table 7.1. The table includes six categories: basic water quality parameters, metals, PAHs, herbicides, miscellaneous organic compounds (Eriksson et al. 2007) and bacterial indicators (Oliveri et al. 1977; Characklis et al. 2005). The presence of microorganisms in stormwater and wet discharges is not specified in the list derived from the DayWater project but bacterial indicators such as faecal coliforms and pathogenic organisms are also identified as stormwater priority pollutants and hence given due consideration and are also presented in Table 7.1. (Oliveri et al. 1977; Characklis et al. 2005).

Table 7.1 List of selected stormwater priority pollutants (indicator parameters)

Type	Name
Basic parameters	Biochemical oxygen demand,
	Chemical oxygen demand
	Suspended solids, Phosphorus
	Nitrogen, pH
Metals	Chromium as chromate
	Cadmium, Platinum, Copper, Nickel, Lead, Zinc
PAHs	Benzo[a]pyrene, Naphthalene, Pyrene
Herbicides	Phenmedipham, Pendimethalin
	Terbutylazine, Glyphosphate
Miscellaneous	Nonylphenol ethoxylates and degradation products (e.g., nonylphenol), 2,4,4-Trichlorobiphenyl (Polychlorinated biphenyl 28), Di(2-ethylhexyl) phthalate, Methyl *tert*-butyl ether, Pentachlorophenol
Bacterial indicators	Fecal coliforms (*Escherichia coli*,[b] *Enterococci* sp.[b])
	pathogens (*Pseudomonas aeruginosa*,[a]
	Staphylococcus aureus,[a] *Clostridium perfringens*[b])

From [a]Oliveri et al. (1977), [b]Characklis et al. (2005), Eriksson et al. (2007)

7.3 Uses of Urban Stormwater

On one side, stormwater is creating multiple problems but on the other side, reclaimed stormwater can be used for several non-potable purposes. Stormwater requires less treatment than municipal wastewaters. However, the perceived risks, particularly those associated with public health should be properly addressed before using the treated stormwater for any purpose. It is advisable to manage the stormwater runoff for beneficial purposes like reducing pollution and erosion issues in receiving water bodies. Potential applications of reclaimed stormwater are given below (Lundy et al. 2017).

Non irrigational use: Toilet flushing, vehicle washing, street cleaning.

Irrigational use: Lawns/flowers and shrubs Parks, playground, public open spaces, sports ground, nurseries, agricultural crops, orchards.

Habitats, aesthetics and recreation: Ornamental/recreational waterbodies, detention/retention basins, wetlands.

Water supply and recharge: Surface reservoirs and ground recharge.

7.4 Best Management Practices (BMPs) for Stormwater Treatment

The degree of the stormwater pollution is expected to worsen further due to the growing population, swing of the world's population to urban locales and landscape alteration that takes place there. There is urgent need of adopting proper management practices to lower down the concentration of pollutants in stormwater before they are discharged into the receiving waters (Malaviya and Singh 2012). Stormwater control measures (SCMs) or integrated structural and non-structural BMPs needs to be designed as an efficient system to tackle with this issue of stormwater pollution. BMPs must incorporate watershed goals, site characteristics, land use development, erosion and sedimentation controls, aesthetics, monitoring and maintenance.

Non-structural SCMs or BMPs includes product substitution, better site design, conservation of natural areas, and watershed and land-use planning which can radically reduce the volume of surface runoff and the pollutant load that it carries. Ion exchange, electrolyte or liquid extraction, electrodialysis, precipitation, reverse osmosis, etc. are few conventional and emerging methods that can be applied to reduce contaminants to acceptable level before discharging stormwater into water bodies. Structural BMPs or sustainable urban drainage systems (SUDs) like filter strip, swales, infiltration trench, infiltration and detention basins, sedimentation tanks, rain gardens, constructed wetlands, lagoons, etc. are also widely used these days to reduce the problems arising due to stormwater runoff (Eriksson et al. 2007). Different best management practices for stormwater have been discussed along with the short descriptions in Table 7.2 (Scholes et al. 2005).

The major limitation of stormwater runoff is its diffuse delivery, which requires extensive regional infrastructure in union with a system that allows degradation as well as removal of pollutants present in it (Malaviya and Singh 2012). One of the BMPs, rain gardens, can overcome this problem by allowing infiltration of diffused runoff from sealed surfaces or roofs into themselves. Rain gardens are also known as bioretention cells, bioretention systems, biofilters, bioswales. Contrary to the all of the above-mentioned solutions, rain gardens provide a cost-effective, environmentally sound solution (Sweets 2013). This is an important BMP and is commonly adopted in commercial and residential areas (Davis et al. 2006; Dietz 2007; Boivin et al. 2008; Piguet et al. 2008).

Rain gardens have been recommended as a best management practice (BMP) to reduce NPS pollution from urban areas (Prince George's County 1993; US EPA 2000). Rain gardens are shallow depressions in the landscape planted with trees and/or shrubs and covered with a bark mulch layer or ground cover. They allow stormwater to creep down into the soil structure and replenish aquifers, reduce peak flows, effectively remove sediments, heavy metals, phosphorus, nitrogen, hydrocarbons and pathogenic bacteria from stormwater and hence also improve the quality of ecosystem. Rain gardens make use of various processes like evapotranspiration, infiltration, overflow, exfiltration to native soils, underdrain discharge, sorption, vegetative uptake and particulate capture to manage stormwater runoff (Roy-Poirier 2009).

Table 7.2 List of different types of best management practices (BMP) (Scholes et al. 2005)

BMP type	Description
Filter strip	Grassed or vegetated strip of ground that stormwater flows across
Swales	Vegetated broad shallow channels for transporting Stormwater
Soakaways	Underground chamber or rock-filled volume; stormwater soaks into the ground via the base and sides; unplanted but host to algal growth
Infiltration trench	A long thin soakaway; unplanted but host to algal growth
Infiltration basin	Detains stormwater above ground, which then soaks away into the ground through a vegetated or rock base
Detention basin	Dry most of the time and able to store rainwater during wet conditions; often possess a grassed surface
Retention pond	Contain some water at all times and retains incoming stormwater; frequently with vegetated margins
Sedimentation tank	Symmetrical concrete structure containing appropriate depth of water to assist the settling of suspended solids under quiescent conditions
Extended detention basin	Dry most of the time and able to store rainwater during wet conditions for up to 24 h; grassed surface and may have a low basal marsh
Filter drains	Gravelled trench systems where stormwater can drain through the gravel to be collected in a pipe; unplanted but host to algal growth
Lagoons	Pond designed for the settlement of suspended solids; fringing vegetation can sometimes occur
Constructed wetlands	Vegetated system with extended retention time
(a) Subsurface flow	Typically contain a gravel substrate, planted with reeds, through which the water flows
(b) Surface flow	Typically contain a soil substrate, planted with reeds, over which the water flows
Porous asphalt	Open graded powdered/crushed stone with binder: high void ratio; no geotextile liner present
Porous paving	Continuous surface with high void content, porous blocks or solid blocks with adjoining infiltration spaces; an associated reservoir structure provides storage; no geotextile liner present; host to algal growth

7.5 Rain Gardens: An Eco-technological Advancement as Structural Best Management Practice

Rain gardens (bioretention systems) are usually built close to cemented areas such as roads and parking lots to reduce peak runoff flow and improve water quality. A rain garden is a small depression constructed in residential lawn to temporarily hold and soak rainwater coming from a house roof, driveway or other open area. The collected rainwater allows groundwater to replenish and the plants to retain some of the stormwater. Rain gardens are typically planted with ornamental grass, perennial flowers and woody shrubs that are well adapted to wet and dry conditions. The rain garden aims at recovery of water quality in nearby water bodies and storage of rainwater for plant use which is otherwise sent through drains straight into sea (Prince George's Country 1993; Alyaseri et al. 2017; Kluge et al. 2018).

Rain gardens (also known as bioretention systems) are a key piece of green infrastructure, an infrastructure that simulates natural processes to vaporize, reuse, or return stormwater to groundwater (Sweets 2013). Rain gardens are the most commonly used GIs (Green infrastructures) because of their versatile nature and better performance capability. Rain gardens make use of evapotranspiration and infiltration processes to restore natural hydrology of the site. A rain garden is a depression of depth of about 0.8–0.9 m and backfilled with highly pervious medium to increase infiltration. The filling media comprised of sand, top soil, organic matter (Alyaseri et al. 2017).

The first rain gardens were constructed primarily to impersonate the natural water retention areas that occurred before development of an area. The rain gardens for residential use were firstly developed by Dick Brinker in 1990 in Prince George's County, Maryland, when he had the idea to replace the traditional pond with a bioretention area. He discussed his idea with Larry Coffman, an environmental engineer and the county's Associate Director for Programs and Planning in the Department of Environmental Resources (US EPA 1995a, b). That idea resulted in the construction of rain gardens on each house's property in Somerset, a residential subdivision (Wisconsin Natural Resources (magazine) 2003). This way, concept of rain gardens came into practical utilisation and quiet common nowadays.

7.5.1 Structure of Rain Gardens

Rain gardens normally look like common gardens but there are some specific features that distinguish them from ordinary gardens. These specific features favour increased runoff infiltration and short-time storage to underlying soil layers and contribute to total runoff reduction as well as its peak (Katsifarakis et al. 2015). A typical rain garden mainly consists of the three parts; ponding area, inflow and overflow structures (Basdeki et al. 2016):

(a) **Ponding area:** It is a naturally occurring or artificially constructed depression in the ground. Bottom of the ponding area is filled with mulch layer and then the top soil is added. Gravel layer needs to be constructed on bottom, in case the water infiltration rate in the underlying strata is not adequate. A punctured under-drain pipe can also be used in same situation. Surfaces with large slope are not much suitable for rain garden construction, therefore, at such landscapes, it is formed by soil excavation and building an earth berm at the downslope side.

(b) **Inflow structure:** Runoff from downspouts or surrounding impermeable areas (streets, sidewalks) is directed by inflow structure into the ponding area.

(c) **Overflow structure:** When the ponding area is full, collected water leaves the rain garden through overflow structure and is directed towards the desired place (usually the sewer network). This structure helps to reduce erosion risk. The detailed structure of a rain garden is shown in Fig. 7.1.

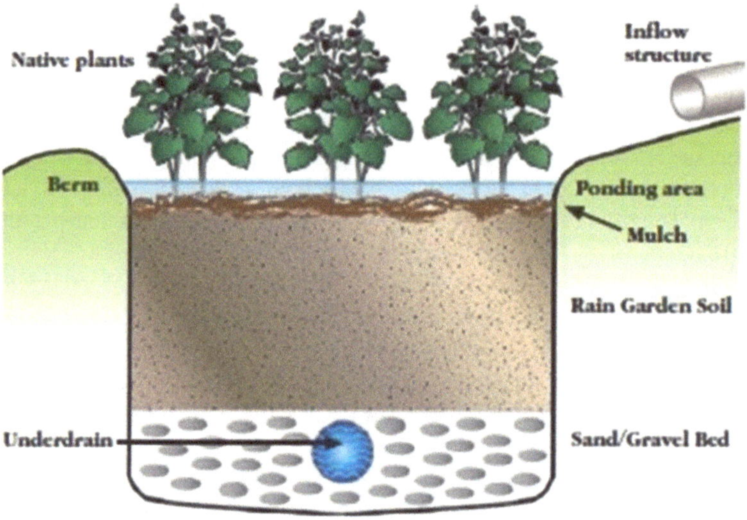

Fig. 7.1 Rain garden with ponding area, inflow and underdrain structure, berm, garden bed and plants growing in it. (Source: https://www.fairfaxcounty.gov/soil-water-conservation/sites/soil-water conservation/files/Assets/images/rgcross_thumb.jpg)

7.5.2 Benefits of Planting Rain Gardens

Rain gardens have a number of benefits associated with them like they reduce total runoff, store water for future use, replenish groundwater recharge, mitigate pollution, etc. (Katsifarakis et al. 2015). Various advantages of rain gardens are given below.

- Reduces total stormwater runoff and peak flow.
- Stores rainwater which can be used for irrigation of gardens.
- Increases groundwater recharge and helps in local aquifer replenishment.
- Reduces property loss and activity disruption that is caused by excessive runoff.
- Make use of processes like pollutant retention, filtration, decomposition, plant uptake, etc. to improve the quality of urban runoff.
- Reduces pollutant load of runoff receiving water bodies.
- Plantings of flowers, grasses and ornamental plants, berms adding height, contrast, and texture to level areas adds to the beauty of place and provides a pleasing garden view.
- Used as a stormwater retrofit by altering existing landscaped areas or resurfacing parking lot.

7.5.3 *Drawbacks of Rain Gardens*

Although rain gardens have a number of advantages but few drawbacks are also associated with them like stagnant water in them may attract pests, can pose accidental drowning, contaminate groundwater in case they are not isolated from ground water table, etc.

- Rodents, raccoons, opossums, insects and many other pests are attracted towards the rain gardens.
- If not drained properly, they create stagnant pools of water that act as breeding grounds for pests.
- There is also the possibility of drowning hazard in rain garden when it is amply pooled.
- They can be placed over utility crossings only if the trench dams are installed.
- The ground water table can intersect with the bed of the bioretention facility, if rain garden is not isolated from it.

7.5.4 *Types of Rain Gardens*

Depending upon the function, rain gardens are grouped into detention basins, filtration basins, infiltration basins, ephemeral ponds and sedge meadows. (https://www.slideshare.net/SarinaLotlikar/raingarden-education)

- **Detention basins:** artificial ponds built adjacent to rivers, lakes, streams and ponds to protect against flooding, erosion and hold water for short period of time.
- **Filtration basins:** shallow artificial ponds that infiltrate water and protect water quality in lakes, streams, rivers and ponds.
- **Infiltration swales:** shallow channels constructed adjacent to streets or residential areas that collect and dispose stormwater runoff.
- **Ephemeral ponds:** natural or artificial pools that provide habitat for some animals.
- **Sedge meadows:** large natural or artificial wetland areas planted mostly with sedges and other native wildflowers. Sedge meadow infiltrates water and also protects shorelines from erosion.

Depending upon the soil type and extent of infiltration of inflow, rain gardens are of full infiltration, full infiltration with reservoir, partial infiltration and partial infiltration with reservoir types (https://www.crd.bc.ca/education/green-stormwater-infrastructure/rain-gardens).

- **Full Infiltration type:** These rain gardens have soil permeability greater than 30 mm/hr. In this type of rain gardens, inflow completely infiltrates into the underlying subsoil.

- **Full Infiltration with Reservoir type:** These rain gardens have subsoil permeability greater than 15 mm/h and show full infiltration of water. They have a drain rock reservoir that causes surface water to move swiftly through the installed growing medium but infiltrate slowly into subsoils from the reservoir below.
- **Partial Infiltration type:** In these rain gardens, most of the surface water infiltrates into the underlying soil and surplus overflow is drained by perforated pipes. Such pipes are placed near the top of the drain rock reservoir. These rain gardens have subsoil permeability ranging between 1–15 mm/h.
- **Partial Infiltration with Flow Restriction type:** These are a kind of rain gardens which have subsoil permeability less than 1 mm/h, act like a small detention facility, offer water quality treatment and allow some infiltration. Additionally, there is a flow restrictor assembly with a small outlet which slowly empties the top portion of the water reservoir.

7.6 Construction and Design Considerations

Design Considerations Before designing rain gardens, various important factors should be kept in mind like type of soil and plants, location where it is to be constructed, appropriate depth, size and shape of garden, etc. (http://extensionpublications.unl.edu/assets/pdf/g1758.pdf).

- A rain garden should be kept at least 10 feet downslope from a house.
- Depth of rain garden should be 2–6 inch and approximately 18 inches, if standing water is not desired and desired, respectively.
- Size of rain garden should be 70 square feet approximately. Shape or design should follow the drainage system of the landscape.
- Water mains, electrical lines and all other utilities should be well marked to avoid digging up or over them during rain garden construction.
- Rain gardens should be built away from the septic drain fields.
- Construction of gardens should be avoided in right of way areas (e.g. phone lines, adjacent public roads) unless specific permission is granted by the utility that owns the right of way.
- Plants adaptable to rain garden conditions, regionally native or adapted plants found in natural landscapes should be planted in rain gardens.
- Soils with infiltration rate >0.25 inches/h are believed to be effective for rain gardens.
- Insolation should be adequately available for rain garden plants.

Construction Selection of location, size and depth of rain gardens are foremost points that are to be kept in mind before designing rain gardens. After that garden bed is prepared to grow and support suitable plants. Timely maintenance is also needed for long life of rain gardens. Following steps are involved in constructing rain garden. (http://www.montgomeryconservation.org/wp-content/uploads/2012/11/01_raingardenbrochure.pdf)

Selection of location of garden

↓

Determining the appropriate size of garden

↓

Preparation of garden bed

↓

Selection of plants to be grown

↓

Timely care and maintenance

- Appropriate location of rain garden is selected. Garden is located near downspouts and/or in a low spot where water collects.
- After the location has been decided, next step is to determine the proper size of rain garden. Size is not a critical factor for a residential rain garden. Rain gardens are typically 100 square-feet or less in size and 6–8 inch deep.
- For preparation of garden bed, ditch is dug out and bottom of the bed is levelled to the depth of 6–8 inch. A berm (mound) is built at the low end of the garden. If soil is clayey or compacted, the garden is over-dug 12 inch below the calculated rain garden depth, removing some soil and replacing with a mix of sand and compost. A path from the downspout to the garden is constructed using a grass swale, gravel, or plastic pipe/extension. Stones are arranged at the inlet of water to the garden to slow the flow and prevent erosion.
- Selection of the plants which are to be grown in the rain garden is made. Native plants are best suited. Soil type (sand, silt, loam or clay) is also taken into consideration during plant selection. Taller plants are planted in the middle or back of the garden and drier species on the berm and perimeter of the garden.
- Care and maintenance is necessarily required for efficient construction and working of rain gardens. Plants are watered regularly until they become established. Mulching with hardwood and removal of weeds is done. It is advisable to prune and remove dead vegetation in spring. Plants can also be moved around if it appears that they would do better in drier or wetter part of the garden.

7.7 Material Specifications for Rain Gardens

Rain gardens are basically constructed by using materials like mulch, gravels, stones, manure and soil. There are some specifications associated with each of the material used. Appropriate size and quantity of the materials and the way to use

Table 7.3 List of materials used in rain gardens and their specifications

Material	Specifications	Size/depth	Notes
Mulch	Double-shredded hardwood or approved pine straw substitute	NA/2–3 inches	Aged minimum of 6 months.
Cobble/Stone	Washed river rock, large gravel, or small rip-rap	3–5 inch diameter stone/ 1 or 2 layers deep	Use at downspouts, inlets, outlets, and along hardscape edges as needed to dissipate flow and prevent soil erosion. Use filter fabric under stone.
Compost *(if not replacing the soil)*	The material should be well composted and free of viable weed seeds.	NA/Add 2 inches of compost across rain garden surface area and incorporate into top 6 inches of soil	Follow recommendations from soil test if compost amendments are suggested.
	Fresh manure should not be used because of its high bacteria and nutrient levels.		
Filter bed soil mix *(if replacing the soil)*	Bioretention soil mix from an approved Arlington County vendor OR mixture of approximately:	NA/18–36 inches	Plan for a volume of soil based on 110% of the design volume, to account for settling or compaction.
	80–90% sand		
	10–20% topsoil		
	3–5% organic matter (compost)		

https://arlingtonva.s3.dualstack.us-east-1.amazonaws.com/wp-content/uploads/sites/13/2016/05/Rain-Garden-Spec-FINAL.pdf

them plays an important role in efficient functioning of rain gardens in stormwater management. Table 7.3 shows various materials which are used in rain gardens along with their specifications, required size/depth, etc. (https://arlingtonva.s3.dualstack.us-east-1.amazonaws.com/wp-content/uploads/sites/13/2016/05/Rain-Garden-Spec-FINAL.pdf)

7.7.1 Plants Used

Plants are integral part of rain gardens. Plants remove nutrients and pollutants from stormwater through vegetative uptake and microbial community support. Plants remove surplus water through evapotranspiration and also create pathways for infiltration through root development. Different types of trees, shrubs, perennials, ferns, grasses and groundcover species can be used in rain gardens but preferably native plants are used in rain gardens. Plants that are tolerant to periodic inundation are most suitable. Different plants which are suitable for rain gardens are listed below with their common as well as scientific names. (https://pubs.ext.vt.edu/content/dam/pubs)

Trees

Bald Cypress	*Taxodium distichum*
Black Gum	*Nyssa sylvatica*
Green Ash	*Fraxinus pennsylvanica*
Hackberry	*Celtis occidentalis*
Katsura Tree	*Cercidiphyllum japonicum*
Lacebark Elm	*Ulmus parvifolia*
Red Maple	*Acer rubrum*
River Birch	*Betula nigra*
Sweetgum	*Liquidambar styraciflua*
Water Oak	*Quercus nigra*
Willow Oak	*Quercus phellos*

Shrubs

Arrowwood	*Viburnum dentatum*
Buttonbush	*Cephalanthus occidentalis*
Chokeberry	*Aronia arbutifolia*
Devilwood	*Osmanthus americana*
False Indigo	*Amorpha fruticosa*
Inkberry	*Ilex glabra*
Spicebush	*Lindera benzoin*
Steeplebush	*Spiraea tomentosa*
Swamp Rose	*Rosa palustris*
Winterberry	*Ilex verticillata*

Perrenials

Arrowhead	*Sagittaria latifola*
Beardtongue	*Penstemon digitalis*
Beebalm	*Monarda didyma*
Calla Lily	*Zantedeschia* spp.
Canna Lily	*Canna* spp.
Crinum Lily	*Crinum* spp.
Daylilies	*Hemerocallis* spp.
Gingers	*Hedychium* spp.
Lizard Tail	*Saururus cernuus*
Lungwort	*Pulmonaria* spp.
Rain Lilies	*Zephyranthes* spp.
Spiderwort	*Tradescantia* spp.
Turtleheads	*Chelone lyonii/obliqua*
Wild Ginger	*Asarum canadense*

Ferns

Holly Fern	*Cyrtomium falcatum*
Lady Fern	*Athyrium felix-femina*
Royal Fern	*Osmunda regalis*
Wood Ferns	*Dryopteris* spp.

Grasses and grass-like

Broom Sedge	*Andropogon virginicus*
Foxtail Grass	*Alopecurus pratensis*
Sweetflag	*Acorus* spp.
Switchgrass	*Panicum virgatum*
Sedges	*Carex* spp.

Ground covers

Bugleweed	*Ajuga* spp.
Foamflower	*Tiarella cordifolia*
Lilyturf	*Liriope spicata*
Mazus	*Mazus reptans*

7.8 Rain Gardens as Stormwater Management Tool

The use of rain gardens epitomizes a potentially cost-effective, practical and a relatively novel approach for treatment of stormwater. The idea of commissioning rain gardens for capturing stormwater runoff and pollutants has been extracted from an understanding of the role they play in urban areas like controlling peaks and volumes of surface runoff, maintaining evapotranspiration and groundwater recharge and reducing pollution of water bodies (Zhang and Guo 2014). Rain gardens have been recognized as an effective low impact development (LID) practice, comprehensively advocated to be built with urban landscaping for reducing stormwater runoff through the retention and infiltration processes. However, the field performance and regional effect of rain gardens have not been exhaustively studied so far.

Rain gardens have been found to be effective in flow retention and pollutants removal in several studies (Zhang and Guo 2014). Study by Hunt et al. (2008) reported about 46–100% flow reduction with rain gardens. This range was supposed to be affected by seasonal difference and stormwater characteristics. In another study by Hatt et al. (2009), peak flow reductions between 49% and 80% were observed from different rain gardens. Brown and Hunt (2011) used two bioretention cells having variable media depth in their study to find the extent of stormwater retention and their study concluded that larger depth resulted in more flow volume reduction and vice versa. Similarly, Davis (2008) studied on flow reduction with the help of two bioretention cells for 49 runoff events in 2 years and the flow peak reductions ranging from 44% to 63% were observed. This percentage of peak reductions satisfied the LID goals set for stormwater management.

Infiltration rate of a functional rain garden must be necessarily sustained because holding capability of a rain garden partially depends on infiltration capacity of its soil media. Regular influxes of stormwater deposits fine particles which cause soil clogging and reduce infiltration rate of rain gardens. Choking of the filter media by fine particles occurs with time, depends upon the characteristics of stormwater and consequently poses threat to the infiltration rate (Siriwardene et al. 2007). In few field studies, it has been found that the accumulation of fine particles do not always result in decrease of infiltration rates of rain gardens. It is because of the root development of plants and resultant macro pores in the soil medium (Jenkins et al. 2010).

Available studies on rain gardens are mostly engrossed on the performance of individual facilities which accomplish hydrological benefits through storage and permeation of stormwater runoff and are mostly conducted over a short period of time. Little attention is paid towards the regional effect of rain gardens in retaining rain runoff and their longstanding use. Research by Endreny and Collins (2009) showed that the regional effect of rain gardens and distribution of such facilities are closely linked. For proper evaluation of the system performance and validation of hydrological models, the long term monitoring of the rain garden flow processes is important, which can be then used for envisaging hydrological effect of rain garden constructions on urban stormwater reduction.

7.9 Rain Gardens for Treatment of Various SSPSs

Rain gardens/ bioretention systems are management practices designed to remove pollutants from urban runoff. This relatively new BMP is effective at removing sediments, heavy metals, pathogens, phosphorus, nitrogen and hydrocarbons from stormwater (Ning-Yuan and Tian 2016). Various studies regarding treatment of different SSPPs using rain gardens (bioretention systems) are discussed in following subsections:

7.9.1 Nitrogen Removal

Urban stormwater contains nitrogen (N) and carries it to receiving waters. The major nitrogen sources are fertiliser application and deposition of vehicle emissions. Addition of excess of N to aquatic ecosystems cause eutrophication, which alters community structure, degrades habitat quality and gives rise to harmful algal blooms. Ammonium (NH_3-N), nitrate (NO_3-N), nitrite (NO_2-N), dissolved organic nitrogen (DON) and particulate organic nitrogen (PON) are the different chemical forms in which nitrogen is present in stormwater. The nitrogen composition depends on land use and hydrologic conditions (Li and Davis 2014).

Several studies have reported poor removal of NO_3-N by rain gardens. Efforts have been made by the researchers to boost the NO_3–N removal performances of rain gardens with main focus on the development of different soil media composition and variable design configurations. Kim et al. (2003) and Dietz and Clausen (2006) in their studies found that the removal efficiency of NO_3-N from synthetic runoff can be improved by creating a submerged zone through elevating the outlet. Hsieh et al. (2007a) also reported improved NO_3-N removal by extending the contact time by using less pervious soil.

Ning-Yuan and Tian (2016) used three bioretention systems to study nitrogen removal under wetting and drying regimes: conventional bioretention column (T1) with only one layer of media; a three layered bioretention column (T2) with less pervious soil layer, planting soil layer and a sand filter layer and two- layered bioretention column (T3) composed of the planting soil layer and sand filter layer. Behaviours of inorganic N species in T1, T2 and T3 under an intermittently wetting regime were analysed. The mean NH_4^+N, NO_3-N and total nitrogen removal efficiencies for T1 were found to be 71%, 1% and 41%, respectively; for T2, respective efficiencies were 83%, 84% and 82%, and the values for T3 were found to be 63%, 31% and 53%, respectively. Important timeframe for the removal of NO_3-N was draining period.

In another study, Wan et al. (2018) reported a wood-chip bioretention system that receives the runoff from an elevated highway. Different nitrogen species were transformed which indicated that the field-scale bioretention system with wood-chips also inhibited leaching of N. Hydraulic loading rate of the bioretention system affected nitrate removal and removal was mainly observed during the wet period.

7.9.2 Phosphorus Removal

Phosphorus, besides being a macronutrient for all life-forms, is also a water pollutant. The major phosphorus sources are fertiliser application and deposition of vehicle emissions. High concentrations of phosphorus in the water bodies can lead to algal blooms and eutrophication, which has noteworthy ecological, environmental and economical impacts. Total phosphorus (TP) removals ranging between 70–85% with an average of 82% were noted by Davis et al. (2006) in bioretention box experiments.

Two rain gardens were constructed by Dietz and Clausen (2005) in Haddam, Connecticut. An increase in TP concentration was consistently observed in the underdrain of both rain gardens, over a monitoring period of 1 year. An exponentially declining trend in underdrain phosphorus concentrations and linearly decreasing trend in the inlet was observed over time. Similarly, monitoring of two field-scale bioretention cells in the University of Maryland campus, showed TP removals of 79% and 77% on mass basis for two respective cells, one of the cells was designed to include an anoxic layer (Davis 2007).

The short-term effectiveness of bioretention soil media in retaining phosphorus was studied through lab-scale column studies by Hsieh et al. (2007b). Their study reported total phosphorus removal of 85% (on mass basis) for high conductivity soil filtration media. Low conductivity media showed lower phosphorus retention potential (65% on mass basis).

Zhang et al. (2008) studied the potential of different materials used as soil media in bioretention systems to remove phosphorus. Fly ash was found to have significant potential for phosphorus sorption after performing sorption tests. Since fly ash has low permeability, therefore, sand amended with 5% of fly ash can better improve phosphorus retention in bioretention systems. They also carried out desorption tests which showed insignificant amounts of phosphorus leaching from the mixture under low concentration influents, while 42% of previously sorbed phosphorus leached from a non-amended sand sample.

7.9.3 Heavy Metal Removal

Washout from tyres, vehicular exhausts, road asphalt, fuel burning, parking dirt, and recreational land are the sources of heavy metals present in stormwater. Discharge of heavy metals into surface and subsurface water bodies can badly affect public health and also cause environmental hazards. Rain gardens are proposed as BMPs to treat stormwater contaminated with heavy metals (Reddy et al. 2014).

Glass and Bissouma (2005) measured inlet and outlet concentrations of a variety of heavy metals in a parking lot bioretention cell over a period of 90 days and reported removal efficiencies of 81%, 66%, 79%, 53%, 75%, 17%, 11% and 53% for Cu, Cd, Zn, Cr, Pb, Al, Ar and Fe, respectively.

Efficiency of bioretention systems in removing heavy metals from synthetic runoff under cold climatic conditions was examined by the researchers of Norwegian University of Science and Technology and Lulea University of Technology in Sweden (Muthanna et al. 2007a). Experiments were performed in April and August. Metal retention was good during both months leading to 90% removal of Zn in the study period. Removals were favoured by warmer temperatures. Increase from 83% to 89% was seen for lead and removal of copper from 60% to 75% was observed.

Muthanna et al. (2007b) performed a series of bioretention box experiments to test the capacity of bioretention cells for heavy metal removal from snowmelt. They used snow collected from three sites with different pollutant loadings resulting from roadway traffic. In all the cases, Zn, Cu, Pb and Cd removals in the range of 81–99% (on a mass basis) were reported.

Li and Davis (2008), in their study investigated heavy metal capture and accumulation in bioretention media through the use of one-dimensional filtration equation for particulate metals, advection/dispersion/adsorption transport equations for dissolved metals and successive extractions. Zinc, copper and lead profiles showed highest accumulation at surface and decreasing accumulation trend with increasing media depth. Highest concentrations of heavy metals and runoff particles were shown in surface street particle- enriched areas. Soluble-exchangeable bound metals from the sequential extraction allied well with predicted water dissolved metals; the more strongly connected metal fractions correlated with modelled runoff and media particulate metals.

7.9.4 *Total Suspended Solids (TSS) Removal*

The presence of suspended solids in water indicates the presence of organic matter, particulate nutrients, heavy metals and several other pollutants. Hsieh and Davis (2005) used synthetic runoff and carried out a series of twelve bioretention column tests. During the first 6-h run, sediment washout was observed and total TSS removal of 91% was noted at the end of experiment.

Field monitoring of two bioretention facilities by Davis (2007), over a period of 14 months, showed low levels of TSS removal. 54% and 59% (on a mass basis) of influent TSS was retained, respectively, in the first and second bioretention cells. Event mean concentrations (EMCs) for 23 storm events were decreased by 41% and 22% by first and second bioretention cells, respectively. Anoxic zone is included at the base of second cell. Sediment washout was also observed as TSS effluent concentrations exceeded influent concentrations, mostly during first storm events following the initiation of the system.

Li and Davis (2008) investigated the filtration mechanism of bioretention cells by performing a series of bioretention column tests and field observations. In their study, they found that stratification tends to occur in the media as particulate matter is captured by bioretention media. They also observed that finer soil particles were present in the upper media and suspended solids were trapped inside the media which altered the characteristics of media and thus reduces its hydraulic conductivity. Their study also concluded that clogging may limit the useful lifespan of bioretention filter media.

7.9.5 *BOD and Pathogens Removal*

When organic matter rich stormwater enters the water bodies, it leads to oxygen depletion because aerobic digestion occurs. The presence of pathogens in stormwater is of great concern because they pose health hazards to both human and aquatic species.

Column tests performed by Rusciano and Obropta (2007) showed that bioretention systems have the ability to reduce pathogen count in stormwater to a significant level. For the 13 experiments comprising a range of fecal coliform counts, an average reduction of 91.6% and removal efficiencies ranging between 54.5–99.8% was obtained.

The results of a 2-year monitoring project on bioretention cell constructed by Hunt et al. (2008) showed an average reduction of 69% in fecal coliform count and 71% reduction in *E. coli*. They confirmed that the bioretention cells have ability to reduce pathogens from stromwater.

7.9.6 Removal of Polycyclic Aromatic Hydrocarbons (PAHs)

Stormwater runoff containing soil and sediments eroded from roadsides often contain particles contaminated with polycyclic aromatic hydrocarbons (PAHs). The sources of PAH compounds in the soil and sediments adjacent to roads are atmospheric depositions from vehicular emissions, coal-tar based sealcot and abraded bituminous asphalt particles from paved roads (Azah et al. 2015).

Diblasi et al. (2009) installed a bioretention cell and conducted a study on 16 polyaromatic hydrocarbons (PAHs). Their study reported that mean total PAH concentration was reduced from 2.08 ug/L in the inflow of bioretention system to 0.22 ug/L in the outflow, corresponding to 90% mean reduction in concentration. The study showed that concentrations of PAHs in the outflow were same for all the events. In events with high influent concentrations, higher removal percentages were noted which suggested that bioretention cells can reduce PAHs concentrations to a specific threshold, independent of influent concentrations. Core analysis of soil media suggested that lower media depths may be enough to reduce PAHs loads because PAHs concentrations were higher in the top 10 cm of the soil media than at greater depths.

7.10 Removal of Organic Contaminants by Rain Gardens

Rain gardens permits water runoff to soak into the ground through soil media, reduce peak flows and remove particulate and dissolved pollutants. Soil organic matter (SOM) has been found to efficiently sorb many pollutants, amend the bioretention medium with highly effective adsorbents and extend lifetime of bioretention system. To compare the removal of organic compounds by rain gardens having SOM and rain gardens having soil amended with SBAC (sewage sludge-based activated carbon), few researchers have carried out studies.

One such study was carried out by Bjorklund and Li (2017). They explored the role of activated carbon in increasing the efficiency of rain gardens to remove hydrophobic organic compounds (HOCs) from stormwater. In their study, they packed three lab-scale columns with soil collected from a rain garden receiving runoff from roof and a grassed area receiving runoff from surface. Out of the three columns, activated carbon made from the sewage sludge (0.5% w/w) was added to only two. Leaching tests were performed to ensure that the soil did not release even a single studied HOCs in a detectable amount. Compounds studied were fluorene, anthracene, pyrene, 4-nonlyphenol, 4-tert-octylphenol, dibutyl phthalate and di(2-ethylhexyl) phthalate. Activated carbon (SBAC) showed considerably more adsorption capacity for tested compounds and also showed the potential to extend lifetime of bioretention medium by approximately 10–20 years before saturation of these pollutants occurs.

7.11 Performance of Bioretention Systems After Long-Term Operation

Bioretention systems are broadly used to treat stormwater pollutants and allow infiltration of diffused runoff from concrete surfaces and roof tops. However, the question arises about the long term hydraulic performance and accumulation of pollutants in these systems. Rain gardens have a predicted design life of about 25 years but currently limited data is available on metal accumulation of these systems with operational time >10 years.

Study carried out by Kluge et al. (2018) dealt with the metal accumulation in a variety of long term operational bioretention systems (11–22 years) to derive further operation recommendations for the authorities. Hydraulic performance of most bioretention systems was observed to meet the technical guidelines even after long-standing operation; significant metal accumulation was seen in the top layer of soil (0–20 cm), median concentrations of all metals were highest at the soil surface (0–10 cm) but followed declining pattern with increasing depth. High concentrations of metals were found at the inflow points of runoff water. At a distance more than 1.5 m from the inflow point, metal concentrations were increased slightly as compared to the initial soil concentrations. They concluded from the leachability tests that most of the metals deposited in bioretention soil- media were slightly soluble in water.

To determine initial flow reduction, total suspended solids, and nutrient removal, Willard (2014) evaluated the effectiveness of a bioretention cell for 5 months in Blacksburg, Virginia, USA. Results of the study indicated that the bioretention medium did not accumulate nitrogen and phosphorus since installation, and the bioretention system remained effective at reducing flow volume and peak flow rates, as well as TSS, TN, TP, total coliforms, *E. coli*, and enterococci loads.

Thus, it can be deduced that even after the long-term operations, the hydraulic conductivity of the bioretention systems show good performance. Various researches also underline the strong retaining ability of bioretention systems after long-term operation. The research on long-term operations suggested that to enhance the working quality of bioretention systems, regular replacement of the accumulated sediments, fine particles or soils at the inflow points after 20–25 years should be made. Technical optimization of the bioretention systems also leads to more homogeneous infiltration of stormwater.

7.12 Case study

Rain Gardens as an Integral Parts of Urban Sewage Systems- A Case Study in Thessaloniki (Greece)

Rain gardens cannot replace sewer network but can be used as their integral part and that too cost- effectively. Integration of rain gardens with sewage systems have been studied with the help of a case study, in Saranta Ekklisies, a neighbourhood close to the centre of Thessaloniki, Greece (Basdeki et al. 2016). Five rain gardens were constructed at different sites in the study area. The selected rain garden sites are shown in Fig. 7.2 and described hereunder.

1. Location K1 (Agiou Dimitriou Street): The area of proposed rain garden was 2415 m^2.
2. Location K2 (at the junction of Raidestou and Anaktoriou Streets): The area of the proposed rain garden was 33 m^2.
3. Location K3, K4 and K5 (parks around the A and C student dormitories) with proposed area equal to 273 m^2, 360 m^2 and 1543 m^2, respectively.

Study took into account a 2-year return period, namely the worst rain event that was likely to happen every 2 years. To calculate the rain-runoff quantities, a diagram was used in which x- and y- axis showed the duration and intensity of the rainfall, respectively while the different curves represented the worst rainfall likely to happen once in every 2, 5, 10, 20 and 50 years (Fig. 7.3). It was concluded that proposed rain gardens were able to mitigate the rain runoff in the study area. Also, the rain garden construction was calculated to be less than 50 Euros per m^2 and could be combined with rehabilitation of green areas.

Fig. 7.2 Map of the study area showing selected rain garden sites (based on EYATH map). (Basdeki et al. 2016)

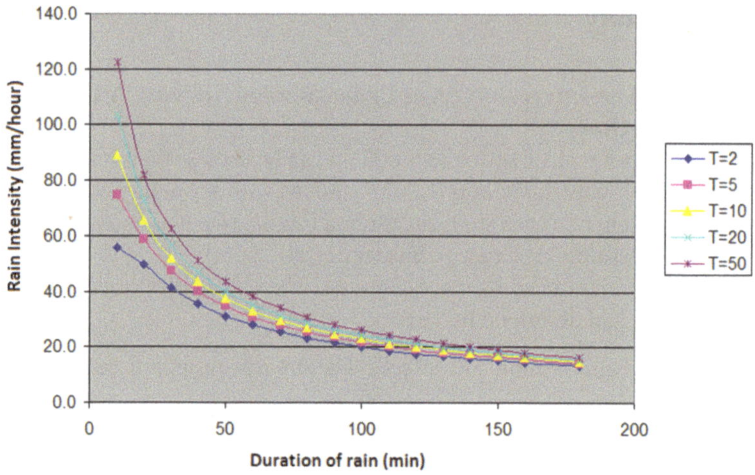

Fig. 7.3 Rain curves for the study period for different time periods. (Basdeki et al. 2016)

7.13 Conclusion

Urbanization has resulted in reduction of green grounds. Due to the replacement of urban permeable land surfaces with impermeable concrete made surfaces, soil layers have been completely sealed and blocked the way of groundwater recharge and infiltration of surface runoff. This has resulted in higher flood peaks in urban streams and rivers, pollution, reduced ground water table, etc. We need to bring drastic changes in already existing stormwater management practices so as to protect human health, urban infrastructure and natural ecosystems from stormwater hazards. Emphasis should be laid on source-control stormwater management techniques.

Rain gardens (bioretention systems), one of the Green infrastructure (GI) technologies, play an important role for urban environments to reduce impacts of stormwater runoff. In the present chapter, types, design and construction, advantages and disadvantages, composition materials, suitable plants, various functions and features of rain gardens were discussed. Rain gardens can reduce surface runoff, property damage and flood peaks, decrease pollutant load of the runoff receiving water bodies, and replenish groundwater resources. This chapter also discussed about different models of rain gardens that are constructed with different types of fill media, different area ratios, varied depression depths, different levels of imperviousness, different types of soil, etc. to treat various pollutants like nitrogen, phosphorus, TSS, BOD, PAHs, pathogens, heavy metals, etc. The rain gardens having soil amended with activated carbon were found to be more efficient in removing the organic contaminants than those having simple soil media.

In general, the environmental benefits of a green infrastructure (GI) approach to surface water drainage provision not only include reduced surface water runoff and associated pollution, but also serve to absorb carbon and link niche habitat sites within the wider urban area. Rain gardens can offer cost-effective and a low environmental impact alternative to expensive reconstruction of sewer networks. Moreover, hydraulic performance and the accumulation of the pollutants in the bioretention systems remain quite satisfactory even after the long term. Rain gardens, however, could not be considered as panacea for stormwater pollution but when combined with other structural and non-structural BMPs, effective performance could certainly be obtained.

Acknowledgements The first author (PM) acknowledges the financial support provided by University Grants Commission (UGC) in the form of UGC Research Award.

References

Alyaseri, I., Zhou, J., Morgan, S. M., & Bartlett, A. (2017). Initial impacts of rain gardens' application on water quality and quantity in combined sewer: Field-scale experiment. *Frontiers of Environmental Science & Engineering, 11*(4), 19. https://doi.org/10.1007/s11783-017-0988-5.

Azah, E., Kim, H., & Townsend, T. (2015). Source of polycyclic aromatic hydrocarbon in roadway and stormwater system maintenance residues. *Environment and Earth Science, 74*(4), 3029–3039.

Basdeki, A., Katsifarakis, L., & Katsifarakis, L. A. (2016). Rain gardens as an integral parts of urban sewage system- A case study in Thessaliniki, Greece. *Procedia Engineering, 162*, 426–432.

Bjorklund, K., & Li, L. (2017). Removal of organic contaminants in bioretention medium amended with activated carbon from sewage sludge. *Environmental Science and Pollution Research, 24*, 19167–19180.

Boivin, P., Saade, M., Pfeiffer, H. R., Hammecker, C., & Degoumois, Y. (2008). Depuration of highway runoff water into grass-covered embankments. *Environmental Technology, 29*, 709–720.

Brown, R. A., & Hunt, W. F. (2011). Impacts of media depth on effluent water quality and hydrologic performance of undersized bioretention cells. *Journal of Irrigation and Drainage Engineering, 137*(3), 132–143.

Characklis, G. W., Dilts, M. J., Simmons, O. D., Likirdopulos, C. A., Krometis, L. A. H., & Sobsey, M. D. (2005). Microbial partitioning to settleable particles in stormwater. *Water Research, 39*, 1773–1782.

Davis, A. P. (2007). Field performance of bioretention: Water quality. *Environmental Engineering Science, 24*(8), 1048–1064.

Davis, A. P. (2008). Field performance of bioretention: Hydrology impacts. *Journal of Hydrologic Engineering, 13*(2), 90–95.

Davis, A. P., Shokouhian, M., Sharma, H., & Minami, C. (2006). Water quality improvement through bioretention media: Nitrogen and phosphorus removal. *Water Environment Research, 78*(3), 284–293.

Diblasi, C. J., Li, H., Davis, A. P., & Ghosh, U. (2009). Removal and fate of polycyclic aromatic hydrocarbon pollutants in an urban stormwater bioretention facility. *Environmental Science & Technology, 43*(2), 494–502.

Dietz, M. E. (2007). Low impact development practices: A review of current research and recommendations for future directions. *Water, Air, and Soil Pollution, 186*(1–4), 351–363.

Dietz, M. E., & Clausen, J. C. (2005). A field evaluation of rain garden flow and pollutant treatment. *Water, Air, and Soil Pollution, 167*(1–4), 123–138.
Dietz, M. E., & Clausen, J. C. (2006). Saturation to improve pollutants retention in a rain garden. *Environmental Science & Technology, 40*(4), 1335–1340.
Endreny, T., & Collins, V. (2009). Implications of bioretention basin spatial arrangements on stormwater recharge and groundwater mounding. *Ecological Engineering, 35*(5), 670–677.
Eriksson, E., Baun, A., Scholes, L., Ledin, A., Ahlman, S., Revitt, M., Noutsopoulos, C., & Mikkelsen, P. S. (2007). Selected stormwater priority pollutants: A European perspective. *Science of the Total Environment, 383*, 41–51.
Fu, B., Zhao, W., Chen, L., Liu, Z., & Lu, Y. (2005). Eco-hydrological effects of landscape pattern change. *Landscape and Ecological Engineering, 1*, 25–32.
Gallagher, M. T., Snodgrass, J. W., Ownby, D. R., Brand, A. B., Casey, R. E., & Lev, S. (2011). Watershed-scale analysis of pollutant distributions in stormwater management ponds. *Urban Ecosystems, 14*, 469–484.
Glass, C., & Bissouma, S. (2005). Evaluation of a parking lot bioretention cell for removal of stormwater pollutants. *WIT Transactions on Ecology and the Environment, 81*, 699–708.
Hatt, B. E., Fletcher, T. D., & Deletic, A. (2009). Hydrologic and pollutant removal performance of stormwater biofiltration systems at the field scale. *Journal of Hydrology, 365*(3), 310–321.
Hostetler, M. (2009). *Conserving biodiversity in subdivision development* (pp. 71–80). Gainesville: University of Florida.
Hsieh, C., & Davis, A. (2005). Evaluation and optimization of bioretention media for treatment of urban storm water runoff. *Journal of Environmental Engineering, 131*, 1521–1531. https://doi.org/10.1061/(asce)0733-9372(2005)131:11(1521.
Hsieh, C. H., Davis, A. P., & Needelman, B. A. (2007a). Nitrogen removal from urban stormwater runoff through layered bioretention columns. *Water Environment Research, 79*(12), 2404–2411.
Hsieh, C., Davis, A. P., & Needelman, B. A. (2007b). Bioretention column studies of phosphorus removal from urban stormwater runoff. *Water Environment Research, 79*(2), 177–184.
http://extensionpublications.unl.edu/assets/pdf/g1758.pdf. Accessed on 10 May 2018.
http://www.montgomeryconservation.org/wpcontent/uploads/2012/11/01_raingardenbrochure.pdf. Accessed on 14 May 2018.
https://arlingtonva.s3.dualstack.us-east-1.amazonaws.com/wp-content/uploads/sites/13/2016/05/Rain-Garden-Spec-FINAL.pdf. Accessed on 15 May 2018.
https://pubs.ext.vt.edu/content/dam/pubs_ext_vt_edu/426/426-043/426-043_pdf.pdf. Accessed on 11 May 2018.
https://www.crd.bc.ca/education/green-stormwater-infrastructure/rain-gardens. Accessed on 15 May 2018.
https://www.fairfaxcounty.gov/soil-water-conservation/sites/soil-waterconservation/files/Assets/images/rgcross_thumb.jpg. Accessed on 14 May 2018.
https://www.slideshare.net/SarinaLotlikar/raingarden-education. Accessed on 11 May 2018.
Hunt, W. F., Smith, J. T., Jadlocki, S. J., Hathaway, J. M., & Eubanks, P. R. (2008). Pollutant removal and peak flow mitigation by a bioretention cell in Urban Charlotte, N.C. *Journal of Environmental Engineering, 134*(5), 403–408.
Ishimatsu, K., Ito, K., Mitani, Y., Tanaka, Y., Sugahara, T., & Naka, Y. (2017). Use of rain gardens for stormwater management in urban design and planning. *Landscape and Ecological Engineering, 13*, 205–212.
Jenkins, J. K. G., Wadzuk, B. M., & Welker, A. L. (2010). Fines accumulation and distribution in a storm-water rain garden nine years post construction. *Journal of Irrigation and Drainage Engineering, 136*(12), 862–869.
Katsifarakis, K. L., Vafeiadis, M., & Theodossiou, N. (2015). Sustainable drainage and urban landscape upgrading using rain gardens. Site Selection in Thessaloniki, Greece. *Agriculture and Agricultural Science Procedia, 4*, 338–347.
Kim, H., Seagren, E. A., & Davis, A. P. (2003). Engineered bioretention for removal of nitrate from stormwater runoff. *Water Environment Research, 75*(4), 355–367.

Kluge, B., Markert, A., Facklam, M., Sommer, H., Kaiser, M., Pallasch, M., & Wessolek, G. (2018). Metal accumulation and hydraulic performance of bioretention systems after long-term operation. *Journal of Soils and Sediments, 18*, 431–441.

Li, H., & Davis, A. P. (2008). Heavy metal capture and accumulation in bioretention media. *Environmental Science & Technology, 42*(14), 5247–5253.

Li, L., & Davis, A. P. (2014). Urban stormwater runoff Nitrogen composition and fate in bioretention systems. *Environmental Science & Technology, 48*(6), 3403–3410.

Lundy, L., Revitt, M., & Bryan, E. B. (2017). An impact assessment for urban stormwater use. *Environmental Science and Pollution Research, 25*, 1–12. https://doi.org/10.1007/s11356-017-0547-4.

Malaviya, P., & Singh, A. (2012). Constructed wetlands for management of urban stormwater runoff. *Critical Reviews in Environmental Science and Technology, 42*, 2153–2214.

Muthanna, T. M., Viklander, M., Gjesdahl, N., & Thorolfsson, S. T. (2007a). Heavy metal removal in cold climate bioretention. *Water, Air, and Soil Pollution, 183*(1–4), 391–402.

Muthanna, T. M., Viklander, M., Blecken, G., & Thorolfsson, S. T. (2007b). Snowmelt pollutant removal in bioretention areas. *Water Research, 41*(18), 4061–4072.

Ning-Yuan, T., & Tian, L. (2016). Nitrogen removal by three types of bioretention columns under wetting and drying regimes. *Journal of Central South University, 23*, 324–332.

NRC National Research Council. (2008). *Urban stormwater management in the United States*. Washington, DC: National Academy Press.

Oliveri, V. P., Kruse, C. W., Kawata, K., & Smith, J. E. (1977). *Microorganisms in urban stormwater*. USEPA Report no EPA-600/2-77-087.

Piguet, P., Parriaux, A., & Bensimon, M. (2008). The diffuse infiltration of road runoff: An environmental improvement. *Science of the Total Environment, 397*, 13–23.

Prince George's County. (1993). *Design manual for use of bioretention in stormwater management, Prince George's County (MD) Government, department of environmental protection*. Watershed Protection Branch, Landover, MD.

Reddy, K. R., Xie, T., & Dastgheibi, S. (2014). Removal of heavy metals from urban stormwater runoff using different filter materials. *Journal of Environmental Chemical Engineering, 2*(1), 282–292.

Roy-Poirier, A. (2009). *Bioretention for phosphorus removal: Modelling stormwater quality improvements*. Master of Science in Engineering thesis, Department of Civil Engineering, Queen's University, Ontario, Canada 272 p.

Rusciano, G. M., & Obropta, C. C. (2007). Bioretention column study: Fecal coliform and total suspended solids reductions. *Transactions of the ASABE, 50*(4), 1261–1269.

Scholes, L., Revitt, D. M., & Ellis, J. B. (2005). *The fate of stormwater priority pollutants in BMPs*. Public report in the DayWater project. https://www.leesu.fr/daywater/REPORT/D5-3-BMP-Benchmark-2005-04-07.pdf

Siriwardene, N. R., Deletic, A., & Fletcher, T. D. (2007). Clogging of stormwater gravel infiltration systems and filters: Insights from a laboratory study. *Water Research, 41*(7), 1433–1440.

Steuer, J., Selbig, H. W. N., & Prey, J. (1997). *Sources of contamination in an urban basin in Marquette, Michigan and an analysis of concentrations, loads, and data quality*. WRI Report 97-4242. Middleton: U.S. Geological Survey.

Sweets, P. R. (2013). Assessing the hydrology of indianapolis rain gardens. In *Proceedings of the National Conference on Undergraduate Research (NCUR) at University of Wisconsin La Crosse, WI*.

Taebi, A., & Droste, R. (2004). Pollution loads in urban runoff and sanitary wastewater. *Science of the Total Environment, 327*, 175–184.

U.S. Environmental Protection Agency. (1995a). *National water quality inventory: 1994 report to congress*. EPA/841/R-95/005. Washington, DC: U.S. EPA.

U.S. Environmental Protection Agency (EPA). (1995b). *Urban runoff: Nonpoint source newsnotes*. Washington, DC: U.S. EPA.

U.S. Environmental Protection Agency. (2002). *National management measures to control nonpoint source pollution from urban areas-draft*. EPA 8421-B-02–003. Washington, DC: U.S. Office of Water.
United States Environmental Protection Agency. (2000). *Low Impact Development (LID), a literature review*. EPA-841-B-00-005, Office of Water, Washington, DC, 20460.
Van-Meter, R. J., Swan, C. M., & Snodgrass, J. W. (2011). Salinisation alters ecosystem structure in urban stormwater detention ponds. *Urban Ecosystems, 14*, 723–736.
Wan, Z., Li, T., & Liu, Y. (2018). Effective nitrogen removal during different periods of a field-scale bioretention system. *Environmental Science and Pollution Research, 25*, 17855–17861. https://doi.org/10.1007/s11356-018-1954-x.
Willard, L. L. (2014). *Does it pay to be mature? Assessing the performance of a bioretention cell seven years post-construction*. Master of Science in Biological Systems Engineering thesis, Faculty of Virginia Polytechnic Institute, USA 127 p.
Wisconsin Natural Resources Magazine. (2003). *Rain gardens made one Maryland community famous*. https://dnr.wi.gov/wnrmag/html/supps/2003/feb 03/run.htm
Zhang, S., & Guo, Y. (2014). Stormwater capture efficiency of bioretention systems. *Water Resources Management, 28*, 149–168.
Zhang, W., Brown, G. O., Storm, D. E., & Zhang, H. (2008). Fly-ash-amended sand as filter media in bioretention cells to improve phosphorus removal. *Water Environment Research, 80*(6), 507–516.

Chapter 8
Application of Geospatial Techniques in Hydrological Modelling

Dharmaveer Singh, S. P. Rai, and Dinesh Rai

Abstract Water management planners face considerable uncertainty in determining water availability, as change in global climate and land use/land cover patterns have altered hydrologic conditions and led to change in the dynamics of the water cycle. Therefore, it calls for having a better understanding of hydrologic processes and surface/groundwater dynamics in hydrologic and water resources management studies. The water budget analysis of a region can be done with the help of hydrological models and geospatial techniques. In the present chapter, applicability of geospatial techniques in hydrological modeling is described in detail.

Keywords Geographic information systems (GIS) · Remote sensing · Hydrological modelling · Water resource management

8.1 Introduction

Water is an important natural resource and it is essential to sustain life over the Earth' surface. Availability of abundant water on time is a prime concern for agriculture, energy and industrial sectors. The world has viewed an increase in demand of water due to rapid growth in population, urbanization and industrialization. However, usable water is reducing and becoming scarce day by day due to changed pattern of climate, reduced storage areas and increased anthropogenic disturbances leading to the changed land use/land cover patterns. These, in combined, have altered the hydrologic conditions and led to the reduced availability of water from various surface and sub-surface sources. The Intergovernmental Panel on Climate Change (IPCC) in its Fifth Assessment Report (AR5) have expressed concern for global

D. Singh (✉)
Symbiosis Institute of Geo-informatics, Symbiosis International (Deemed University), Pune, India

GIS Cell, Motilal Nehru National Institute of Technology, Allahabad, India

S. P. Rai · D. Rai
Hydrological Investigation Division, National Institute of Hydrology, Roorkee, India

S. Shah et al. (eds.), *Sustainable Green Technologies for Environmental Management*, https://doi.org/10.1007/978-981-13-2772-8_8

warming and extreme climate events as these are anticipated to become more persistent in the future (IPCC 2013). In consonance with, changed in patterns of rainfall is also predicted, which may affect the spatial and temporal availability of regional water. This may, further, aggravate crisis in water sector. Therefore, supplying water for various sectors such as agricultural, industrial, energy and domestic is one of the great challenges for twenty-first century. This suggests for adopting some efficient approaches for analyzing the uses, depletion, and productivity of water to maintain continuous water supply (Qin et al. 2008). In this respect, geospatial techniques, combined with hydrological models, can be of great importance, and effectively be employed in hydrologic and water resources management studies.

8.2 Hydrological Models

Hydrological or watershed models, developed to simulate the catchment behavior, are complex mathematical models (Xu et al. 2001). These models use complex mathematical equations to explain how climatological forcings (precipitation and moisture) are converted into the watershed response in the form of runoff through processes such as interception, evaporation, transpiration and infiltration etc. (Shelton 2009). These models are being used as state-of-the-art tools due to their economic feasibility in water resource management since many years. These assist in better understanding of hydrological phenomena functioning in the basin and also show how changes in the basin parameters affect these phenomena. Hydrological models, applied to predict/simulate hydrological behaviour of the basin under changing climate conditions and land use/land cover patterns, provide synthetic time series of hydrological data for facility design (Greene and Cruise 1995).

Different hydrological models have been developed for particular applications considering underlying difficulty of quantifying a basin-scale response to small-scale spatial complexity of physical processes (Shelton 2009). Based on levels of sophistication and complexity, these are categorized into a number of groups. Xu (2002) discussed the chronological development of various hydrological models as well as their applications in detail. These models can be grouped into stochastic, deterministic, empirical, physical, conceptual, lumped, distributed and semi-distributed models based on randomness, purpose and characteristics of model structure and spatial variations (Fig. 8.1) (Chow et al. 1988; Sui and Maggio 1999; Thompson 1999; Xu 2002).

Stochastic models deal with large random variables to represent process uncertainty, and use probability distributions of hydro-climatic variables to simulate the conversion of basin precipitation into runoff. On the other hand, *deterministic models* are based on real physical processes that are involved in transforming precipitation into runoff. In case of *empirical models*, also known as black-box models, hydrological output is directly linked with climatic inputs through some empirical equations without defining the physical processes involved in modelling. These do not aid in physical understanding due to ignorance of internal structure and resultant response of the basin (Shelton 2009).

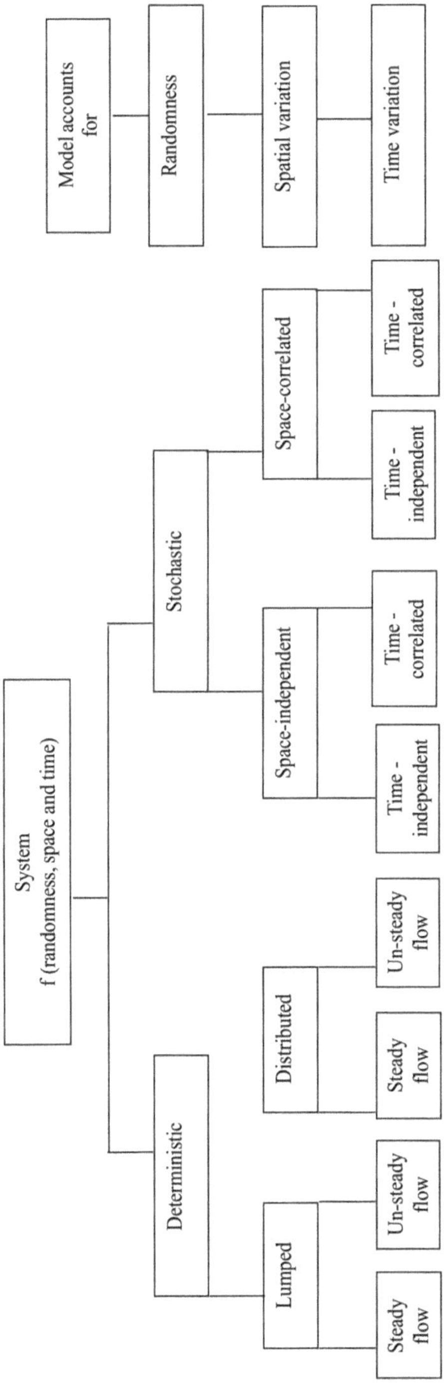

Fig. 8.1 Groups of hydrological models. (From Sui and Maggio 1999)

Conversely, *physical models* incorporate complex systems of equations based on physical laws and theoretical concepts that govern hydrological processes such as evapotranspiration, infiltration, overflow and saturated and unsaturated zone flow. These models have a logical structure similar to the real-world, and are data intensive and operate at fine temporal scales (Arnell 1996; Xu 2002). *Conceptual models* are between empirical and physical models. They take into account physical laws but in very simplified form and characterise the catchment as an idealized representation of storages.

Lumped models consider the complete basin as a homogeneous whole that can be characterised by a single set of parameters. Generally, these are represented by a set of differential or empirical algebraic equations without considering spatial variability of processes, inputs, boundary conditions (Singh and Valiron 1995). Lumped models ignore spatial traits of the basin. They incorporate average values of the catchment characteristics *i.e.* vegetation, soils, geology or topography. In *distributed models*, the basin is divided into separate units or grids and flows are passed from one node to another as water drains through the basin (Singh 1988). Their ability in representing spatial heterogeneity of the catchment and their realistic nature make them applicable over wide range of physical environment including gauged and ungauged watersheds.

Semi-distributed models use a combination of lumped and distributed model attributes by assigning separate units as homogeneous areas to simulate runoff. The basin is divided into number of small sub-basins where model parameters are allowed to vary partially in space. The major advantage of semi distributed models is that their structure is more physically based than the structure of lumped models and that these are computationally less demanding than fully distributed models (Cunderlik 2003; Gassman et al. 2007). Soil and Water Assessment Tool (SWAT) is an example of this category (Arnold et al. 1993).

8.3 Role of Geospatial Techniques in Hydrological Modelling

The ability of distributed and semi-distributed hydrological models in integrating with spatial distribution of various inputs and boundary conditions has raised the demand for spatial data (Qin et al. 2008). The spatial data applied in hydrological modelling can be divided into two classes, *viz.*, topographical data and topological data (De Vantlier and Feldman 1993). Elevation property of the terrain is defined by the topographical data whereas topological data represents spatial distribution of terrain attributes. This mainly involves catchment area, land use and land cover classes, flow length, surface roughness and soil types. These attributes help to describe the ability of a region to store and transmit water (Singh and Fiorentino 1996).

Recently, remote sensing techniques have emerged as an advanced tool in deriving spatial data required in hydrological modelling, and in some extreme cases it is the only source of getting data needed for hydrological models (Brivio et al. 2002).

Remotely sensed data, including topography, land use/land cover, precipitation, snow, climatic variables, ET, and vegetation characteristics, may be used directly as inputs in hydrological models. These data are available at different temporal and spatial resolutions. This makes remote sensing techniques as an indispensable and decisive tool for successful hydrological model analysis, prediction and validation (Jain et al. 1998).

The proper handling of large volumes of spatial and temporal datasets as well as their integration on a common spatial platform has always been a serious issue in hydrological modelling. This has also been acknowledged in past research studies (Kopp 1996; Al-Sabhan et al. 2003). However, to the larger extent, this problem has been resolved with advent of Geographic Information System (GIS) and its integration with hydrological models (Chairat and Delleur 1993; Singh and Fiorentino 1996; Xu et al. 2001). GIS is a system that facilitates the preparation and analysis of georeferenced data. The term GIS is spelled out in many ways. Burrough (1996) defined GIS as a tool that can be used in coding, storing and retrieving geographic information. Further, Morris (2006) defined GIS as a computer based system that can be used to store, manage, manipulate, analyses and retrieve large volumes of georeferenced spatial data and associated attributes collected from a variety of sources. The most obvious feature of GIS is its capability of performing spatial analysis (Jain et al. 1998). The ability of GIS in processing Digital Elevation Model (DEM) data has offered modelers with new platforms for data handling and visualization (DeVantier and Feldman 1993). Grid-based GIS is a very suitable tool for hydrologic modeling due to its capability in processing DEM (Ashour 2000).

The coupling of Geographic Information System with hydrological models can be established using four approaches. These are embedding GIS like capabilities in hydrological modelling package, embedding hydrological modelling in GIS package, loosely coupled integration and tightly coupled integration (Fig. 8.2) (Kopp 1996; Sui and Maggio 1999). In this way, distributed hydrological models, together with remote sensing and GIS, can be effectively applied in water resource management practices. The integration of SWAT model with GIS is an example of loose coupling approach where model input and output parameters are facilitated directly by GIS.

8.4 Conclusion

Water is a precious natural resource essential for existence of living organisms. Timely water availability is growing concern for agriculture, energy and industrial sectors. Increasing population in tandem with urbanization and industrialization increases the demand for water necessitating an in-depth understanding of hydrologic processes and spatio-temporal variability of water resources. In this respect, geospatial techniques, combined with hydrological models, have shown significant potential in hydrologic and water resource management studies. Hydrologic models, developed to simulate the catchment behaviour, are grouped categorically into

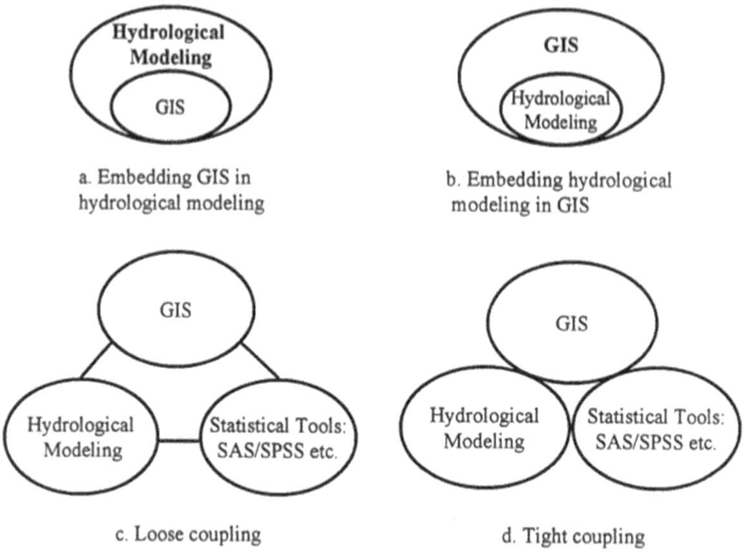

Fig. 8.2 Approaches to integration of GIS with hydrological modelling. (**a**) Embedding GIS in hydrological modelling; (**b**) Embedding hydrological modelling in GIS package; (**c**) Loosely coupled integration, and (**d**) Tight coupled integration. (From Sui and Maggio 1999)

stochastic, deterministic, empirical, physical, conceptual, lumped, distributed and semi-distributed models based on the level of sophistication and complexity. Further, to augment the performance of distributed and semi-distributed models, these models are integrated with geospatial techniques. The chapter endeavoured to dwell on the applicability of geospatial techniques in hydrological modeling.

Acknowledgement The chapter presented here is part of Ph.D. thesis of Dr. Dharma veer Singh. Authors acknowledge Director, Motilal Nehru National Institute of Technology Allahabad, India for providing necessary administrative facilities for carrying out this research work.

References

Al-Sabhan, W., Mulligan, M., & Blackburn, G. A. (2003). A real-time hydrological model for flood prediction using GIS and the WWW. *Computers Environment and Urban Systems, 27*(1), 9–32.

Arnell, N. W. (1996). *Global warming, river flows and water resources*. Chichester: Wiley.

Arnold, J. G., Allen, P. M., & Bernhardt, G. (1993). A comprehensive surface-groundwater flow model. *Journal of Hydrology, 142*, 47–69.

Ashour, R. A. (2000). *Description of a simplified GIS-based surface water model for an arid catchment in Jordan*. In Proceedings of 20th annual international ESRI user conference, held in San Diengo, CA, USA, on June 26–30.

Brivio, P. A., Colombo, R., Maggi, M., & Tomasoni, R. (2002). Integration of remote sensing data and GIS for accurate mapping of flooded areas. *International Journal of Remote Sensing, 23*(3), 429–441.
Burrough, P. A. (1996). Opportunities and limitations of GIS-based modeling of solute transport at the regional scale. In D. L. Cowrin & K. Loague (Eds.), *Applications of GIS to the modeling of non-point source pollutants in the vadose zone, SSSA Special Publication No. 48* (pp. 19–38). Madison: SSSA.
Chairat, S., & Delleur, J. W. (1993). Effects of the TOPOGRAPHIC index distribution on predicted runoff using GRASS1. *Journal of the American Water Resources Association, 29*, 1029–1034. https://doi.org/10.1111/j.1752-1688.1993.tb03266.x.
Chow, V. T., Maidment, D. R., & Mays, L. W. (1988). *Applied hydrology*. New York: McGraw Hill.
Cunderlik, J. (2003). *Hydrologic model selection for the CFCAS project: Assessment of water resources risk and vulnerability to changing climatic conditions*. Department of Civil and Environmental Engineering, The University of Western Ontario.
De Vantlier, B. A., & Feldman, A. D. (1993). Review of GIS applications in hydrologic modeling. *Journal of Water Research Pl-ASCE, 119*(2), 246–261.
Gassman, P. W., Reyes, M. R., Green, C. H., & Arnold, J. G. (2007). The soil and water assessment tool: Historical development, applications, and future research directions. *American Society of Agricultural Biological Engineers, 50*(4), 1211–1250.
Greene, R. G., & Cruise, J. F. (1995). Urban watershed modeling using geographic information system. *Journal of Water Research Pl-ASCE, 121*(4), 318–325.
IPCC. (2013). *Climate change 2013 the physical science basis*. Contribution of working group I to the fifth assessment report of the intergovernmental panel on climate change. Edited by Stocker et al. Cambridge/New York: Cambridge University Press.
Jain, S. K., Kumar, N., Ahmad, T., & Kite, G. W. (1998). SLURP model and GIS for estimation of runoff in a part of Satluj catchment, India. *Hydrological Sciences Journal, 43*(6), 875–884.
Kopp, S. M. (1996). Linking GIS and hydrological models: Where we have been, where we are going? In K. Kovar & H. P. Natchnebel (Eds.), *Application of geographic information systems in hydrology and water resources* (Vol. 235, pp. 133–139). Wallingford: IAHS Publishing.
Morris Steven, P. (2006). Geospatial web services and geo archiving: New opportunities and challenges in geographic information services. *Library Trends, 55*(2), 285–303.
Qin, C., Jia, Y., Su, Z. Z., Zhou, Z., Qiu, Y., & Suhui, S. (2008). Integrating remote sensing information into a distributed hydrological model for improving water budget predictions in large-scale basins through data assimilation. *Sensors, 8*, 4441–4465. https://doi.org/10.3390/s8074441.
Shelton, M. L. (2009). *Hydrology (perspective and applications)*. New York: Cambridge University Press.
Singh, V. P. (1988). *Hydrologic systems, Volume 1: Rainfall-runoff modelling*. New Jersey: Prentice Hall.
Singh, V., & Fiorentino, M. (1996). *Geographical information systems in hydrology*. Dordrecht: Springer.
Singh, V. P., & Valiron, F. (1995). *Computer models of watershed hydrology*. Colorado: Water Resources Publications.
Sui, D. Z., & Maggio, R. C. (1999). Integrating GIS with hydrological modeling: Practices, problems, and prospects. *Computers Environment and Urban Systems, 23*(1), 33–51.
Thompson, S. A. (1999). *Hydrology for water management*. Rotterdam: AA Balkema.
Xu, C-yu. (2002). Chapter1: Modelling in hydrology. In *Hydrologic models*. Uppsala University Department of Earth Sciences Hydrology.
Xu, Z. X., Ito, K., Schultz, G. A., & Li, J. Y. (2001). Integrated hydrologic modeling and GIS in water resources management. *Journal of Computing in Civil Engineering, 15*, 217–223.

Chapter 9
Sustainable Energy: Challenges and Perspectives

S. Prasad, K. R. Sheetal, V. Venkatramanan, S. Kumar, and S. Kannojia

Abstract Currently, energy security, sustainable development and wellbeing are the energy policy drivers throughout the world. India has made significant progress, far more rapidly in the past 2 years, increasing the installed capacity of sustainable energy, and potentially this upward drift is anticipated to persist. The innovation in new and advanced technologies, aggressive energy policies, action, and planning activities has enabled India to resolve the barriers of commercial production of sustainable energy. The domestic production and use of renewable energy, such as off-grid power sources, i.e., solar power, wind power, small hydropower, biofuels and bioenergy from new biomass will help to reduce the fossil fuel use and its imports from other countries.

Sustainable economic and industrial growth also requires safe and sustainable energy resources. The use of sustainable energy will help in strengthening low-carbon energy in India and providing a clean environment through reduction of pollutants and greenhouse gas emissions. Prospective attention to financial and development needs by the use of sustainable energy will also improve the living standards of society with equity and economic sustainability. There is a strong need to extensively adopt and use sustainable energy technologies to supply off-grid power, especially in the areas with difficulty in accessing the central grid power, such as un-electrified villages, remote areas, and hilly terrains. Finally, these sustainable energy sources offer massive benefits and can contribute significantly to ensuring a secure energy future for India.

S. Prasad (✉) · S. Kumar · S. Kannojia
Centre for Environment Science and Climate Resilient Agriculture,
Indian Agricultural Research Institute, New Delhi, India

K. R. Sheetal
Central Arid Zone Research Institute, Regional Research Station,
Bikaner, Rajasthan, India

V. Venkatramanan
School of Interdisciplinary and Transdisciplinary Studies,
Indira Gandhi National Open University, New Delhi, India

S. Shah et al. (eds.), *Sustainable Green Technologies for Environmental Management*, https://doi.org/10.1007/978-981-13-2772-8_9

Keywords Renewable energy · Sustainable energy · Solar energy · Biomass energy · Fossil fuels

9.1 Introduction

Energy is a key driver for agriculture, industries, and service sectors that influence economic development, but today's rising concern over its sustainability has put India in a critical position. The burning of fossil fuels causes multiple environmental problems such as air pollution and global climate change (Zidanšek et al. 2009). Global change threatens the very existence of life. Transforming the present coal dominated energy mix to renewable and sustainable energy dominated energy use is one of the herculean tasks facing India (Institute for Energy Research 2014). A fine balance of environmental sustainability with necessary economic development is required. On the other hand, the transition to sustainable and renewable energy technologies provides an opportunity to address not only the environmental problems but also overall economic and developmental needs to improve the living standards of people with equity and economic sustainability (IAC 2007).

The global community is becoming increasingly clear that management of the considerable risks due to air pollution and climate change, necessitates reduction in fossil fuel utilization, as has been explained in several reports including the fifth assessment report released by IPCC. It is of utmost importance, urgency and priority that a policy objective of finding ways and methods to generate and use energy that helps to preserve the integrity of elementary natural systems, arrest environmental degradation and sustain development toward a more sustainable, fair and civilized world (IAC 2007). Solar, wind and water-based energy and biofuels are a few examples of sustainable energy (IAC 2007; Steven and Majumdar 2012). The innovation in new and advanced energy technologies has enabled India to resolve the barriers of commercial production of sustainable energy. Therefore, sustainable energy and low-carbon technologies could provide ways for development opportunities that result in a world that is healthy, viable and has a secure energy source (Steven and Majumdar 2012).

9.2 Energy Consumption Pattern: Global Scenario

According to 5th Assessment Report of IPCC, the world energy consumption is rising at the rate of 2.3% annually (IPCC 2014). The global energy mix in 2012 was dominated by coal, natural gas, and oil with 87%. However, the dominance changes between these three sources with year, although only marginally. In the same year, while the contribution of natural gas and coal increased 0.1% and 0.2%, that of oil reduced by 0.3%, in the global energy consumption. This trend was predicted to lead to replacement of oil by coal by the year 2017 by the International Energy Agency (IEA) (World Watch Institute 2015). The IEA also expects an increase in global energy requirement under the existing scenario, from 12 to approximately 17 billion tonne oil equivalents (t.o.e.) in the time period from 2009 to 2035 (WEO 2009).

9.3 Energy Consumption Pattern: Indian Scenario

India was the 3rd largest consumer of petroleum products and crude oil in 2015, and it was also the 3rd largest net importer of crude oil and petroleum products after the United States and China in 2016 (IBEF 2017). The Indian energy demand is rising at a rate of 6.5% $year^{-1}$. With energy consumption of 0.55 tonnes of oil equivalent per capita, India's consumption is far below the international average of 1.9 tonnes; but the consumption is expected to almost double by 2035 (The Hans India 2017). With a gap of more than four times between the demand (4 million barrels per day) and production (1 million barrels per day) of oil as of 2015, the gap between India's oil demand and supply is widening (Dunn 2016). India's consumption was nearly 194 million tonnes in 2016–2017 as against 148 million tonnes in 2012 fiscal. By 2040, the Indian oil production predicted to remain steady while the oil demand is predicted by IEA to increase up to 7.5 million barrels per day. Energy companies in India are hence turning to other energy sources to reduce the country's dependency on oil imports.

The net oil import dependency of India rose from 43% in 1990 to 75% in 2015 that resulted in a massive strain on the current account as well as the government exchequer (US EIA 2014; Dunn 2016). The transport sector accounts for the most significant share in terms of utilization of petroleum products in India, consuming nearly 70% of diesel and 99.6% petroleum, and the demand is anticipated to show a growth of 6–8% in the coming time in tandem with the rapidly growing vehicle ownership. This implies that imports will also rise to 92% by 2030 (WEO 2009). With domestic oil production providing only one-fourth of the national demand, energy security has emerged as a vital question for India. This situation necessitates the production of alternate energy from readily available resources (IEC 2015).

9.4 Energy and Environmental Quality

The advancements in scientific monitoring capabilities and increased awareness in recent past has shed more light on the more understated effects associated with energy production, conversion, and use, on ecology and environment like being the cause for pollution and climate (Levine 1991; World Watch Institute 2015).

9.4.1 Fossil Fuels and Air Pollution

Present energy and transportation systems, mainly based on fossil fuels, are the most significant contributor to air pollution throughout the world. The burning of fossil fuel releases carbon monoxide (CO), Volatile Organic Compounds (VOCs), nitrogen oxides (NOx) and particulate matter (PM). In the presence of sunlight, the VOC and NOx mixture results in ozone (O_3) formation in the troposphere, the leading constituent of photochemical smog (Levine 1991). Coal-based power plants are known to have higher CO_2 emissions and other pollutants per kWh electricity generation (IEA 2004).

The combustion of coal also severely affects local air quality by emitting sulfur dioxide (Institute for Energy Research 2014). Contributing about 40% and 38% of gross CO_2 and ozone, burning of biomass is also a grave source of gaseous emissions (Levine 1991). In addition, combustion of biomass as fuels in conventional systems also contribute to 1.4 million tonnes (MT) of methane (CH_4). The emissions are not the only points of ecological concern while using energy (Prasad et al. 2012). Hence, this energy system is turning out to be unsustainable, and sustainability is an essential criterion for energy in this century (Steven and Majumdar 2012).

9.4.2 Fossil Fuels and Global Climate Change

One of the significant threats facing the world today is climate change (Zidansek et al. 2009). According to IPCC 5th Assessment Report, the major anthropogenic contributor to climate change is the use of fossil fuels (coal, gas, oil) which came into practice at the dawn of the industrial era, which amplified the atmospheric concentration of heat-trapping greenhouse gases. Atmospheric greenhouse gas concentration since the pre-industrial revolution to the recent time and global warming potential is given in Table 9.1 (Blasing 2014). The report also reveals that of the 10 $GtCO_2eq$ increase in annual anthropogenic GHG emission that occurred in the last decade (2001–2010), energy supply was the major contributor with 47%, followed by industry, transport and building sectors with 30%, 11% and 3% respectively (IPCC 2014).

Table 9.1 Recent tropospheric greenhouse gas (GHG) concentrations

GHGs	Pre-1750 level	Recent level	GWP(100-year time horizon)	Increased radiative forcing (W/m^2)
CO_2	280	395.4 ppm	1	1.88
CH_4	722	1893/1762 ppb	28	0.49
N_2O	270	326/324 ppb	265	0.17
O_3	237	337 ppb	n.a.	0.40
CFC-11	zero	236/234 ppt	4660	0.061

Source: Blasing (2014)

According to IPCC 5th Assessment Report, more than three-fourth of the increase in greenhouse gas emissions in the 40 years since 1970 was mainly from industry and burning fossil fuels. CO_2 emissions from fossil fuel use have touched 32 (±2.7) $GtCO_2$/year, in 2010, and increased further by about 3% during the next year and approximately 1–2% in the year after (IPCC 2014). Global climate change is predicted to cause irreversible and irreparable adverse impacts on agriculture, health sector and on the earth and the ecosystem as a whole (IPCC 2014).

The herculean and insurmountable task of reducing GHG emissions by 60–80% by 2050 compared to 1990s, necessary to limit temperature rise to 2 °C above pre-industrial levels and to stabilize CO_2 concentrations below 550 ppm, requires a total shift to low-carbon emission technologies that can effectively tackle climate change (IPCC 2014). It was observed that global CO_2 emissions from fossil fuels and industry were almost stable in the years 2014 to 2016, increasing only 0.2%, against the 2.2% average rise during the previous decade. This decline was primarily due to decreasing in global coal use and improvement in efficiency of energy use and increased utilization of renewable alternative energy sources (REN21 2017). India is also developing its various renewable energy sources, especially power generation from solar, wind, biomass, and other viable alternative sources (Zidansek et al. 2009). From 2012 onwards, India has started reforms in oil and gas pricing to promote sustainable investment and to lower subsidy costs (Dani 2014), which could efficiently augment energy supply, at the same time improving energy efficiency in India (Prasad et al. 2014).

9.5 Renewable and Sustainable Energy

Solar energy, wind power, and biofuels are examples of renewable energy which refers to the energy obtained from natural resources that will not deplete over time (Farrel and Gopal 2008). Non-renewable energy sources like fossil fuels form over millions of years and hence do not regenerate easily. The technologies developed to exploit renewable energies are known as renewable energy technologies (RET) or clean technologies or green energy. Sustainable energy goes one step further in terms of energy efficiency than renewable energy. Optimizing energy supply and use leading to minimum wastage leads to higher energy efficiency (Beckett 2012). Renewable energy base and energy efficiency are the twin pillars of sustainable energy.

Sustainable energy systems, technologies or resources support economic and human development needs, at the same time conserving the environment, reducing climate change risks by reducing GHG emissions, giving equal chances for all people to access energy and also improving energy security (IAC 2007; Steven and Majumdar 2012; Beckett 2012).

9.6 Renewable Energy: Global Landscape

Today, the scope of renewable energy lies beyond providing the viable future energy sources (IAC 2007). It is a tool to deal with problems caused by fossil fuel use on environmental and human health and promote energy security, economic and social welfare (REN21 2014). The contribution of renewable energy to global energy production is increasing gradually. In 2015, renewable energy provided 19.3% of global energy production with modern renewables contributing around 10%, and rest by biomass (Fig. 9.1). Out of total energy use, heat energy from modern renewable sources contributed to 4.2%; hydropower accounted for about 3.6% and 2.4% from other renewable sources (REN21 2017).

As per International Renewable Energy Agency's (IRENA) estimate, the global share of renewable energy can exceed 30% by 2030 for which technologies are already on hand. This global share can be further enhanced by 6% by improving energy efficiency and upgrading energy access (MNRE 2009). This trend is visualized through the fast growth rate of renewable installed capacity and production during past years, predominantly in the power sector (Fig. 9.2). Renewables provided 24.5% of electric power worldwide by latter part of 2016. The highest improvement in renewable sector was found in solar energy which accounted for almost half of the newly installed power, followed by wind and hydropower (REN21 2017). Similarly, growth rate of biofuel production, mainly for transport sector, also picked up pace in 2013, after a slow pace from 2010 to 2012. Although biofuel blends remain the primary focus of the policy support for alternate energy in the

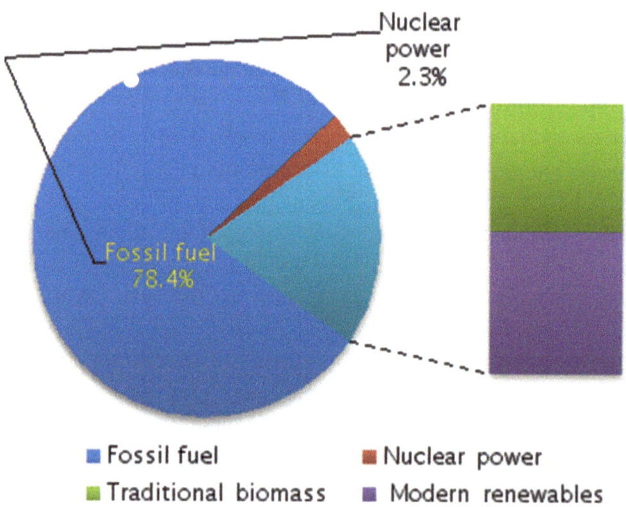

Fig. 9.1 Renewable energy share of global final energy consumption in 2016. (From REN21 2017)

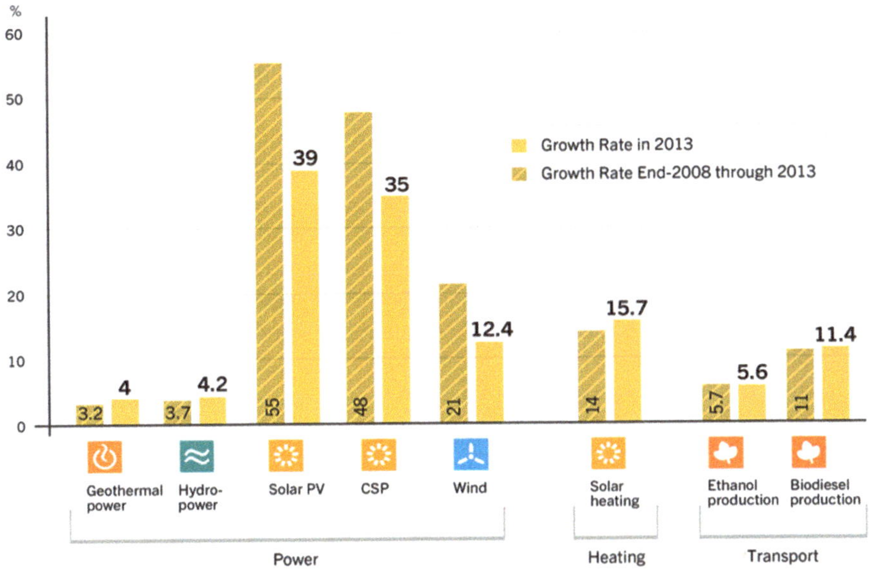

Fig. 9.2 Average annual growth rates of renewable energy capacity and biofuels production. (From REN21 2014)

transport segment, policies encouraging the exploitation of electric vehicles (EVs) are emerging (REN21 2017).

9.7 Sustainable Energy Policies, Institutions, and Programs in India

Global policy decisions are at present being greatly influenced by problems of environmental pollution, energy security and climate change (Planning Commission 2003). The Public Utilities Regulatory Policy Act, 1978 forced the purchase of electricity from independent power generators at reasonable rates in the US. As a result, by 1990s, similar rules were brought in other countries as biomass electricity production grew substantially (Larson and Kartha 2000).

India is one country to have a separate 'Ministry of New and Renewable Energy' to address issues of development of renewable energy sources along with biofuels. Of the total installed power, renewable power (14.8%) has secured the second position after thermal (69.4%) and is spreading its wings rapidly in India (MNRE 2017).

Numerous initiatives were taken up by the Government of India in the past few years including the idea of solar park and Green Energy Corridor concept, rooftop solar program initiation, increase in clean environment cess, solar pump scheme, making purchase of energy from waste to energy plants, etc. attempt has been made to create 50,000 people trained in solar photovoltaic systems under the Surya Mitra

Scheme by March 2020. Advancing towards improved cook-stoves initiatives; commencing coordinated R & D activities in solar thermal and photovoltaic systems; second-generation biofuels, hydrogen energy, and fuel cells, etc. are the other noteworthy schemes.

Since 1993, India has had a fixed purchase price for biomass-generated electricity to encourage expansion of biomass-generating capacity. A task force was constituted by MNRE leading to development of a National Programme on Biomass-based co-generation (Shukla 2000), which recognized bagasse waste as a potential energy resource and suggested initial thrust on bagasse co-generation in the sugar industry. The Government of India (GOI) is enthusiastic in increasing contributions of sustainable energy (resources as a low-carbon generation), undertaking necessary planning and policies to ensure the use of renewable energy in all sectors. The success of this goal is ensured by national-level institutes such as National Institute of Solar Energy (NISE), National Institute of Wind Energy (NIWE), and Alternate Hydro Energy Center (AHEC) (MNRE 2014), under the government's supervision.

Financial support for renewable energy development and improving energy efficiency projects is provided by the Indian Renewable Energy Development Agency (IREDA), also supervising the renewable energy incentive programs. Funding is also provided by GOI and multilateral lending agencies. Ministry of Power, Planning Commission and Prime Minister's Council on climate change are few other government institutions responsible for developing renewable energy (REN21 2014).

The GOI in 1981 launched a national project on biogas development. The National Biogas and Manure Management Program (NBMMP) is implemented by MNRE in the country. It is executed by the state nodal departments/state nodal agencies and Khadi and Village Industries Commission (KVIC), and Biogas Development and Training Centers (BDTCs). MNRE is also funding research projects on different aspects of hydrogen energy, geothermal energy technology development. These projects are assisting in the development of indigenous research and industrial base, proficiency, trained workforce and models/devices/systems in the country (MNRE 2014).

In a bid to decrease dependency on imported fuels, a notification was made by GOI in Sept 2003 making blending of petrol with 5% ethanol compulsory initially in 9 states and 3 union territories, which was to be extended to the whole of India later on. In the case of biodiesel also some steps were taken such as the identification of *Jatropha curcas* as the highly likely candidate for biodiesel production through the 'National Mission on Biodiesel, 2003' and emphasis on wasteland plantations of this tree-borne oilseed (MNRE 2009, 2014). Further, a 'biodiesel purchase policy' was brought into action in 2005 to enable oil companies to purchase biodiesel for 5% blending with diesel by the Ministry of Petroleum and Natural Gas (MOP&NG). In another step forward, the National Biofuels Policy (NBP), 2009 was launched to ensure prevention of debates between food and fuel and the use of only wastelands for biofuel crop cultivation (MNRE 2009). It also gave attention to issues like Minimum Support Prices (MSPs), subsidies for biofuel crops growers, marketing, subsidies for the biofuel industry, mandatory blending of with ethanol and biodiesel, testing and certification of biofuels (Planning Commission 2003; MNRE 2014).

9.8 Renewable Energy Installed Capacity of India

With a noteworthy growth in clean energy over the past years, our country is catching up fast on the determination to becoming a global leader in production of renewable energy (MNRE 2014; IRENA 2014). India has tremendous growth potential in the energy production from different renewable sources, with a potential of about 900 GW (Table 9.2). Table 9.3 shows India's total renewable energy installed capacity (in MW). As per MNRE reports, solar, wind, small hydropower and biomass together contribute to 16% of total electrical installed capacity by the end of 2016 (MNRE 2017), with wind energy contributing more than half of the installed capacity among renewable (GWEC 2017).

India has set a target of 16,660 MW of renewable installed energy for the year 2016–2017, including solar (12,000 MW), wind (4000 MW), SHP (250 MW), bio-energy (400 MW) and energy from waste (10 MW). Other than this, solar photovoltaic energy, cogeneration from non-bagasse biomass, energy from waste, gasification of biomass, small wind/hybrid systems, micro hydel systems are also targeted to produce about 100, 60, 15, 10, 1, 1 MW eq. of renewable energy as off-grid energy, along with large number of biogas plants for the same financial year 2016–2017 (MNRE 2017). For the future, GOI also hopes to achieve establishment 100, 60, 10 and 5 MW of solar, wind, biomass and small hydropower leading to reaching the target of 175 GW of clean energy by 2022, through the work of several organizations and provisions dedicated towards this goal. Advancing the national goal of the creation of renewable energy systems not only benefits ecology and environment in the long term, but also helps to secure energy requirements, employment creation, achieve financial development and reduce dependency upon the exhaustible energy resources (IRENA 2014; Kumar et al. 2015).

Table 9.2 Total renewable energy potential from various sources in India

Sectors	Potential (in MW)
Wind Energy	102,788
Solar Energy	748,990
Small Hydro Power (SHP)	19,749
Biomass	17,538
Bagasse Cogeneration	5000
Waste to Energy	2556
Total	896,621

Source: Energy Statistics (2017), Govt. of India

9.9 Renewable and Sustainable Energy Technologies: The Path Forward

The challenge of sustainable development can be realized through development of renewable energy technologies (RETs) (Make in India 2015). Many scientific and technological obstacles need to be overcome in the next 5–10 years to meet the ambitious RETs capacity addition goal and biofuel mandates for sustainable energy supply. However, the current innovations in RETs are capable of resolving the barriers of commercial production.

9.9.1 Wind Power

Among renewable energy sources, wind energy is becoming one of the essential sources of power generation. In India, wind turbines are generally used for off-grid mechanical power or electricity generation. The wind energy plants comprise of a wind turbine and an electrical generator connected using a gearbox. The wind turbine converts wind kinetic energy into mechanical power. It can be used for specific tasks such as grinding grain or pumping water, or an electrical generator can convert this mechanical power into electrical energy (Ahmad et al. 2014). The rate of wind energy is now waning due to noteworthy advancements in wind mill technology, as well as raise in the height of the wind towers.

Industry associations of India emphasize the high potential for wind energy from 65 to 242 GW, if tower heights are kept more than 50 m, energy conversion efficiencies are improved and necessary policy initiatives are adopted (Ramasamy et al. 2015). At a hub height of 100 m, India's wind power potential is approximated to be 302 GW as per NIWE. Presently, with 31 GW installed capacity in the first quarter of 2017 and contributing to 57% of total renewable energy, India is at the 4th position globally, with respect to wind power (Table 9.3) (MNRE 2014; GWEC 2017). This energy is mainly generated in the states of Tamil Nadu, Maharashtra, Gujarat, Rajasthan, and Karnataka, together supplying 94% of total wind energy generated. The Wind Resource Assessment is led by the coastal state of Gujarat with 84.4 GW estimated potential, followed by the other states mentioned earlier (GWEC 2017).

Table 9.3 Total renewable energy installed capacity (as on 31 Dec 2016)

Renewable energy (RE) sources	Power (MW)	% Contribution to Installed energy
Wind Power	28700.44	57.3
Solar Power	9012.66	18.0
Small Hydro Power	4333.85	8.6
Biomass energy	7907.34	15.8
Waste to Power	114.08	0.2
Total	50068.37	

Source: MNRE (2017)

Table 9.4 Installed capacity of wind power in various states of India (as on 31 Dec 2016)

S. No.	States of India	Installed capacity of wind power (in MW)	% Wind power potential used
1.	Tamil Nadu	7613.86	22.53
2.	Karnataka	2869.15	5.14
3.	Maharashtra	4653.83	10.25
4.	Rajasthan	3993.95	21.28
5.	Andhra Pradesh	1431.45	3.24
6.	Madhya Pradesh	2141.1	20.42
7.	Gujarat	3948.61	4.68
8.	Kerala	43.5	2.56
9.	Telangana	77.7	1.83
10.	Others	4.3	0.13

Source: MNRE (2017)

To fulfill the rising demand for energy, other states are also following in the footsteps to increase wind power production (Ahmad et al. 2014).

Installed wind power capacity of various Indian states till March 2016 is shown in Table 9.4. Tamil Nadu, the state with maximum wind energy installed, increased its capacity from over a period of 5 years from 2009, with a result of 69% increase (Make in India 2015). Apart from the on-shore potential, India also has high off-shore wind power potential owing to its long coastline (7600 km) and wind patterns. The states of Tamil Nadu and Gujarat will play a major role in this addition of energy production capacity with proper planning and estimation (Ramasamy et al. 2015). Increasing the energy conversion efficiencies can help elevate Indian power position to that of the global wind energy leaders, which requires technological and market interventions (Ramasamy et al. 2015).

9.9.2 *Solar Energy*

There are two techniques of using solar energy: thermal and photovoltaic way. The thermal route employs the heat from solar energy for cooking, water heating and purification, drying, etc. The solar photovoltaic (SPV) method transforms the light in solar energy into electricity via the use of solar cell installed in a solar panel, which can then be used for lighting, pumping, communications, and off-grid power supply in un-electrified areas (Indian Renewable Energy Status Report 2010). Being a tropical country, India is bestowed with abundant number of clear sunny days, obtaining 4–7 kilowatt-hour per square meter per day radiation on an average and as a result, shows good potential for development of solar power (Ramasamy et al. 2015). The desert state of Rajasthan receives the highest annual solar radiation in India.

The solar energy plans are executed by the MNRE, considered one of the most extensive programs globally. India has an assessed solar power potential of around 748,990 MW, out of which the solar grid had a cumulative capacity as of October 2017 of 15.60 GW (Energy statistics 2017; MNRE 2017). Increase in solar capacity has led to decline in solar electricity rates, which has gone below that of coal based electricity. Out of the total commissioned solar power capacity in the country as of March 2017, the share of Andhra Pradesh is maximum, closely followed by Rajasthan Tamil Nadu, Telangana, and Gujarat (Bridge to India 2017).

The state and central governments combined initiation of the Jawaharlal Nehru National Solar Mission (JNNSM) in 2010, is a momentous stride towards the support of sustainable energy in India. The objectives of the mission are (i) to establish 20,000 MW of grid-connected solar power (ii) to develop 20 million solar lights which will meet the target of 2000 MW of off-grid power (iii) to cover an area of 20 million meter squares with the solar thermal collector; by the year 2022. The tie-up of the MNRE with IREDA aims to encourage use of solar energy and in general, increase clean energy capacity in India, through location-specific and need-based research, demonstrations, financial support and in league with private sector projects (MNRE 2014; Ramasamy et al. 2015).

9.9.3 Small Hydropower (Less than 25 MW)

When economic feasibility is taken into consideration, small hydro projects were found better than other sources (Balachandar 2014). However, presently only 5% of this source's capacity has been exploited on a global scale, out of 150–200 GW potential as per the International Energy Agency. Another advantage is the lower and comparable rates of this small hydropower to traditional thermal power of around Rs 2–3/kWh, without the addition of any associated variable fuel cost (i.e. natural gas or coal) as in thermal power (World Bank 2010). In India also, this source can play a major role in supplying energy to remote and inaccessible regions, carving its niche in the Indian energy sector. Small hydropower plants are categorized in the following segments presented in Table 9.5.

India has an estimated small hydropower (less than 25 MW) potential of approximately 20,000 MW out of which the collective installed capacity as of 31st December, 2016 was 4333.85 MW counting both off-grid and grid-connected

Table 9.5 Classification of Small Hydro Power Project (SHP)

S. No.	Category	Station capacity in kW
1.	Micro Hydro	Up to 100
2.	Mini Hydro	101–2000
3.	Small Hydro	2001–25,000

Source: MNRE (2014), Annual Report, Govt. of India

sources (MNRE 2017). Irrigation channels in plains and rivers in hilly areas are the areas with high potential to generate this power. To successfully tap into this power, the government has set subsidies for new installations as well as to maintain existing hydropower plants (Indian Renewable Energy Status Report 2010). Further incentives for establishment of this source is provided through loans at low interest rates and income tax exemptions, all of which has culminated in its significant share in India's electrification efforts in rural areas without access to electricity from the central grid (Ramasamy et al. 2015; World Bank 2010).

9.9.4 Geothermal Energy

Thermal springs have drawn consideration, being the surface manifestation of the vast resources of energy at depth in the form of geothermal reservoirs (Mittra 2011). The geothermal energy is now better known for electricity generation. It is derived from the Earth's core, mantle, and crust heat, and has been helping meet both industrial as well as household energy needs with great success. Geothermal energy holds the key to solve the power problem of India. It is an ideal alternative energy resource meeting all requirements as a clean, non-exhaustive energy (Razdan et al. 2008). The global leader in this sector is the United States with 3.6 GW installed capacity followed by Philippines and Indonesia in 2016, with global production being a projected 78 terawatt-hours (REN21 2017).

India is at the budding stages of development of geothermal energy, chiefly due to the lower rates of coal. Reports from the Geological Survey of India (GSI) reveal the existence of about 340 hot springs, which are distributed in seven geothermal provinces of the country (Aggarwal 2016), i.e. Himalayan (Puga, Chhumathang), Sahara Valley, Cambay Basin, Son-Narmada-Tapi (SONATA) lineament belt, West Coast, Godavari and Mahanadi basin. Most of them are in the low surface temperature range from 37 to 90 °C, which is suitable for direct heat and power applications (Kakkar et al. 2012). At present, the geothermal resource is tiny, but the GOI has an ambitious plan to produce 10,000 MW of geothermal energy by 2030 (Aggarwal 2016). According to International Sustainable Energy Agency, geothermal solutions can provide not only sustainable but also economical and safe energy for all by 2050.

9.9.5 Biomass-Derived Bioenergy and Biofuels

Plants capture sunlight and carbon dioxide from the atmosphere for growth, and this basic mechanism leads to the classification of biomass as a renewable source (Larson and Kartha 2000). Tropical countries, like India have enormous potential for energy generation through biomass and its residues. It is available in plenty in the form of agricultural waste (crop residue and cattle dung, etc.), urban waste (municipal solid waste, etc.) and industrial waste. India has about 500 million metric tons of biomass availability annually, which can be converted to bioenergy (by the thermo-chemical and biochemical process) or directly used as a heat and power source (IRENA 2014).

Deploying bioenergy with carbon capture and sequestration results in a net decline in atmospheric carbon. The vast potential of biofuels becomes more noticeable as an alternative to the depleting fossil fuels. Moreover, they can be locally produced, requiring very less modification before final-use. Biomass energy also supports rural economy, requires lesser capital investment than other renewable, leading to lesser unit cost (Sharma and Trikha 2013).

9.9.5.1 Biomass to Bioenergy by Thermo-chemical Process

The conversion of biomass to an alternative form of modern bio-energy can be achieved primarily through thermo-chemical conversion of biomass. The end-products of the thermo-chemical processes such as combustion, pyrolysis, and gasification, include heat, electricity, or gaseous or liquid precursors which may be modified into biofuels. With biomass serving as an energy source for more than three-fourth of the country and producing 32% of the total primary energy use, biomass occupies a central position in our energy security (Biomass Knowledge Portal 2017), with over 30,000 MW estimated power production of which only a meager 8% has been exploited as yet (Schroder et al. 2008). Currently, India has over 5940 MW power capacity from biomass based plants of which only 4946 MW is grid-connected (mainly from bagasse cogeneration and waste to energy plants) and 994 MW off-grid power plants (mainly from non-bagasse cogeneration and biomass gasifier systems) (Biomass Knowledge Portal 2017).

Biomass energy along with other renewable, can play a major role in the planned electrification of about 24,500 remote identified villages of India where the extension of grid electricity is not possible or very expensive (EAI report 2012; Thomas et al. 2015). The uniform availability of biomass throughout India, particularly in the rural areas, increases the importance of biomass as an energy source for electrification. This abundant resource worth annual investments of crores of rupees with >700 billion electricity units production and also employing several people, has the potential to form a significant element of India's energy mix in upcoming future (Thomas et al. 2015).

9.9.5.1.1 Direct Combustion of Biomass

The direct combustion of biomass has been carried out worldwide since ancient times. Since ancient times, firewood, field level residues and cow dung, have been a crucial segment of energy supply and these traditional biomass still has its followers particularly in rural areas and remote location. In complete combustion, biomass is oxidized giving out carbon dioxide, water and heat energy (Farrell and Gopal 2008). A popular 'biomass combustion technology' or the traditional cooking method in developing countries involves burning biomass under a pot supported by three stones. Despite its huge popularity, energy efficiency of this method is very less (15%), simultaneously exposing users to air pollutants such as methane (CH_4) carbon monoxide (CO), particulates (PM), nitrogen oxides (NO_X), and tars (Farrell and Gopal 2008).

9.9.5.1.2 Biomass Pyrolysis

Pyrolysis is essentially the thermal breakdown of organic components in waste to energy under inert atmospheric conditions or in limited air supply, at temperatures ranging from 350–550 °C up to 700–800 °C (EAI report 2012). During the process, the long carbon, hydrogen and oxygen compound chains break down into smaller molecules in the form of highly heterogeneous gaseous, liquid, and solid by-products (Jahirul et al. 2012). The pyrolysis oil, a liquid product is a heterogeneous blend constituted of high oxygen content and resembling a very viscous tar, from which fuels or chemicals may be produced (Fisher et al. 2002). Char is the intermediate solid residue which is formed in reactors during pyrolysis processes. This residual char finds use as a fuel or soil amendment (Jahirul et al. 2012).

9.9.5.1.3 Biomass Gasification

Gasification is the efficient means of producing green power. It is incomplete oxidation of biomass under controlled conditions aimed at getting peak yields of gaseous compounds rich in carbon and hydrogen compounds such as CO, CH_4, H_2, CO_2 and heat (Fisher et al. 2002). These by-products are referred to as syngas or producer gas can be used with minor changes in air inlet, to operate a diesel engine in dual fuel mode. A drawback of this process is due to one of its condensate product 'tar', an environmental pollutant, formed from high molecular weight volatiles (EAI report 2012; Fisher et al. 2002).

The gas can be put into use directly for electricity generation via a spark-ignited gas engine incorporated with an alternator (Tanger et al. 2013). This electricity can be used locally or off-grid or can be made on-grid, providing enough flexibility for small or large scale production for remote and difficult areas. These features also ensure the sustainability and feasibility of biomass gasification in India.

9.9.5.2 Biomass to Liquid Biofuels by Biochemical Process

Liquid biofuels like ethanol and biodiesel are the most feasible green alternatives to fossil fuel-based transportation fuels. Molasses, a waste product of sugar industry is the main feedstock for bioethanol production in India, while biodiesel is made by the transesterification of non-edible oilseeds. Indian government has brought forth several missions and policies in support of biofuels, along with loans at lower interest rates for establishing ethanol production plants, encouraging cultivation of non-edible and tree-borne oilseeds and tax exemptions for biodiesel use.

9.9.5.2.1 Ethanol as Biofuel

Production of energy efficient technologies offers the world the safest and the most environmentally benign way to sustainability. In the past few decades, great attention has been focused on the development of alternative fuel sources with particular reference to ethanol. Raw materials containing sugars, or resources which can be converted into sugars, can be used as fermentation feedstocks for ethanol production (Meshram and Mohan 2007). The fermentable raw materials can be grouped into three (a) directly fermentable sugar containing materials such as starch (b) lignocellulosic biomass (c) urban/industrial organic residues. Reports are also available on direct fermentation of sugarcane, sugar beet and sweet sorghum to ethanol (Prasad et al. 2009), which require the least costly pretreatment, where starchy, lignocellulosic resources and other wastes need expensive pretreatment, to change into fermentable substrates (Rao et al. 2013). However, since maintaining the food security is given higher priority in our agrarian economy faced with rising population pressure, it is not possible to divert a fraction of the sugary or starchy food sources towards biofuel production.

Recently, lignocellulosic biomass, as the most abundant renewable resource has been widely considered for ethanol production. It gives us an opportunity to look forward to the day when all fuel needs can be met from the utilization of agricultural residues, as it has been reported that use of currently available 74 Tg of agricultural residues across the earth could produce 16 times higher ethanol production over present level (Prasad et al. 2007). The ethanol production potential from biomass is high enough to replace 32% of the global petrol consumption (Meshram and Mohan 2007). With respect to India, the entire annual petrol consumption can be met by utilizing jut one-third of available surplus biomass (Ramasamy et al. 2015).

Ethanol from biological systems is considered of particular importance because it can be readily used as a fuel for spark ignition engines without necessitating any modifications. Ethanol contains high oxygen percent of 34.7% by weight, aiding complete fuel burning and a marked drop in emissions. It is usually used as an additive in petrol to increase octane number, and improve the type of emissions. Ethanol, being the low-carbon fuel, appears to have enormous potential benefits to minimize the risk of greenhouse gas emissions. It also reduces carbon monoxide emissions and smog; usually caused due to sulfur deposits in gasoline (Kim and Dale 2004).

9.9.5.2.2 Biodiesel as Biofuel

Biodiesel is defined as ‘monoalkyl esters of long chain fatty acids obtained from renewable lipid feedstock, such as vegetable oil or animal fat’ by the American Society for Testing and Materials (ASTM) (Prasad et al. 2012). As the best diesel alternate, it may be used directly or after blending. The 'B20' blend which consists of biodiesel and diesel in 1:4 ratio by volume is most common. Biodiesel fuel blends cut down emissions of particulate matter (PM), carbon monoxide (CO), hydrocarbon (HC) and sulfur oxides (SOx), thus less toxic to humans and environment. Also with 80–95% lesser carbon dioxide emissions over its life cycle, biodiesel has very little share in climate change, as compared to fossil fuels (Karmakar et al. 2010).

Substantial developmental actions have been done regarding the production and use of biodiesel. A 1000 liters/day capacity biodiesel production plant was established by the Aatmiya Biofuels Pvt. Ltd. at Por-Vadodara, Gujarat, using *jatropha* seeds as feedstock (Marina et al. 2002). Another commercial level plant with an expandable daily biodiesel production capacity up to 100 tonnes was proposed by Southern Online Biotechnologies (P) Ltd., in Andhra Pradesh. Lurgi Life Science Engineering, Germany is providing the technology for this unit, along with their Delhi-based associates (Biswas et al. 2006).

9.9.6 Hydrogen Energy

Hydrogen (H_2) is a sustainable energy, which is produced from available sources and used in every application where fossil fuels are being used. H_2 is the fuel of the future, largely due to its high conversion efficiency, non-polluting nature, and recyclability, yielding only water after combustion (Francis et al. 2005). There are many hydrogen production processes, including electrolysis of water, biomass gasification, and biological processes. Currently, steam reformation of methane or electrolysis of water is employed almost exclusively to produce H_2. Currently, H_2 is produced, almost exclusively, by steam reformation of methane or by water electrolysis. Supercritical water partial oxidation, steam reforming, and gasification are the thermo-catalytic processes employed for H_2 production (Gupta et al. 2013). In the last decade, the important problems associated with H_2 economics have changed dramatically. Nevertheless, refineries now have turned into net consumers of H_2 to cut pollution and meet environmental rules and regulations.

In India, hydrogen production from renewable resources is at the dawn of development. The Many research projects focusing on the development of different aspects of hydrogen production technology is being undertaken, funded by Government of India. The Indian Institute of Science, Bangalore, is researching on employing an open top downdraft gasification system to enhance the performance of the oxy-steam gasification unit for hydrogen production at a rate of about 0.1 kg/kg biomass at various steam-to-biomass ratios. The National Institute of Technology, Rourkela is working on the development of a 5 kW capacity bench scale fluidized bed gasifier for hydrogen production rate of about 0.09 kg/kg of feedstock (Sherif et al. 2003). The roadmap envisions initiating research and development activities in various sectors and aspects of hydrogen energy technologies and envisaged goals of aggregate hydrogen-based power generation capacity of 1000 MW and one million hydrogen-fuelled vehicles in the country by 2020 (Nouni 2012).

9.9.7 Biogas

Biogas production is a sustainable energy technology which not only efficiently manages and converts organic wastes into clean bioenergy but also allows the best possibilities to cut tropospheric emissions of greenhouse gases (Kumar et al. 2010). Biogas is produced via the anaerobic digestion of organic and bio-degradable resources such as agricultural waste, food waste, municipal waste, sewage etc. The slurry produced after digestion is rich in macro and micronutrients and can be used directly as valuable organic fertilizer. Biogas is a mixture of different gases primarily methane (CH_4) and carbon dioxide (CO_2) and may have small quantities of hydrogen sulphide (H_2S) and moisture. Methane is the only combustible constituent of biogas (Pathak et al. 2009).On account of methane's low specific gravity, methane biogas will rise on escaping, thus dissipating from the site of a leak which makes it safer than other fuels like petrol and Liquefied Petroleum Gas. One cubic meter of biogas generates 1.5 kWh which is equivalent to 1 lb of LPG, 0.54 L of petrol, 0.52 L of diesel and 4 kWh of electricity in terms of calorific value.

Biogas finds its use both as a source of heat for cooking by direct combustion and for mechanical or electrical power applications with internal combustion engines (Pathak et al. 2009; Rao et al. 2010). Biogas plant helps to attain self- sufficiency for cooking gas and highly nutrient- rich organic fertilizer for households having feed material (Horst 2000). Further, this technology can solve the problems of indoor air pollution in households and while saving on the cost of refilling of LPG cylinders. The Government of India offers a subsidy for family type biogas plants as per the rates are given in Table 9.6.

Table 9.6 Subsidy for setting up of biogas plants under national biogas and manure management programme

	Particulars of central financial assistance	Family type biogas plants size cubic meter (M^3) capacity day^{-1}	
S. No.	Central subsidy rates applicable (In Rs.)	1 M^3	2–6 M^3
1.	North Eastern Region States, Sikkim (except plain areas of Assam) and including SC & ST Categories of NE Region States	15,000	17,000
2.	Plain areas of Assam	10,000	11,000
3.	Jammu & Kashmir, Himachal Pradesh, Uttarakhand, Niligiri of Tamil Nadu, Sadar Kurseong & Kalimpong Sub-Divisions of Darjeeling, Sunderbans (West Bengal) and Andaman & Nicobar Islands	7000	11,000
4.	Scheduled Castes/Scheduled Tribes of all other States except North Eastern Region States (including Sikkim)	7000	11,000
5.	All Others	5500	9000

Source: MNRE (2014), Annual Report, India

9.10 Contribution of Sustainable Energy to CDM

Despite all challenges, the clean development mechanism (CDM) saw rapid growth in the past decade, quickly becoming the most significant carbon offsetting mechanism in the world. Currently, India accounts for approximately 4% of global greenhouse gases emissions. However, on per capita basis, its emissions are only one-fourth of the global average and less than 10% of those of most developed nations. India has committed to reduce its economy's emission intensity to 20–25% below 2005 levels by 2020 and has pledged that per capita GHG emissions will not surpass those of industrialized countries (Ngumah et al. 2013).

In 2008, NAPCC (National Action Plan on Climate Change) was launched to promote development goals while addressing GHG mitigation and climate change adaptation. As per NAPCC, renewable sources have the potential to meet 15% of India's energy demand by 2020 (Ramasamy et al. 2015). The NAPCC includes eight dedicated missions, out of which one is devoted to solar energy (Sood et al. 2014). India is observed to be one of the most attractive Non-annexe-I countries for CDM project development.

The Clean Development Mechanism (CDM) of the Kyoto Protocol extends supporting hands for the development of renewable energy projects in India. The fundamental economic hurdle is the relatively higher costs of electricity generation through Renewable Energy Technology (RET), although the scale of difference varies with each technology. This price difference can be compensated to some extent by the revenues attained from selling CERs from a CDM project (Sood et al. 2014).

9.11 Conclusion

Sustainable energy is the need of the time and it is gaining currency due to its low–carbon generation potential, often accompanied by its co–benefits. The transition of the fossil fuel based conventional energy to sustainable energy will help in effective management of air pollution, climate change mitigation, and spur sustainable economic growth and improve living standards of people with equity and environmental quality. The IPCC Fifth Assessment Report validates the promising role of bioenergy in encouraging economic development by increasing and diversifying farm incomes and providing rural employment and in offering cheap alternatives for mitigating climate change. There is a need to adopt and use renewable and sustainable energy technologies for mitigating of environmental challenges across all sectors.

Acknowledgment Authors wish to thank Indian Agricultural Research Institute, New Delhi, India for providing financial support.

References

Aggarwal, M. (2016). *India looks to harness 10 GW geothermal power by 2030*. Available at: http://www.livemint.com/Industry/qFWtAToDvYCYqtKjTcfixH/India-looks-to-harness-10-GW-geothermal-power-by-2030.html

Ahmad, M. A., Khalid, S., Qaisar, S. H., & Tiwari, G. (2014). The contemporary scenario of Indian renewable energy sector. *International Research Journal of Environmental Sciences, 3*(11), 82–89.

Balachandar, G. (2014, April 5). Solar capacity in India crosses 2,500 MW. *The Hindu*. http://www.thehindu.com/business/Industry/solar-capacity-in-india-crosses-2500-mw/article5876808.ece

Beckett, M. (2012). *Biotechnology's role in sustainable energy for all*. Available at: http://www.pub.ac.za/files/Biotech%20and%20Sustainable%20energy.pdf

Biomass Knowledge Portal. (2017). *Overview of biomass power sector in India*. Available at: http://biomasspower.gov.in/About-us-3-Biomass%20Energy%20scenario-4.php

Biswas, S., Kaushik, N., & Srikanth, G. (2006). *Biodiesel: Fuel for the future*. Available online at: http://tifac.org.in/index.php?option=com_content&view=article&id=776:biodiesel-fuel-for-the-future&catid=120:publication-bioprocess-a-bioroducts&Itemid=1380

Blasing , T. J. (2014). *Recent greenhouse gas concentrations, February 2014*. Carbon Dioxide Information Analysis Center (CDIAC). https://doi.org/10.3334/CDIAC/atg.032 Available at: http://cdiac.ornl.gov/pns/current_ghg.html

Bridge to India. (2017). *India solar handbook 2017*. Bridge to India Energy Private Limited, Gurgaon, India. Available at: http://www.bridgetoindia.com/wp-content/uploads/2017/05/BRIDGE-TO-INDIA_India-Solar-Handbook_2017-1.pdf

Dani, S. (2014). Escalating petrol prices in India: Repulsive government policies and turbulence in public. *IOSR Journal of Economics and Finance, 4*(6), 31–35.

Dunn, C. (2016). *India continues developing its strategic petroleum reserve as its oil imports grow*. US EIA. Available at: https://www.eia.gov/todayinenergy/detail.php?id=27132

EAI Report. (2012). *India biomass gasification power production*. Available electronically at: http://www.eai.in/ref/reports/biomass_gasification.html

Energy Statistics. (2017). *Central statistics office*. New Delhi: Ministry of Statistics and Programme Implementation. Government of India.

Farrell, A. E., & Gopal, A. R. (2008). Bioenergy research needs for heat, electricity, and liquid fuels. *MRS Bulletin, 33*, 373–380.

Fisher, T., Hajaligol, M., Waymack, B., & Kellogg, D. (2002). Pyrolysis behavior and kinetics of biomass-derived materials. *Journal of Applied Pyrolysis, 62*, 331–349.

Francis, G., Edinger, R., & Becker, K. (2005). A concept for simultaneous wasteland reclamation, fuel production, socio-economic development in degraded areas in India- need, potential, and perspectives of Jatropha plantations. *Natural Resources Forum, 29*, 12–24.

Gupta, S. K., Kumari, S., Reddy, K., & Bux, F. (2013). Trends in hydrogen production: Major challenges and state-of-the-art developments. *Environmental Technology, 34*(13–16), 1653–1670.

GWEC. (2017). *Indian wind energy – A brief outlook 2016*. http://www.gwec.net/wp-content/uploads/vip/GWEC_IWEO_2016_LR.pdf

Horst, W. D. (2000). *Biofuel generation: Proceedings of national workshop on integrated food production and resource management*, pp. 101–105.

IAC. (2007). *Inter academy council. Lighting the way: Toward a sustainable energy future; 2007*. Royal Netherlands Academy of Arts and Sciences, P.O. Box 19121, NL-1000 GC Amsterdam. http://www.interacademycouncil.net/24026/25142/24333.aspx

IBEF. (2017). *Oil and gas*. Available at: https://www.ibef.org/download/Oil-and-Gas-February-2017.pdf

IEA. (2004). *World energy outlook 2004*. International Energy Agency, Available at http://www.worldenergyoutlook.org/media/weowebsite/2008-1994/weo2004.pdf

IEC. (2015). Integrated energy planning, 6th edition of India energy congress: Securing India's energy future: Integration and action on 28–29, January 2015, in New Delhi.

Indian Renewable Energy Status Report. (2010). *Background report for DIREC 2010*. Available electronically at: http://www.osti.gov/bridge

Institute for Energy Research. (2014). *BP energy review coal use on the rise globally*. Available at: http://instituteforenergyresearch.org/analysis/coal-use-rise-globally-bp-energy-review-2014/

IPCC. (2014). *Summary for policymakers*. In Climate change 2014, mitigation of climate change: Contribution of working group III to the 5th Assessment report of the Intergovernmental Panel on Climate Change (IPCC). Available electronically at http://www.climatechange2013.org/images/report/WG1AR5_SPM_FINAL.pdf

IRENA. (2014). *Global bioenergy supply and demand projections: A working paper for REmap 2030*. http://irena.org/remap/IRENA_REmap_2030_Biomass_paper_2014.pdf

Jahirul Md, I., Rasul Md, G., Chowdhury, A. A., & Nanjappa, A. (2012). Biofuels production through biomass pyrolysis. *Technological Review Energies, 5*, 4952–5001.

Kakkar, V., Agarwal, N. K., & Kumar, N. (2012). Geothermal energy: New prospects. *International Journal of Advances in Engineering & Technology, 4*(2), 333–340.

Karmakar, A. A., Karmakar, S. B., & Mukherjee, S. (2010). Properties of various plants and animals feedstock for biodiesel production. *Bioresource Technology, 101*, 7201–7210.

Kim, S., & Dale, B. E. (2004). Global potential bioethanol production from wasted crops and crop residues. *Biomass and Bioenergy, 26*, 361–375.

Kumar, A., Kumar, K., Kaushik, N., Sharma, S., & Mishra, S. (2010). Renewable energy in India: Current status and future potentials. *Renewable and Sustainable Energy Reviews, 142*, 434–2442.

Kumar, A., Kumar, N., Baredar, P., & Shukla, A. (2015). A review on biomass energy resources, potential, conversion and policy in India. *Renewable and Sustainable Energy Reviews, 45*, 530–539.

Larson, E. D., & Kartha, S. (2000). Expanding roles for modernized biomass energy. *Energy for Sustainable Development, 4*(3), 15–25.

Levine, J. S. (1991). *Global biomass burning: Atmospheric, climatic, and biospheric implications* (pp. 25–30). Cambridge, MA: The MIT Press.

Make in India. (2015). *Renewable energy statistics*. Available electronically at: http://makeinindia.com/sector/renewable-energy/

Marina, E., Antonio, S., Boyan, K., & Peder, J. (2002). *Techno-economic analysis of Biodiesel production in the EU: A short summary for decision-makers*. European communities Report EUR 20279 EN.

Meshram, J. R., & Mohan, S. (2007). Biomass power and its role in distributed power generation in India. In *25 years of renewable energy in India* (pp. 109–134). New Delhi: Ministry of New and Renewable Energy.

Mittra, S. (2011). Small hydro: Too small for a national mission? *ORF Issue Brief, 34*, 1–6. Available online at: www.orfonline.org.

MNRE. (2009). *GOI national policy on biofuels*. Available electronically at: http://mnre.gov.in/file-manager/UserFiles/biofuel_policy.pdf

MNRE. (2014). *Annual report 2013–14*. Ministry of New and Renewable Energy (MNRE), Government of India.

MNRE. (2017). *New and renewable energy sector achievements report*. Department of Industrial Policy and Promotion.
Ngumah, C., Ogbulie, J., Orji, J., & Amadi, E. (2013). Potential of organic waste for biogas and biofertilizer production in Nigeria. *Environmental Research, Engineering and Management, 1*(63), 60–66.
Nouni, M. R. (2012). Hydrogen energy and fuel cell technologies, Ministry of New and Renewable Energy, GOI. *Akshay Urja, 5*(5), 10–15.
Pathak, H., Jain, N., Bhatia, A., Mohanty, S., & Gupta, N. (2009). Global warming mitigation potential of biogas plants in India. *Environmental Monitoring and Assessment, 157*, 407–418.
Planning Commission, GOI. (2003). *Report of the committee on development of biofuels*. Available at : http://planningcommission.nic.in/reports/genrep/cmtt_bio.pdf
Prasad, S., Singh, A., & Joshi, H. C. (2007). Ethanol as an alternative fuel from agricultural, industrial and urban residues. *Resources, Conservation and Recycling, 1854*, 1–39.
Prasad, S., Lata, J. H. C., & Pathak, H. (2009). Selection of efficient *Saccharomyces cerevisiae* strain for ethanol production from sorghum stalk juice. *Current Advances Agricultural Sciences, 1*(2), 70–72.
Prasad, S., Dhanya, M. S., Gupta, N., & Kumar, A. (2012). Biofuels from biomass: A sustainable alternative to energy and environment. *Biochemical and Cellular Archives, 12*(2), 255–260.
Prasad, S., Kumar, A., & Muralikrishna, K. S. (2014). Biofuels production: A sustainable solution to combat climate change. *Indian Journal of Agricultural Sciences, 84*(12), 1443–1452.
Ramasamy, P., Chandel, S. S., & Yadav, A. K. (2015). Wind speed prediction in the mountainous region of India using an artificial neural network model. *Renewable Energy, 80*, 338–347.
Rao, P. V., Baral, S. S., Dey, R., & Mutnuri, S. (2010). Biogas generation potential by anaerobic digestion for sustainable energy development in India. *Renewable and Sustainable Energy Reviews, 14*, 2086–2094.
Rao, S. S., Patil, J. V., Umakanth, A. V., Mishra, J. S., Ratnavathi, C. V., Shyam, P. G., & Rao, B. D. (2013). Comparative performance of sweet sorghum hybrids and open-pollinated varieties for millable stalk yield, biomass, sugar quality traits, grain yield and bioethanol production in tropical Indian condition. *Sugar Tech, 15*(3), 250–257.
Razdan, P. N., Agarwal, R. K., & Singh, R. (2008). Geothermal energy resources and its potential in India. *E-Journal Earth Science India, 1*(1), 30–42.
REN21. (2014). *Renewables global status report* (Paris: REN21 Secretariat). Available electronically at http://www.ren21.net/Portals/0/documents/e-paper/GSR2014/index.html
REN21. (2017). *Renewables 2017*. Global status report (Paris: REN21 Secretariat). ISBN 978-3-9818107-6-9 Available at: http://www.ren21.net/wp-content/uploads/2017/06/17-8399_GSR_2017_Full_Report_0621_Opt.pdf
Schroder, P., Herzig, R., Bojinov, B., Ruttens, A., Nehnevajova, E., & Stamatiadis, S. (2008). Bioenergy to save the world producing novel energy plants for growth on abandoned land. *Environmental Science and Pollution Research, 15*, 196–204.
Sharma, O. P., & Trikha, P. (2013). Geothermal energy and its potential in India, Ministry of New and Renewable Energy, GOI. *Akshay Urja, 7*(1), 14–18.
Sherif, S. A., Barbir, F., & Veziroglu, T. N. (2003). Principles of hydrogen energy production, storage and utilization. *Journal of Scientific and Industrial Research, 62*(01–02), 46–63.
Shukla, P. R. (2000, February 3–5). *Biomass energy in India: Policies and prospects*. Workshop on biomass energy: Key issues and priority needs. Paris: International Energy Agency, p. 20. Available at: http://www.e2analytics.com
Sood, R., Aggarwal, R. K., & Mahajan, P. K. (2014). Estimation of consumption and carbon emission from household sector in mid-Himalaya, region of Himachal Pradesh, India. *Research Journal of Forest Environmental Protection, 1*(2), 112–116.
Steven, C., & Majumdar, A. (2012). Opportunities and challenges for a sustainable energy future. *Nature, 488*, 294–303.

Tanger, P., John, L. F., Jahn, C. E., Morgan, W. D., & Jan, E. L. (2013). Biomass for thermochemical conversion: Targets and challenges. *Frontiers in Plant Science, 4*, 218.
The Hans India. (2017). *India's crude facts*. Available at: http://www.thehansindia.com/posts/index/Hans-Classroom/2017-05-24/Indias-crude-facts/302061
Thomas, J., Ashok, S., & Jose, T. L. (2015). A pricing model for biomass-based electricity. *Energy Sources, Part B: Economics, Planning and Policy, 10*, 103–110.
US EIA. (2014). *Energy information administration countries analysis briefs*. Available at: http://www.eia.gov/countries/analysisbriefs/India/india.pdf
WEO. (2009). International Energy Agency in World Energy Outlook IEA Publications, Paris, http://www.worldenergyoutlook.org
World Bank. (2010). *Unleashing the potential of renewable energy in India*. Energy sector management assistance Program, South Asia energy unit, World Bank http://siteresources.worldbank.org/EXTENERGY2/Resources/Unleashing_potential_of_renewables_in_India.pdf
World Watch Institute. (2015). *Fossil fuels dominate primary energy consumption*. Available electronically at: http://www.worldwatch.org/fossil-fuels-dominate-primary-energy-consumption-1
Zidansek, A., Blinc, R., Jeglic, A., Kabashi, S., Bekteshi, S., & Šlause, I. (2009). Climate changes, biofuels and the sustainable future. *International Journal of Hydrogen Energy, 34*(16), 6980–6983.

Chapter 10
Microbial Fuel Cell: Sustainable Green Technology for Bioelectricity Generation and Wastewater Treatment

Shachi Shah, V. Venkatramanan, and Ram Prasad

Abstract Global Environmental Change and the rapid exhaustion of non-renewable energy sources like coal, and petroleum products have kindled the necessity of humankind for the invention of viable and efficient sustainable green technologies for harvesting energy resources. Microbial Fuel Cell (MFC) technology offers an effective carbon neutral alternative for bioelectricity generation. This environmentally benign technology capitalizes the ability of electrogenic bacteria to produce electricity from chemical energy produced from the degradation of organic substrates including wastewater. Bioelectricity generation by electrogenic bacteria is influenced by factors like nature/type of substrate, concentration of substrate, hydrogen ion concentration, organic loading rate and internal resistance. Of late, studies in the field of substrates for MFCs demonstrate that a diverse group of organic sources can be used as a substrate for microbes and consequently, sustainable energy can be produced. MFC-based systems found applications in hydrogen production, environmental sensors, seawater desalination, bioremediation, and microbial electro-synthesis and energy recovery. In this chapter, an insight has been given to the principles, components, upscaling, and potential applications of MFC as green and clean technology for bioelectricity generation and waste reduction.

Keywords Microbial fuel cells · Bioremediation · Bioelectricity generation · Biohydrogen · Biosensor

S. Shah (✉) · V. Venkatramanan
School of Interdisciplinary and Transdisciplinary Studies, Indira Gandhi National Open University, New Delhi, India
e-mail: sshah@ignou.ac.in

R. Prasad
School of Environmental Science and Engineering, Sun Yat-Sen University, Guangzhou, China

S. Shah et al. (eds.), *Sustainable Green Technologies for Environmental Management*, https://doi.org/10.1007/978-981-13-2772-8_10

10.1 Introduction

Microbial fuel cells technology gained currency in the backdrop of energy crisis, widening demand -supply gap, and global climate change. Microbial fuel cells technology (MFC) is promising, efficient and environmentally benign approach in sustainable wastewater treatment and bioelectricity generation. Microbial fuel cells technology takes advantage of the metabolic activity of electrogenic bacteria which generate bioelectricity through degradation of organic substrates. As the technology entails renewable methods and biological processes devoid of toxic outputs, this green microbial fuel cells technology can be construed as an alternative and sustainable method for power generation that can bridge the gap between energy demand and supply at the local level (Logan 2004; Mohan et al. 2008; Chaturvedi and Verma 2016; Roy et al. 2017). MFCs are bioelectrochemical devices which are capable to thrive on diverse biodegradable substrates. Fundamentally, the MFCs contain electrodes namely anode and cathode. A proton exchange membrane separates both electrodes. In the anode compartment, oxidation of substrates like organic substance, wastewater, contaminants, etc., results in the production of electrons and protons. While, the protons through the proton exchange membrane reaches the cathode, the electrons from the anode through an external circuit or load reaches the cathode compartment, only to be accepted by the electron acceptor like oxygen and reduces to form water in the presence of a proton (Ucar et al. 2017). In the recent past, MFCs have found application in wastewater treatment, microbial solar cells, bioelectricity generation, industrial chemicals recovery, pollutant removal, microbial desalination cells, sensors, hydrogen production, bioremediation, and energy recovery (Zhang and Angelidaki 2011; Chandrasekhar and Venkata Mohan 2012; Roy et al. 2017; Ucar et al. 2017). As regards the MFCs in wastewater treatment, the pollutants are removed either at cathode as an electron acceptor or as an electron mediators, or at anode as an electron donors (Ucar et al. 2017).

10.2 Microbes as Energy Producer

Biologist Luigi Galvani way back in 1780 exhibited the connections existing between biological organisms and bioelectricity generation. Nevertheless, in the early twentieth century, potential of microbes in electricity generation was demonstrated vividly by Potter (Potter 1911) and eventually extensive studies were conducted to know the role of micro-organisms in the degradation of organic materials, and to gauge the capability of biological organisms to generate bioelectricity. Potter demonstrated using *Saccharomyces cerevisiae* as a bioelectrogenic bacteria, glucose as a carbon source, and Pt as an electrode, bioelectricity generation of 0.3–0.5 V (Potter 1911). MFCs tap the cellular respiration of microbes for electricity generation. The products of cellular respiration in addition to carbon dioxide are protons and electrons. The electrons released on account of bacterial respiration reaches the anode in the MFC either in the presence of mediators or absence of mediators. Mediators like thionine, neutral red, etc., enable transfer of electrons from electrogenic bacteria to anode. Interestingly, studies undertaken by Kim and co-workers in the late 1990s was path breaking as it led to the development of mediatorless MFC (Allen and Bennetto 1993; Park and Zeikus 1999; Kim et al. 2002).

10.3 Fundamentals of MFC

Microbial fuel cells enable generation of bioelectricity through the redox reactions of electrogenic bacteria. Fundamentally, microbial fuel cells are made of anode compartment with an anode electrode and anolyte; Cathode compartment with cathode and catholyte; and Ion exchange membrane between anode and cathode compartment (Logan et al. 2006; Ishii et al. 2013). While the anode electrode is placed in the anoxic environment, the cathode is placed in aerobic condition which is made possible by supply of oxygen (Kim and Lee 2010; Nastro 2014). The electrogenic bacteria in the process of degradation or oxidation of carbon rich organic matter gain electrons due to cellular respiration and the electrons are then transferred to the anode electrode. From the anode, the electrons are transferred to the cathode through an resistor or load which generates power. On reaching the cathode, the electrons are accepted by terminal electron acceptor and the terminal electron acceptor gets reduced. This completes the circuit (Fig. 10.1) (He and Angenent 2006; Logan et al.

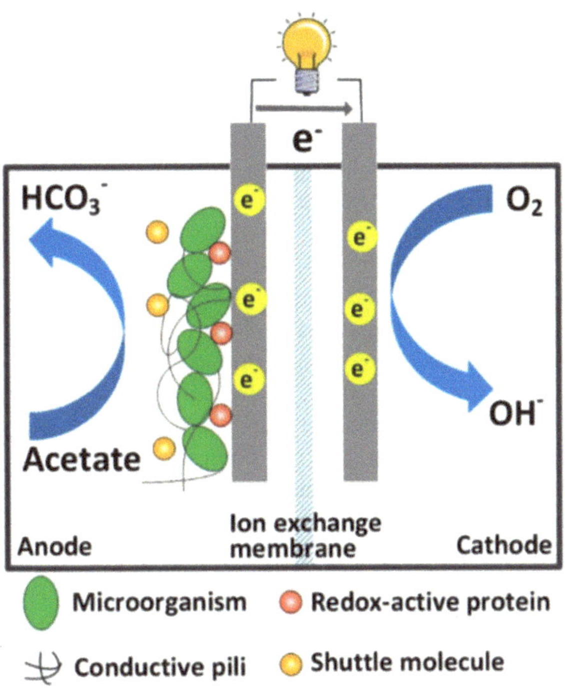

Fig. 10.1 A typical two-chamber microbial fuel cells. (Yuan and He 2015)

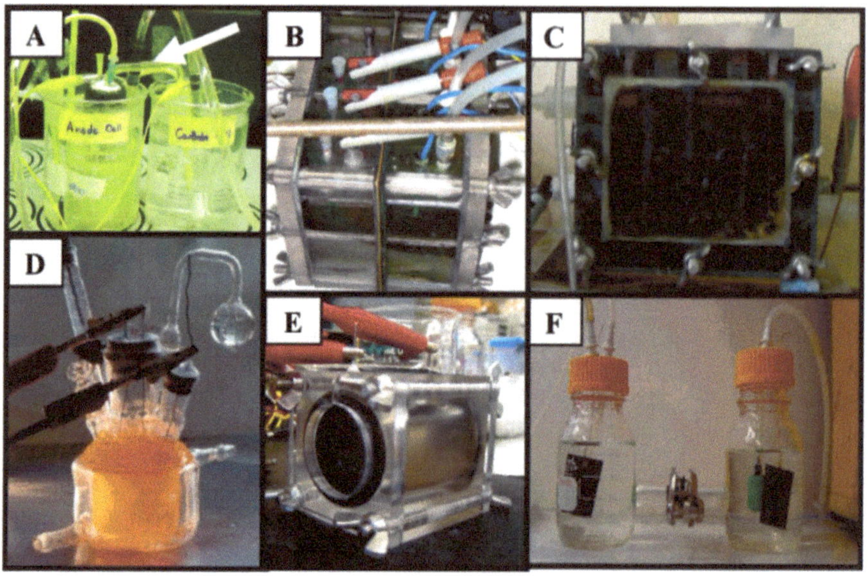

Fig. 10.2 Types of MFC (With permission from Logan et al. 2006): (**a**) Easily constructed system containing a salt bridge (shown by arrow) (Min et al. 2005); (**b**) four batch-type MFCs where the chambers are separated by the membrane (without a tube) and held together by bolts (Rabaey et al. 2005); (**c**) same as B but with a continuous flow-through anode (granular graphite matrix) and close anode-cathode placement (Rabaey et al. 2005); (**d**) photoheterotrophic type MFC (Rosenbaum et al. 2005); (**e**) single-chamber, air-cathode system in a simple "tube" arrangement (Liu and Logan 2004); (**f**) two-chamber H-type system showing anode and cathode chambers equipped for gas sparging (Logan et al. 2005)

2006; Logan 2008; Roy et al. 2017). It was observed that the anodic compartment in the absence of oxygen are efficient in power generation. So, the anode compartment is maintained under anaerobic condition. This is made possible by using semipermeable proton exchange membrane which maintains the anode compartment under anaerobic condition. Further, the protons produced at the anode due to oxidation passes through the proton exchange membrane to reach cathode electrode where in the terminal electron acceptor like oxygen accepts the electron and reduces to form water (Logan 2008). Single chamber MFCs and two-chamber MFCs are formed based on the absence and presence of proton exchange membrane respectively (Xia et al. 2018). Many different configurations are possible for MFCs (Fig. 10.2). Two chamber MFC are inexpensive and popular. They are built in the form of "H" shape with two bottles that are well connected by a tube containing cation exchange membrane. The cation exchange membrane (CEM) can be either Nafion, Ultrex or a plain salt bridge (Park and Zeikus 1999; Bond et al. 2002; Rabaey et al. 2003; Min et al. 2005; Logan et al. 2005). Nevertheless, single-compartment MFCs do not require separation (Logan et al. 2006).

The processes occurring at the anode and cathode of MFCs with glucose as organic source are illustrated with the following equations:

$$\text{Anode}: \quad C_6H_{12}O_6 + 6H_2O \rightarrow 6CO_2 + 24H^+ + 24e^- \tag{10.1}$$

$$\text{Cathode}: \quad 24H^+ + 24e^- + 6O_2 \rightarrow 12H_2O \tag{10.2}$$

The efficiency of MFC system depends also on electron transfer mechanisms which entail the "direct" and "mediated electron transfer". As regards the electron transfer through direct mode, "nanowires" and "membrane bound cytochromes" play a vital role. In case of mediated electron transfer, mediators play a prominent role (Feng et al. 2014; Xia et al. 2018). Electrochemically active bacteria are capable of transferring electrons to the anode either through the formation of anode biofilm or electron mediators (Yu et al. 2012). Certain electron mediators like cysteine, flavins (flavin mononucleotide), etc., are synthesized by the electrochemically active microbes in the anode compartment. In certain cases, the e^- mediators like humic substances are present in the anolyte itself (Kumar et al. 2015). Biofilms are polymeric matrix structures produced by certain group of bacteria either in pure culture or consortia. These "electrochemically active biofilms" play an important role in bioelectricity generation as they enable exchange of electrons with electrodes (Babauta et al. 2012). The process of biofilm formation involves a series of stages like movement and attachment of micro-organisms to the electrodes, colonization, and maturation of biofilms (Kumar et al. 2015; Saeed et al. 2015).

As regards the electron acceptors, generally, oxygen is preferred as an electron acceptor as it has an oxidation potential of 1.23 volts (10.3). The cathode compartment of MFCs is normally enriched with electron acceptor-oxygen through bubbling with air.

$$O_2 + 4H^+ + 4e^- \rightarrow 2H_2O \left(E^0 = 1.23v\right) \tag{10.3}$$

However, energy intensive option of using oxygen as an electron acceptor has led to extensive studies on variety of electron acceptors so as to bring down the cost and upscale the bioelectricity generation (Ucar et al. 2017). Interestingly, use of nitrate, iron, copper, and such other electron acceptors has opened up the opportunities to use MFCs in wastewater treatment plants and pollution abatement strategies (Ucar et al. 2017).

10.4 Exoelectrogens in MFC

Exoelectrogens are group of organisms which are capable of transporting electrons directly outside the cell. Exoelectrogens have been found application in MFCs on account of their unique functional characteristics. Extensive studies have been conducted on exoelectrogens like *Geobacter sp.*, *Burkholderia sp.*, *Rhodoferax ferrireducens*, and *Shewanella sp.* to know the electrogenesis process, to understand the diversity of exoelectrogens in the biofilm, and also to decipher the ecology of exoelectrogens (Logan 2008; Feng et al. 2014). Diverse group of micro-organisms and wide variety of substrates are used in MFCs for bioelectricity generation (Table 10.1).

The electron transfer mechanisms adopted by the exoelectrogens can be either through oxidation-reduction active proteins housed on the cell surface or self-produced mediators like flavin nucleotides, etc., or filamentous and conductive structures popularly called nanowires (Fig. 10.3) (Kumar et al. 2015). The performance of MFCs is also dependent on the electron transfer kinetics which can be greatly upscaled by adopting measures like electrode (anode) surface alteration, gene expression enabling the production of e^- mediators like flavin, etc. (Kumar et al. 2016).

Pure culture of electrochemically active bacteria due to the anoxic condition prevailing in anode chamber, are found to grow slowly in the anolyte, resulting in lesser

Table 10.1 Micro-organisms and diverse substrates used in MFCs for bioelectricity generation

Microorganisms	Substrates/ co-substrates	Current density/ power density	References
Pure culture			
Rhodoferax ferrireducens	Glucose, xylose sucrose, maltose	158 mW/m^2	Liu et al. (2006)
Pseudomonas aeruginosa	Pyocyanin	4310 mW/m^2	Rabaey et al. (2004)
Saccharomyces cerevisiae	Glucose	16 mW/m^2	Rahimnejad et al. (2009)
Pseudomonas sp.	Peptone	979 mA/cm^2	Daniel et al. (2009)
Klebsiella pneumoniae strain L17	Glucose	34.77 mW/m^2	Liu et al. (2009)
Shewanella oneidensis strain 14063	Sodium pyruvate	> 40 mW/m^2	Fernando et al. (2012)
Escherichia coli strain K-12	Sucrose	215 mW/m^2	Zheng and Nirmalakhandan (2010)
Cellulose degrading bacteria	Cellulose	188 mW/m^2	Zhou et al. (2013)
Mixed culture			
Thermophilic effluent from anaerobic digestion of brewery wastewater	Acetate	1030 mA/cm^2	Jong et al. (2006)
Gammaproteo and *Shewanella affinis* (KMM3586)	Cyctenin	36 mW/m^2	Zhou et al. (2012)
Desulfobulbus and *Clostridium*	Rice straw hydrolysate	137.6 mA/cm^2	Cui et al. (2014)
Fly ash leachate	Fermentation effluent	85.07 mA/cm^2	Varanasi et al. (2015)

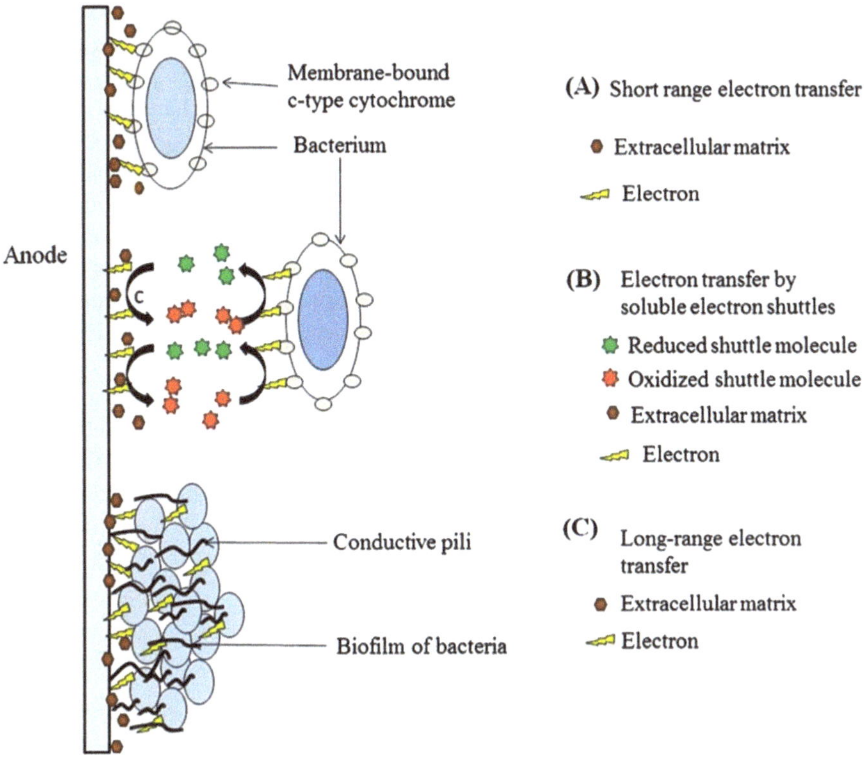

Fig. 10.3 Mechanisms of electron transfers from micro-organisms to electrode. (With permission from Kumar et al. 2015)

bioelectricty generation. Comparatively, a microbial consortia are found to outperform the activity of pure culture in treatment of wastewater, as the diversity in the consortia can widen the biodegradation capabilities. Substrates like sludge, wastewater are preferred to be used in MFCs as they possess rich microbial diversity (Chandrasekhar et al. 2018). Systems operated with mixed cultures or consortia are found to yield higher power densities and the operational cost is much lesser than the pure culture (Rabaey et al. 2005; Ömeroğlu and Sanin 2016).

10.5 Configuration of MFC

Design and configuration of MFCs depend on number of chambers, mode of operations, and other specifications. Basically, MFCs include single chamber and two chambers MFCs. The traditional and classical design of MFCs are made of anode and cathode compartments separated by proton exchange membrane. Such dual chamber MFCs are generally run in batch mode or in the continuous mode depending on the requirements (Du et al. 2007). On the contrary, the single chamber MFCs consist of only anodic chamber which are coupled with air cathode. The MFCs with single chamber which is devoid of a proton exchange membrane, the terminal electron acceptor (O_2) is provided to the cathode compartment directly by the diffusion of air (Liu and Logan 2004; Logan and Regan 2006). Recently, MFCs with triple chambers have been developed to house a desalination chamber in the MFCs. The triple-chamber MFCs consist of desalination chamber sandwiched between the anode and cathode compartments. Triple-chamber MFCs are constructed by using an anion exchange membrane and a proton exchange membrane on both sides of the desalination chamber (Fig. 10.5) (Kim and Logan 2013). The power generation of MFCs are a function of nature of substrate, composition of electrode, exoelectrogens, exoelectrogenic microorganisms, electrode material, electron acceptors, circuit resistance, and the reactor configuration (Kim et al. 2011). As regards the MFC architecture, there is a need for finding suitable electrode composition, augmenting bioelectricity generation and reducing cost.

10.5.1 Configuration of Anode

Configuration of anode entails up-scaling of operating efficiency of anode compartment which squarely depends on the growth of biocatalyst on the anode surface in the form of biofilms. Certain group of bacteria called exoelectrogens which are efficient in electron transfer mechanism are sine quo non for the operation of anode. With respect to the anode material, it should enable the formation of active biofilms which aid in the transfer of electrons from the exoelectrogens to the anode. Further, materials like carbon plates, Pt rods, etc., are used as an anode material. Oxidation of organic wastes, and contaminants in wastewater in the anode compartment releases electrons which are transported to the anode. The electrons passes via an load or the external circuit to the cathode compartment where in the electron is accepted by the electron acceptors like oxygen. In order to increase the O_2 reduction rate at the cathode, the cathode electrode is made of Pt, $MnO_{2,}$ etc. (Pandit and Das 2018). Extensive studies have been initiated to modify the anode electrode using different techniques including nanoengineering methods which can enhance the electron transfer and eventually upgrade the MFC performance (Scott et al. 2007). Carbon nanotubes was found to augment the electron transfer feasibility and incidentally the carbon nanotubes structure provided more electrode surface area (Qiao et al. 2007). It is reported that polymer materials with appreciable conductivity like polyaniline, polytetrafluoroethylene are preferred as anode material (Rahimnejad et al. 2015).

10.5.2 Configuration of Cathode

There is an immense need to up-scale the MFCs and also transfer the technology from pilot scale to industrial scale. However, this proposition demands suitable cathode configuration to produce more bioelectricity. Commonly used cathode material are carbon paper, carbon fibre, Cu, granular graphite, etc. (Niessen et al. 2004; Chen et al. 2008; Ghasemi et al. 2013; Rahimnejad et al. 2015). In the initial period of operation of MFCs with Pt cathodes, bioelectricity generation was much higher than the MFCs with Pt free electrodes. However, after the lapse of few days, there exists no difference in bioelectricity generation between the Pt cathodes and Pt free cathodes (Santoro et al. 2013). Recently, studies have been directed to explore carbon based electrodes for efficient and cost effective functioning of MFCs. Simultaneously, studies have been initiated in right earnest to explore efficient cathodic electron acceptors so that MFCs can find applications in wastewater treatment plants, pollutant removal and at the same time generate increased voltage potential (Ucar et al. 2017). Knowledge of the microbial growth on the cathode surface had given impetus to develop biocathodes which possess the advantage of cost effectiveness and the biofilms formed on the cathode surface aid in enhanced power generation (Niessen et al. 2004).

10.5.3 Membrane Configuration

The selective membrane used in MFCs have a pivotal role to play in electricity generation (Watanabe 2008). The selective membrane placed between the anode and cathode compartment enables movement of protons from the anode compartment to the cathode compartment. Further, the proton selective membrane also negates the movement of oxygen from cathode compartment to anode compartment so as to maintain anoxic condition in anode chamber (Berchmans 2018). The classical MFCs having two chambers essentially require cation selective exchange membrane to set apart the anode and cathode compartments. This may not be the case however in single chamber MFCs. Nowadays, Nafion or perflurosulphonic acid is used as cation exchange membrane (CEM). Research is directed to find alternatives to Pt, Nafion so that cost can be reduced (Logan 2008). As cost reduction is essential for the economic feasibility of MFCs, there is a constant urge to identify suitable alternative membranes which are both cost effective and efficient, for the fabrication of MFCs (Li et al. 2011; Yousefi et al. 2016, 2017).

10.6 Substrates Used in MFC

The activity of electrochemically active bacteria is dependent on the nature and type of substrates. The substrates which are fuel for the biocatalyst varies widely from carbon rich substrates to nitrogen rich substrates; domestic wastewater to petroleum refinery wastewater; and food waste to metal contaminated waste material (Pant et al. 2010; Chandrasekhar and Venkata Mohan 2012). Further, it is reported that certain pollutants play the role of electron shuttle compounds. The role of mediators in electron transfer mechanism cannot be overestimated (Mu et al. 2009; Kumar et al. 2012; Wen et al. 2013). Albeit, the bioelectricity generation is a function of concentration of substrate, increasing the concentration of substrate above a certain threshold have a negative impact on bioelectricity generation. Therefore, it is pertinent to note that in addition to the nature of substrate, the concentration of substrate is equally important warranting optimization so as to obtain maximum bioelectricity output.

10.7 Scaling Up of Technology

Although, MFC technology is gaining currency as a sustainable and an alternate source of electricity generation, the technology faces constraints from the perspective of output, cost, efficiency and up-scalability (Song et al. 2015; Ucar et al. 2017). Indeed, the efficiency of MFC system lies in the power output. Type and nature of substrate, substrate concentration, exoelectrogens, material composition of electrodes, MFC architecture and functioning of electron acceptors greatly influence the power output of MFCs (Ucar et al. 2017). Transfer of MFC technology from pilot scale to industrial scale is need of the hour. In this regard, to attain economic efficiency, measures are warranted in the domain of upscaling MFC reactor size, enhancing the bioelectricity generation, and augmenting the treatment efficiency (Logan 2010). Prospects of MFCs in addition to bioelectricity generation, was identified in wastewater treatment plants. It is essential to develop MFC technology which is sustainable, self-sustained and productive (Zhang et al. 2013). Consequently, the task set for research on MFC is to augment simultaneously the energy generation and upscale of MFC reactor size. In this regard, studies on upscaling of MFC technology highlight the significance of spacing between the electrodes in case of MFC unit containing multiple electrodes and also the need for augmenting the surface area of the electrodes as they determine the power output (Liu et al. 2008; Li et al. 2008). However, factors like MFC systems, high internal resistance, high fabrication and material cost, reduced pollutant degradation dynamics, and voltage losses challenge the upscaling of MFC technology (Fornero et al. 2010; Logan 2010; Gude 2016). Further, granular study of microbial ecology of electrochemical active bacteria and exoelectrogens, and innovations in the application of MFC technology would aid in the development and integration of MFCs in pollution abatement technologies, industrial chemicals recovery plant and wastewater treatment plants (Logan 2008, 2009).

10.8 MFC Applications

MFC technology progressively developed from early twentieth century to late twentieth century. Nevertheless, scientific developments in the field of electron transfer mechanisms and increasing recognition of MFCs as a sustainable and renewable energy source led to extensive studies to map the application of MFC in diverse areas like wastewater treatment, biodegradation, bioelectricity generation, etc. (Fig. 10.4). Studies have reported that diverse waste products ranging from agricultural waste to municipal waste can be used as a substrate for MFCs (Kumar et al. 2012; Chandrasekhar et al. 2015). Erable and co-workers (2010) reported that current focus of research on MFCs is to identify the ways to augment the "power density level" and bioelectricity generation capacity of MFC so that it can be used on industrial scale (Erable et al. 2010). Also, MFCs are used in the areas of biosensors, heavy metal and toxicity detection, hydrogen production, and wastewater quality monitoring (Mook et al. 2013). Further, in the sediment microbial fuel cells (SMFCs), the bioelectricity is generated due to the redox gradient present naturally across the sediment and water interface (De Schamphelaire et al. 2008). SMFCs involve placement of anode electrode in the sediments and the cathode electrode in the water body. SMFCs enable increased oxidation of organics for instance in constructed wetlands, redox control and simultaneous generation of bioelectricity.

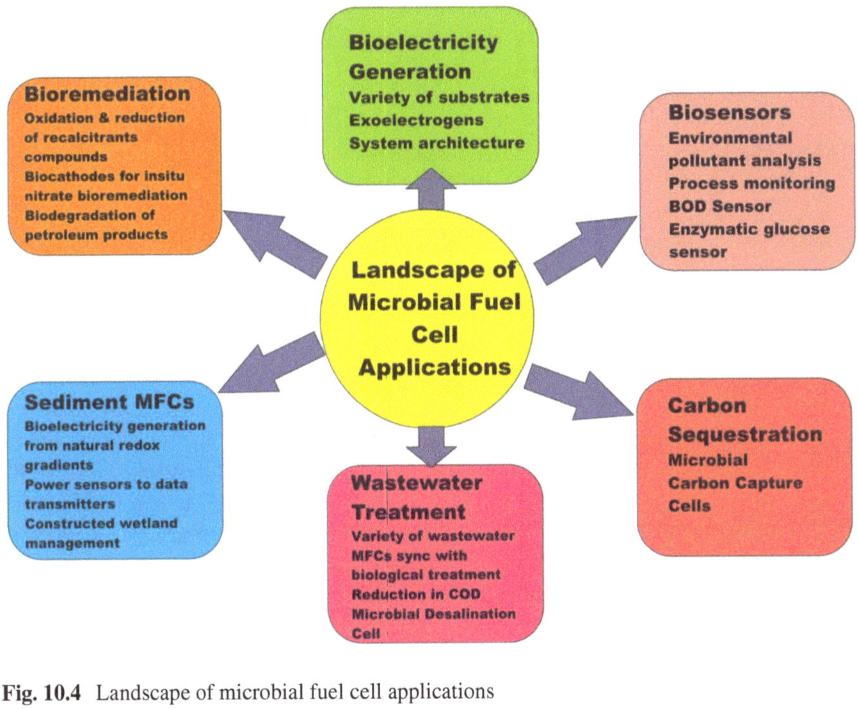

Fig. 10.4 Landscape of microbial fuel cell applications

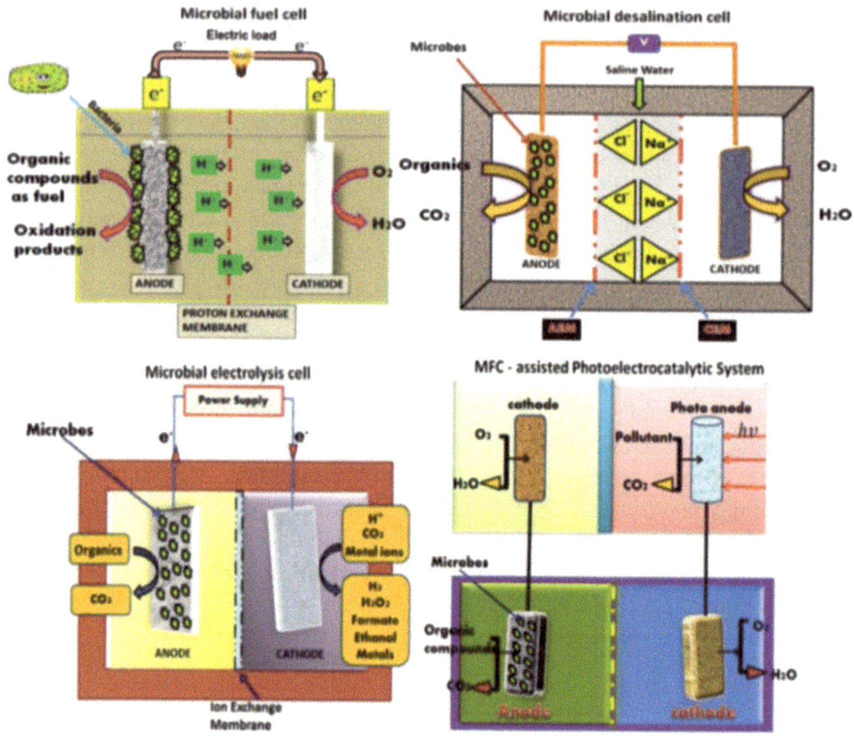

Fig. 10.5 Microbial fuel cells and its analogues. (With permission from Berchmans 2018)

Further, the efficiency of SMFCs can be improved by using carbon nanotube coated cathode and by augmenting the oxygen concentration and rate of reduction of oxygen at the cathode using photosynthetic algae like *chlorella vulgaris* (Wang et al. 2014). With regard to the scale-up of SMFCs, Ewing and co-workers recommended isolation of electrodes through use of a power management system connected to smaller sized SFMCs (Ewing et al. 2014).

Analogous to MFCs, a recent development in the area of bio electricity generation is microbial desalination cell (MDCs), which are capable of generating electricity, treatment of wastewater, desalination and bioremediation of contaminated groundwater (Saeed et al. 2015; Carmalin Sophia et al. 2016). The architecture of MDC is similar to microbial fuel cells but for the addition of desalination chamber which is sandwiched between the anode and cathode compartments of MDC (Fig. 10.5). Structurally, the desalination component of MDC is placed between the two electrode compartments by placing the anion selective membrane and cation selective membrane on both sides of the desalination compartment. Like MFCs, MDCs also generate bioelectricity. Nevertheless, to ensure electro-neutrality of electrode compartments, while anions like chlorine present in the salt water moves across the anion selective membrane to the anode compartment, cations like sodium from salt water moves across the cation selective membrane to the cathode compartment of MDCs. MDCs simultaneously generate bioelectricity and using the self produced energy, removes salts from the salt water (Saeed et al. 2015; Sevda et al. 2015). Intensive study by Sevda and co-workers (2017) highlighted the use of MDCs for simultaneous treatment of wastewater from petroleum refineries and salt removal from the sea water (Sevda et al. 2017). Extensive studies on MDCs led to the novel integration of MDC with microbial electrolysis cells with an aim to remove pollution causing nutrient elements like nitrogen from wastewater; remove salt content from salt water; and to simultaneously generate bioelectricity from the oxidation of organic wastes in wastewater and enable the integration of MDCs and microbial electrolysis cells as a self-sustaining and prospective system (Li et al. 2017).

10.8.1 Wastewater Treatment

Biodegradable organic component of wastewater possess energy which can be tapped using suitable technology (Du et al. 2007). Wastewater treatment plant consists of series of processes aimed at treating the wastewater efficiently, cost-effectively and monitoring the pollution load of the treated effluent. The wastewater treatment plant for domestic wastewater consists of screens and grit chamber (to remove the large sized particles); primary clarifier (to remove the settable particles); secondary or biological treatment stage that includes activated sludge treatment, trickling filters, etc.; and chlorination unit as part of the tertiary treatment of wastewater (Logan 2008). The efficiency of wastewater treatment plants is gauged from point of view of operational feasibility, economic sustainability, and ease of maintenance. As regards the biological wastewater treatment, functionally, it is responsible for the removal of huge quantity of organic matter and nutrients as well (Rahimnejad et al. 2015). Recently, MFCs found recognition on account of their potential to produce bioelectricity from domestic wastewater (Fig. 10.6) (Min and Logan 2004; Logan et al. 2006), cattle manure, brewery wastewater (Feng et al. 2014), and waste sludge (Jiang et al. 2011; Wang et al. 2018). The potential of MFCs for wastewater treatment was explored from the perspective of pollution reduction, cost efficiency, bioelectricity generation and sustainability of the system. Logan 2008 opined that MFCs has potential to be implemented in wastewater treatment either by replacing the traditional biological treatment methods with MFC or as a prelude to the membrane bioreactor or as membrane bioreactor itself (Logan 2008). MFC incorporation in wastewater treatment plant has potential benefits in terms of energy savings, bioelectricity generation and reduced quantum of solids produced from the bioreactor (Logan 2008). Successful integration and functioning of MFCs with wastewater treatment plants depend on oxygen supply to cathode compartment and supply of wastewater as a fuel to the anode compartment. Further, the surface area of the electrodes and internal resistance of the system determines the efficiency of MFCs (Logan 2010). With regard to harvesting of CO_2 which is a greenhouse gas, studies have been conducted to develop "microbial carbon capture

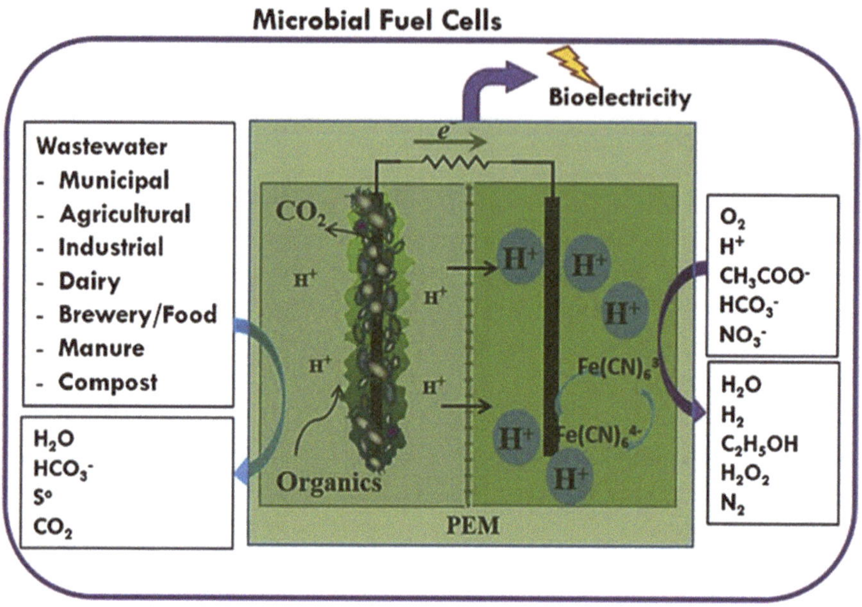

Fig. 10.6 Microbial fuel cells in wastewater treatment. (With permission from Gude 2016)

cells" where in the oxidation product from the anode is transferred to the cathode with photosynthetic *Chlorella vulgaris* which sequesters the CO_2. Wang and co-workers (2010) reported that the microbe enabled CO_2 capture in MFCs was remarkable (about 94%) and further, such analogues of MFCs can be an integral part of the wastewater treatment plants (Wang et al. 2010).

10.9 Conclusion

Biological fuel cells endeavour in the presence of a catalyst to convert the chemical energy in the organic substrates into electrical energy. Based on the biocatalyst used in the fuel cells namely the enzyme or a bacterial cell, biological fuel cells are grouped into enzymatic fuels cells and microbial fuel cells (MFC). The MFCs entail tapping the energy produced from the cellular metabolism of substrates. Interestingly, MFC use wide variety of substrates ranging from agricultural waste, municipal waste, food waste to wastewater. MFCs which are bioelectrochemical devices uses bioelectrogenic micro-organisms to treat different wastewater and other substrates to generate bioelectricity. This virtue of MFCs give a radiant hope to pollution management and bioelectricty generation which are essentially important in the back drop of growing energy crisis and environmental pollution. Biodegradable organic component of wastewater possess energy that can be tapped using MFCs. Generally, wastewater treatment plant aims at treating the wastewater efficiently, cost-effectively and monitoring the pollution load of the treated effluent. Of late, MFCs found recognition in environmental pollution management on account of their potential to produce bioelectricity from domestic wastewater, cattle manure, brewery wastewater, and waste sludge, etc. MFCs prospects for wastewater treatment can be gauged from the perspective of pollution reduction, cost efficiency, bioelectricity generation and sustainability of the system. Further, integration of MFCs in wastewater treatment plants has potential benefits in terms of energy savings, bioelectricity generation and reduced quantum of solids produced from the bioreactor. Granular research is deeply sought for in the areas of reactor design and fabrication, electron transfer mechanisms, microbial ecology of exoelectrogens, and upscaling of MFCs.

References

Allen, R., & Bennetto, H. (1993). Microbial fuel-cells. *Applied Biochemistry and Biotechnology, 39-40*, 27–40. https://doi.org/10.1007/bf02918975.

Babauta, J., Renslow, R., Lewandowski, Z., & Beyenal, H. (2012). Electrochemically active biofilms: Facts and fiction. A review. *Biofouling, 28*, 789–812. https://doi.org/10.1080/08927014.2012.710324.

Berchmans, S. (2018). Microbial fuel cell as alternate power tool: Potential and challenges. In D. Das (Ed.), *Microbial fuel cell*. Cham: Springer.

Bond, D. R., Holmes, D. E., Tender, L. M., & Lovley, D. R. (2002). Electrode-reducing microorganisms that harvest energy from marine sediments. *Science, 295*, 483–485. https://doi.org/10.1126/science.1066771.

Carmalin Sophia, A., Bhalambaal, V., Lima, E., & Thirunavoukkarasu, M. (2016). Microbial desalination cell technology: Contribution to sustainable waste water treatment process, current status and future applications. *Journal of Environmental Chemical Engineering, 4*, 3468–3478. https://doi.org/10.1016/j.jece.2016.07.024.

Chandrasekhar, K., & Venkata Mohan, S. (2012). Bio-electrochemical remediation of real field petroleum sludge as an electron donor with simultaneous power generation facilitates biotransformation of PAH: Effect of substrate concentration. *Bioresource Technology, 110*, 517–525.

Chandrasekhar, K., Amulya, K., & Venkata Mohan, S. (2015). Solid phase bio-electro fermentation of food waste to harvest value-added products associated with waste remediation. *Waste Management, 45*, 57–65.

Chandrasekhar, K., Kadier, A., Kumar, G., Nastro, R. A., & Jeevitha, V. (2018). Challenges in microbial fuel cell and future scope. In D. Das (Ed.), *Microbial fuel cell*. Cham: Springer.

Chaturvedi, V., & Verma, P. (2016). Microbial fuel cell: A green approach for the utilization of waste for the generation of bioelectricity. *Bioresources Bioprocessing, 3*, 38.

Chen, G., Choi, S., Lee, T., Lee, G., Cha, J., & Kim, C. (2008). Application of biocathode in microbial fuel cells: Cell performance and microbial community. *Applied Microbiology and Biotechnology, 79*, 379–388. https://doi.org/10.1007/s00253-008-1451-0.

Cui, Y., Rashid, N., Hu, N., Rehman, M., & Han, J. (2014). Electricity generation and microalgae cultivation in microbial fuel cell using microalgae-enriched anode and bio-cathode. *Energy Conversion and Management, 79*, 674–680. https://doi.org/10.1016/j.enconman.2013.12.032.

Daniel, D., Das Mankidy, B., Ambarish, K., & Manogari, R. (2009). Construction and operation of a microbial fuel cell for electricity generation from wastewater. *International Journal of Hydrogen Energy, 34*, 7555–7560. https://doi.org/10.1016/j.ijhydene.2009.06.012.

De Schamphelaire, L., Rabaey, K., Boeckx, P., Boon, N., & Verstraete, W. (2008). Outlook for benefits of sediment microbial fuel cells with two bio-electrodes. *Microbial Biotechnology, 1*, 446–462. https://doi.org/10.1111/j.1751-7915.2008.00042.x.

Du, Z., Li, H., & Gu, T. (2007). A state of the art review on microbial fuel cells: A promising technology for wastewater treatment and bioenergy. *Biotechnology Advances, 25*, 464–482.

Erable, B., Duteanu, N., Ghangrekar, M., Dumas, C., & Scott, K. (2010). Application of electro-active biofilms. *Biofouling, 26*, 57–71. https://doi.org/10.1080/08927010903161281.

Ewing, T., Ha, P., Babauta, J., Tang, N., Heo, D., & Beyenal, H. (2014). Scale-up of sediment microbial fuel cells. *Journal of Power Sources, 272*, 311–319. https://doi.org/10.1016/j.jpowsour.2014.08.070.

Feng, C., Li, J., Qin, D., Chen, L., Zhao, F., Chen, S., Hu, H., & Yu, C. (2014). Characterization of exoelectrogenic bacteria enterobacter strains isolated from a microbial fuel cell exposed to copper shock load. *PLoS One, 9*, e113379. https://doi.org/10.1371/journal.pone.0113379.

Fernando, E., Keshavarz, T., & Kyazze, G. (2012). Enhanced bio-decolourisation of acid orange by *Shewanella oneidensis* through co-metabolism in a microbial fuel cell. *International Biodeterioration and Biodegradation, 72*, 1–9.

Fornero, J., Rosenbaum, M., & Angenent, L. (2010). Electric power generation from municipal, food, and animal wastewaters using microbial fuel cells. *Electroanalysis, 22*, 832–843. https://doi.org/10.1002/elan.200980011.

Ghasemi, M., Daud, W., Rahimnejad, M., Rezayi, M., Fatemi, A., Jafari, Y., Somalu, M., & Manzour, A. (2013). Copper-phthalocyanine and nickel nanoparticles as novel cathode catalysts in microbial fuel cells. *International Journal of Hydrogen Energy, 38*, 9533–9540. https://doi.org/10.1016/j.ijhydene.2013.01.177.

Gude, V. (2016). Wastewater treatment in microbial fuel cells – An overview. *Journal of Cleaner Production, 122*, 287–307. https://doi.org/10.1016/j.jclepro.2016.02.022.

He, Z., & Angenent, L. (2006). Application of bacterial biocathodes in microbial fuel cells. *Electroanalysis, 18*, 2009–2015. https://doi.org/10.1002/elan.200603628.

Ishii, S., Suzuki, S., Norden-Krichmar, T., Wu, A., Yamanaka, Y., Nealson, K., & Bretschger, O. (2013). Identifying the microbial communities and operational conditions for optimized wastewater treatment in microbial fuel cells. *Water Research, 47*, 7120–7130. https://doi.org/10.1016/j.watres.2013.07.048.

Jiang, D., Curtis, M., Troop, E., Scheible, K., McGrath, J., Hu, B., Suib, S., Raymond, D., & Li, B. (2011). A pilot-scale study on utilizing multi-anode/cathode microbial fuel cells (MAC MFCs) to enhance the power production in wastewater treatment. *International Journal of Hydrogen Energy, 36*, 876–884. https://doi.org/10.1016/j.ijhydene.2010.08.074.

Jong, B., Kim, B., Chang, I., Liew, P., Choo, Y., & Kang, G. (2006). Enrichment, performance, and microbial diversity of a thermophilic mediatorless microbial fuel cell. *Environmental Science & Technology, 40*, 6449–6454. https://doi.org/10.1021/es0613512.

Kim, M., & Lee, Y. (2010). Optimization of culture conditions and electricity generation using *Geobacter sulfurreducens* in a dual-chambered microbial fuel-cell. *International Journal of Hydrogen Energy, 35*, 13028–13034. https://doi.org/10.1016/j.ijhydene.2010.04.061.

Kim, Y., & Logan, B. (2013). Microbial desalination cells for energy production and desalination. *Desalination, 308*, 122–130. https://doi.org/10.1016/j.desal.2012.07.022.

Kim, H., Park, H., Hyun, M., Chang, I., Kim, M., & Kim, B. (2002). A mediator-less microbial fuel cell using a metal reducing bacterium, Shewanella putrefaciens. *Enzyme and Microbial Technology, 30*, 145–152. https://doi.org/10.1016/s0141-0229(01)00478-1.

Kim, J., Han, S., Oh, S., & Park, K. (2011). A non-Pt catalyst for improved oxygen reduction reaction in microbial fuel cells. *Journal of the Korean Electrochemical Society, 14*, 71–76. https://doi.org/10.5229/jkes.2011.14.2.071.

Kumar, A. K., Reddy, M. V., & Chandrasekhar, K. (2012). Endocrine disruptive estrogens role in electron transfer: Bio-electrochemical remediation with microbial mediated electrogenesis. *Bioresource Technology, 104*, 547–556. https://doi.org/10.1016/j.biortech.2011.10.037.

Kumar, R., Singh, L., Wahid, Z., & Din, M. (2015). Exoelectrogens in microbial fuel cells toward bioelectricity generation: A review. *International Journal of Energy Research, 39*, 1048–1067. https://doi.org/10.1002/er.3305.

Kumar, R., Singh, L., & Zularisam, A. (2016). Exoelectrogens: Recent advances in molecular drivers involved in extracellular electron transfer and strategies used to improve it for microbial fuel cell applications. *Renewable and Sustainable Energy Reviews, 56*, 1322–1336. https://doi.org/10.1016/j.rser.2015.12.029.

Li, Z., Yao, L., Kong, L., & Liu, H. (2008). Electricity generation using a baffled microbial fuel cell convenient for stacking. *Bioresource Technology, 99*, 1650–1655. https://doi.org/10.1016/j.biortech.2007.04.003.

Li, W., Sheng, G., Liu, X., & Yu, H. (2011). Recent advances in the separators for microbial fuel cells. *Bioresource Technology, 102*(1), 244–252. https://doi.org/10.1016/j.biortech.2010.03.090.

Li, Y., Styczynski, J., Huang, Y., Xu, Z., McCutcheon, J., & Li, B. (2017). Energy-positive wastewater treatment and desalination in an integrated microbial desalination cell (MDC)-microbial electrolysis cell (MEC). *Journal of Power Sources, 356*, 529–538. https://doi.org/10.1016/j.jpowsour.2017.01.069.

Liu, H., & Logan, B. (2004). Electricity generation using an air-cathode single chamber microbial fuel cell in the presence and absence of a proton exchange membrane. *Environmental Science & Technology, 38*(14), 4040–4046. https://doi.org/10.1021/es0499344.

Liu, Z., Lian, J., Du, Z., & Li, H. (2006). Construction of sugar-based microbial fuel cells by dissimilatory metal reduction bacteria. *Chinese Journal of Biotechnology, 22*, 131–137. https://doi.org/10.1016/s1872-2075(06)60010-1.

Liu, H., Cheng, S., Huang, L., & Logan, B. (2008). Scale-up of membrane-free single-chamber microbial fuel cells. *Journal of Power Sources, 179*(1), 274–279. https://doi.org/10.1016/j.jpowsour.2007.12.120.

Liu, Z., Liu, J., Zhang, S., & Su, Z. (2009). Study of operational performance and electrical response on mediator-less microbial fuel cells fed with carbon- and protein-rich substrates. *Biochemical Engineering Journal, 45*, 185–191. https://doi.org/10.1016/j.bej.2009.03.011.

Logan, B. E. (2004). Extracting hydrogen electricity from renewable resources. *Environmental Science & Technology, 38*(9), 160A–167A. https://doi.org/10.1021/es040468s.

Logan, B. E. (2008). *Microbial fuel cell*. New York: Wiley.

Logan, B. E. (2009). Exoelectrogenic bacteria that power microbial fuel cells. *Nature Reviews Microbiology, 7*, 375–381. https://doi.org/10.1038/nrmicro2113.

Logan, B. E. (2010). Scaling up microbial fuel cells and other bioelectrochemical systems. *Applied Microbiology and Biotechnology, 85*, 1665–1671. https://doi.org/10.1007/s00253-009-2378-9.

Logan, B., & Regan, J. (2006). Microbial fuel cells—challenges and applications. *Environmental Science & Technology, 40*, 5172–5180. https://doi.org/10.1021/es0627592.

Logan, B., Murano, C., Scott, K., Gray, N., & Head, I. (2005). Electricity generation from cysteine in a microbial fuel cell. *Water Research, 39*, 942–952. https://doi.org/10.1016/j.watres.2004.11.019.

Logan, B. E., Hamelers, B., Rozendal, R., Schröder, U., Keller, J., Freguia, S., Aelterman, P., Verstraete, W., & Rabaey, K. (2006). Microbial fuel cells: Methodology and technology. *Environmental Science & Technology, 40*, 5181–5192. https://doi.org/10.1021/es0605016.

Min, B., & Logan, B. (2004). Continuous electricity generation from domestic wastewater and organic substrates in a flat plate microbial fuel cell. *Environmental Science & Technology, 38*, 5809–5814. https://doi.org/10.1021/es0491026.

Min, B., Cheng, S., & Logan, B. (2005). Electricity generation using membrane and salt bridge microbial fuel cells. *Water Research, 39*, 1675–1686. https://doi.org/10.1016/j.watres.2005.02.002.

Mohan, Y., ManojMuthu Kumar, S., & Das, D. (2008). Electricity generation using microbial fuel cells. *International Journal of Hydrogen Energy, 33*, 423–426. https://doi.org/10.1016/j.ijhydene.2007.07.027.

Mook, W., Aroua, M., Chakrabarti, M., Noor, I., Irfan, M., & Low, C. (2013). A review on the effect of bio-electrodes on denitrification and organic matter removal processes in bio-electrochemical systems. *Journal of Industrial and Engineering Chemistry, 19*, 1–13. https://doi.org/10.1016/j.jiec.2012.07.004.

Mu, Y., Rabaey, K., Rozendal, R., Yuan, Z., & Keller, J. (2009). Decolorization of azo dyes in bioelectrochemical systems. *Environmental Science & Technology, 43*(13), 5137–5143. https://doi.org/10.1021/es900057f.

Nastro, R. A. (2014). Microbial fuel cells in waste treatment: Recent advances. *International Journal of Performability Engineering, 10*(4), 367–376.

Niessen, J., Schröder, U., Rosenbaum, M., & Scholz, F. (2004). Fluorinated polyanilines as superior materials for electrocatalytic anodes in bacterial fuel cells. *Electrochemistry Communications, 6*, 571–575. https://doi.org/10.1016/j.elecom.2004.04.006.

Ömeroğlu, S., & Sanin, F. (2016). Bioelectricity generation from wastewater sludge using microbial fuel cells: A critical review. *CLEAN – Soil, Air, Water, 44*, 1225–1233. https://doi.org/10.1002/clen.201500829.

Pandit, S., & Das, D. (2018). Principles of microbial fuel cell for the power generation. In D. Das (Ed.), *Microbial fuel cell* (pp. p21–p41). Cham: Springer.

Pant, D., Van Bogaert, G., Diels, L., & Vanbroekhoven, K. (2010). A review of the substrates used in microbial fuel cells (MFCs) for sustainable energy production. *Bioresource Technology, 101*, 1533–1543. https://doi.org/10.1016/j.biortech.2009.10.017.

Park, D. H., & Zeikus, J. G. (1999). Utilization of electrically reduced neutral red by *Actinobacillus succinogenes*: physiological function of neutral red in membrane-driven fumarate reduction and energy conservation. *Journal of Bacteriology, 181*(8), 2403–2410.

Potter, M. (1911). Electrical effects accompanying the decomposition of organic compounds. *Proceedings of the Royal Society B: Biological Sciences, 84*, 260–276. https://doi.org/10.1098/rspb.1911.0073.

Qiao, Y., Li, C., Bao, S., & Bao, Q. (2007). Carbon nanotube/polyaniline composite as anode material for microbial fuel cells. *Journal of Power Sources, 170*, 79–84. https://doi.org/10.1016/j.jpowsour.2007.03.048.

Rabaey, K., Lissens, G., Siciliano, S., & Verstraete, W. (2003). A microbial fuel cell capable of converting glucose to electricity at high rate and efficiency. *Biotechnology Letters, 25*, 1531–1535. https://doi.org/10.1023/a:1025484009367.

Rabaey, K., Boon, N., Höfte, M., & Verstraete, W. (2005). Microbial phenazine production enhances electron transfer in biofuel cells. *Environmental Science & Technology, 39*, 3401–3408. https://doi.org/10.1021/es048563o.

Rahimnejad, M., Mokhtarian, N., Najafpour, G., Ghoreyshi, A., & Dahud, W. (2009). Effective parameters on performance of microbial fuel cell. 2009 Second International conference on environmental and computer science. *IEEE, 1*, 411–415. https://doi.org/10.1109/icecs.2009.23.

Rahimnejad, M., Adhami, A., Darvari, S., Zirepour, A., & Oh, S. (2015). Microbial fuel cell as new technology for bioelectricity generation: A review. *Alexandria Engineering Journal, 54*, 745–756. https://doi.org/10.1016/j.aej.2015.03.031.

Rosenbaum, M., Schro¨der, U., & Scholz, F. (2005). In situ electro oxidation of photobiological hydrogen in a photo bioelectrochemical fuel cell based on *Rhodobacter sphaeroides*. *Environmental Science & Technology, 39*, 6328–6333. https://doi.org/10.1021/es0505447.

Roy, S., Marzorati, S., Schievano, A., et al. (2017). Microbial fuel cells. In M. A. Abraham (Ed.), *Encyclopedia of sustainable technologies* (pp. 245–259). Amsterdam: Elsevier.

Saeed, H., Husseini, G., Yousef, S., Saif, J., Al-Asheh, S., Abu Fara, A., Azzam, S., Khawaga, R., & Aidan, A. (2015). Microbial desalination cell technology: A review and a case study. *Desalination, 359*, 1–13. https://doi.org/10.1016/j.desal.2014.12.024.

Santoro, C., Ieropoulos, I., Greenman, J., Cristiani, P., Vadas, T., Mackay, A., & Li, B. (2013). Current generation in membraneless single chamber microbial fuel cells (MFCs) treating urine. *Journal of Power Sources, 238*, 190–196. https://doi.org/10.1016/j.jpowsour.2013.03.095.

Scott, K., Rimbu, G., Katuri, K., Prasad, K., & Head, I. (2007). Application of modified carbon anodes in microbial fuel cells. *Process Safety and Environmental Protection, 85*, 481–488. https://doi.org/10.1205/psep07018.

Sevda, S., Yuan, H., He, Z., & Abu-Reesh, I. (2015). Microbial desalination cells as a versatile technology: Functions, optimization and prospective. *Desalination, 371*, 9–17. https://doi.org/10.1016/j.desal.2015.05.021.

Sevda, S., Abu-Reesh, I., Yuan, H., & He, Z. (2017). Bioelectricity generation from treatment of petroleum refinery wastewater with simultaneous seawater desalination in microbial desalination cells. *Energy Conversion and Management, 141*, 101–107. https://doi.org/10.1016/j.enconman.2016.05.050.

Song, H., Zhu, Y., & Li, J. (2015). Electron transfer mechanisms, characteristics and applications of biological cathode microbial fuel cells – A mini review. *Arabian Journal of Chemistry*. https://doi.org/10.1016/j.arabjc.2015.01.008.

Ucar, D., Zhang, Y., & Angelidaki, I. (2017). An overview of electron acceptors in microbial fuel cells. *Frontiers in Microbiology, 8*, 643. https://doi.org/10.3389/fmicb.2017.00643.

Varanasi, J., Roy, S., Pandit, S., & Das, D. (2015). Improvement of energy recovery from cellobiose by thermophillic dark fermentative hydrogen production followed by microbial fuel cell. *International Journal of Hydrogen Energy, 40*, 8311–8321. https://doi.org/10.1016/j.ijhydene.2015.04.124.

Wang, X., Feng, Y., Liu, J., Lee, H., Li, C., Li, N., & Ren, N. (2010). Sequestration of CO_2 discharged from anode by algal cathode in microbial carbon capture cells (MCCs). *Biosensors & Bioelectronics, 25*, 2639–2643. https://doi.org/10.1016/j.bios.2010.04.036.

Wang, D., Song, T., Guo, T., Zeng, Q., & Xie, J. (2014). Electricity generation from sediment microbial fuel cells with algae-assisted cathodes. *International Journal of Hydrogen Energy, 39*, 13224–13230. https://doi.org/10.1016/j.ijhydene.2014.06.141.

Wang, Y., Zhang, H., Li, B., & Feng, Y. (2018). Integrating sludge microbial fuel cell with inclined plate settling and membrane filtration for electricity generation, efficient sludge reduction and high wastewater quality. *Chemical Engineering Journal, 331*, 152–160. https://doi.org/10.1016/j.cej.2017.07.173.

Watanabe, K. (2008). Recent developments in MFC technologies for sustainable bioenergy. *Journal of Bioscience and Bioengineering, 106*, 528–536. https://doi.org/10.1263/jbb.106.528.

Wen, Q., Yang, T., Wang, S., Chen, Y., Cong, L., & Qu, Y. (2013). Dechlorination of 4-chlorophenol to phenol in bioelectrochemical systems. *Journal of Hazardous Materials, 244-245*, 743–749. https://doi.org/10.1016/j.jhazmat.2012.11.002.

Xia, C., Zhang, D., Pedrycz, W., Zhu, Y., & Guo, Y. (2018). Models for microbial fuel cells: A critical review. *Journal of Power Sources, 373*, 119–131. https://doi.org/10.1016/j.jpowsour.2017.11.001.

Yousefi, V., Mohebbi-Kalhori, D., Samimi, A., & Salari, M. (2016). Effect of separator electrode assembly (SEA) design and mode of operation on the performance of continuous tubular microbial fuel cells (MFCs). *International Journal of Hydrogen Energy, 41*(1), 597–606. https://doi.org/10.1016/j.ijhydene.2015.11.018.

Yousefi, V., Mohebbi-Kalhori, D., & Samimi, A. (2017). Ceramic-based microbial fuel cells (MFCs): A review. *International Journal of Hydrogen Energy, 42*(3), 1672–1690. https://doi.org/10.1016/j.ijhydene.2016.06.054.

Yu, J., Cho, S., Kim, S., Cho, H., & Lee, T. (2012). Comparison of exoelectrogenic bacteria detected using two different methods: U-tube microbial fuel cell and plating method. *Microbes and Environments, 27*, 49–53. https://doi.org/10.1264/jsme2.me11205.

Yuan, H., & He, Z. (2015). Graphene-modified electrodes for enhancing the performance of microbial fuel cells. *Nanoscale, 7*, 7022–7029. https://doi.org/10.1039/c4nr05637j.

Zhang, Y., & Angelidaki, I. (2011). Submersible microbial fuel cell sensor for monitoring microbial activity and BOD in groundwater: Focusing on impact of anodic biofilm on sensor applicability. *Biotechnology and Bioengineering, 108*, 2339–2347. https://doi.org/10.1002/bit.23204.

Zhang, X., Liang, P., Shi, J., Wei, J., & Huang, X. (2013). Using a glass fiber separator in a single-chamber air-cathode microbial fuel cell shortens start-up time and improves anode performance at ambient and mesophilic temperatures. *Bioresource Technology, 130*, 529–535. https://doi.org/10.1016/j.biortech.2012.12.091.

Zheng, X., & Nirmalakhandan, N. (2010). Cattle wastes as substrates for bioelectricity production via microbial fuel cells. *Biotechnology Letters, 32*, 1809–1814. https://doi.org/10.1007/s10529-010-0360-3.

Zhou, M., Jin, T., Wu, Z., et al. (2012). Microbial fuel cells for bioenergy and bioproducts. In K. Gopalakrishnan, J. van Leeuwen, & R. Brown (Eds.), *Sustainable bioenergy and bioproducts. Green energy and technology* (pp. 131–171). London: Springer.

Zhou, M., Wang, H., Hassett, D., & Gu, T. (2013). Recent advances in microbial fuel cells (MFCs) and microbial electrolysis cells (MECs) for wastewater treatment, bioenergy and bioproducts. *Journal of Chemical Technology and Biotechnology, 88*, 508–518. https://doi.org/10.1002/jctb.4004.

Chapter 11
Biofuels: A Clean Technology for Environment Management

S. Prasad, V. Venkatramanan, S. Kumar, and K. R. Sheetal

Abstract Air quality (AQ) and greenhouse gas (GHG) emission management are the key drivers related to sustainability goal around the world today because most of the aforementioned issues are attributable to fossil fuels burning. Biofuels are the clean, alternative fuels that are derived from biomass-based resources. It addresses efficient management of air pollution problems and the collective goal of climate change mitigation. Recent times have witnessed accelerated growth in biofuel production worldwide. The Government of India has been very keen on promoting manufacture and blending of ethanol derived from sugarcane molasses, and biodiesel from non-edible and waste oils for mixing with diesel. India's biofuel policy (2003), dealing with bio-ethanol and biodiesel, sights to channelize biofuels into the energy and transport sector to address energy security and improvement in the living standard of rural areas. Biofuels as an alternative energy source can help to lessen the dependency on imported fossil fuel oil, achieve sustainability goal and several other societal requirements.

Keywords Fossil fuels · Biofuels · Air quality · Climate change · Mitigation

S. Prasad (✉) · S. Kumar
Centre for Environment Science and Climate Resilient Agriculture, Indian Agricultural Research Institute, New Delhi, India

V. Venkatramanan
School of Interdisciplinary and Transdisciplinary Studies, Indira Gandhi National Open University, New Delhi, India

K. R. Sheetal
Central Arid Zone Research Institute, Regional Research Station, Bikaner, Rajasthan, India

S. Shah et al. (eds.), *Sustainable Green Technologies for Environmental Management*, https://doi.org/10.1007/978-981-13-2772-8_11

11.1 Introduction

Presently, most of our energy demands are satisfied by fossil fuels based products. Fossil fuels consumption is the well-known contributor to air pollution, (Tyagi et al. 2016), greenhouse gas (GHG) emissions and the environmental consequences with a large endowment of coal and has an energy system that is more carbon intensive (USDoE 2007; Prasad and Dhanya 2011a). The burning of fossil fuels generates almost 21.3 billion tons of carbon dioxide (CO_2) per year, and the natural sources can only absorb nearly half of that amount, so there is a net addition of 10.65 billion tonnes of atmospheric CO_2 (USDoE 2007). CO_2 is one of the principal GHG responsible for the rise in global warming and thus an increase in average surface temperatures of the earth (Prasad et al. 2014).

The Intergovernmental Panel on Climate Change (IPCC) declares that the warming of the climate is unequivocal, with temperatures 0.85 °C higher than those documented since the 1880 baseline. If the current trajectory of emissions continues, the world can expect 3.7 °C to 4.8 °C of warming by 2100. Under all future projections of GHGs, the earth surface temperatures are likely to increase with associated increases in the frequency of heat waves, extreme rainfall, ocean acidification, and global sea levels (IPCC 2014). The burning of fossil fuel also emits a significant amount of pollutant gases such as nitrogen oxides, carbon monoxide, VOCs and particulate matter fractions. The reactions between the combination of VOCs and NOx with sunlight make the tropospheric ozone (O_3), which is the chief constituent of photochemical smog. These gaseous air pollutants have further deteriorated the health of human beings and agricultural productivity (Vinish 2005).

The world is facing a severe environmental crisis due to these critical changes in atmospheric condition. The question now arises how to manage this environment adversity and human well-being sustainably. An essential component of environmental management is promoting pollutants reduction and mitigation of GHGs through available resources, technologies, and policies, which can substantially reduce the possible risks associated with these consequences (IPCC 2014). So, it is essential to look at a broader agenda that promotes energy security, and sustainable economic development without a complete restructuring of the current working energy system (Prasad et al. 2012; Kumar et al. 2016). Over the last few decades, biofuels had come to the forefront of environmental problems management and have the potential to fight the challenges of climate change and air pollution (Sudhakar et al. 2013; Prasad et al. 2014).

The current evidence suggests that the biofuels offer a number of benefits to society such as energy security, pollutant reduction, and sustainable economic development (Sudhakar et al. 2013; Attal et al. 2014). There is a need to manage the environmental issues more sustainably to advance the understanding of energy efficiency, transportation, efficient use of organic materials and sustainability for a low carbon future. Biofuels production practices are considered to be carbon neutral since CO_2 captured by the biomass during its growth and development is only returned to the atmosphere on combustion. The decreased dependency on imported fuels, increased energy efficiency, cut emissions of harmful pollutants, enhanced possibilities for carbon storage and carbon credits are all advantages that can be achieved by the proper management and utilization of biomass resources and biofuels (Lynd et al. 2005; Prasad et al. 2012; Sheetal et al. 2017).

11.2 Biofuels

Biofuels are alternative fuels that originated from biomass resources to meet our growing energy needs. Biofuels have oxygen levels ranging from 10% to 45% while petroleum has essentially none, exhibiting the chemical properties of biofuels quite distinct from those of petroleum (Prasad et al. 2007). Currently, the marketed biofuels are produced from sugar crops (sugarcane, sugar beet) and their by-products like molasses, food grain-starch (corn, potatoes), vegetable oils from soybean, sunflower, and rapeseed, etc., and animal fats like beef tallow. Sugar and grain starch is processed into ethanol by fermentation. Vegetable oils and animal fats are transformed into biodiesel by trans-esterification. Ethanol and biodiesel are the most widely used as an automobile's fuel (Attal et al. 2014; Prasad et al. 2017).

Biofuels are very sought after as an alternative fuel on account of their simplicity, and production through well-known processes and technologies. Biofuels have the capacity for mitigation of global warming without the complete restructuring of the current working energy requirements. Biofuels can be blended with conventional fuels in existing engines and potential to facilitate worldwide mobilization around a standard set of regulations and practices. It is the readily accessible energy source with great public acceptance and ample uniform distribution than fossil fuel and nuclear resources. It also has the potential to have the advantages for the economy especially in rural areas, including employment (Rothengatter 2010).

Bioethanol and biodiesel are the most popular liquid biofuels for global road transport (Attal et al. 2014). Biofuel demand is increased at an average annual growth rate of 1.5, 12.1 and 15.4% for solid biomass, liquid and gaseous biofuels, respectively, between 1990 and 2008. In 2009, biofuels accounted for almost 3% of global road transportation, with the production of bioethanol and biodiesel growing by 10% and 9%, respectively (SRREN 2011). Biofuels, particularly the second and third generation fuels are anticipated to play a significant role in satisfying the requirements for transportation fuel in the future.

11.3 Biofuels as a Leading Alternative to Fossil Fuel

Most of the energy demands are currently supplied by fossil fuels–coal and petroleum-based products and natural gas (MoPNG 2014). Worldwide biofuels have brought considerable attention since they become the first alternative to fossil fuel, are produced domestically by various countries. It requires just modest changes to retailer supply chain and end-use technologies. According to FAO, global ethanol generation is supposed to increase by 70% compared to the average of 2010–2012, and cross 168 billion liters by 2022 (FAO 2013).

The benefit of biofuels is its compatibility with petroleum-fuels to be used as blends in existing petrol engines. Liquid biofuels have similar characteristics to petrol/diesel regarding engine performance and refueling points (Planning commission 2003). On the other hand, limits on the percentage of blends with gasoline (e.g., E85) and the difficulty in its distribution using existing fuel pipelines, because of its tendency to absorb water. Advanced biofuels usually contain higher carbon chains and are considered to avoid the problems (Atsumi and Liao 2008). According to IPCC, biofuels were recognized as a "key mitigation strategy" (IPCC 2007). Nevertheless, the debate especially on biofuel from biomass in the food versus fuel competition, and growing concerns about land use, water, replacement of forests, etc. have acted as catalysts for the promotion and implementation of sustainability criteria, policies, market chain and regulation frameworks (SRREN 2011; Das et al. 2010). Moreover, support for advanced bio-refinery and next-generation biofuel possibilities are driving biofuel to be more sustainable in future (IPCC 2014).

In India, the Biofuel Policy (2003), associated with bioethanol and biodiesel, aims to utilize biofuels in the energy and transportation sector to address issues of food and energy security, climate change mitigation, air pollution abatement and rural development and standard of living (Sheetal et al. 2017). Having sufficient high renewable energy production potential, India is now encouraging the production of bio-ethanol derived from sugarcane molasses and biodiesel derived from non-edible and waste oils for blending with petrol and diesel, respectively (Aradhey 2014).

Though the biomass-derived biofuel industry in the country is still under developmental stages, however extensively used biofuel especially ethanol is being produced from the abundantly available molasses. Currently, in India, there are 140 distilleries, with the potential to distill around 2 billion liters of conventional bioethanol per annum (Aradhey 2014). The production of biodiesel is still commercially negligible and not implemented. Even though India has an installed biodiesel capacity of 1.2 million tons per annum, presently only 7% is utilized because of several other restrictions (Economic Times 2014). Ensuring energy and food security to entire segments of the society, and fulfilling international commitments towards climate change, biofuels provides an opportunity for the strengthening of the existing linkages within society, science, and policy (Das et al. 2010; Sheetal et al. 2017). Besides direct services as alternative fuels to the nation-building, it also has indirect benefits like employment generation, diminishing oil import dependency, sequestration of carbon through biosource utilization. The aim of rural development and an enhanced standard of living can be additionally gained.

11.4 Biofuel Policies

The GOI has undertaken various policy initiatives to augment production and use of biofuels during the last few decades. There are several ministries and departments associated with policymaking, regulation, promotion, and development of the biofuels at the national level (MNRE 2009). The expansion of a domestic biofuels market and supply chain is anticipated to improve rural livelihood by creating employment opportunities (Biswas et al. 2006). Ministry of Petroleum & Natural Gas (MoPNG) announced a notification in September 2002 for mandatory blending of 5% ethanol in major 9 states and 3 Union Territories (Prasad et al. 2007; Prasad et al. 2012). In 2008, the target was tried to be increased to 10% biofuel blends (Bandyopadhyay 2015).

The GOI approved the National Policy on Biofuels (NPB) on December 24, 2009. The Goal of the NPB (2009) is to ensure that a minimum level of biofuels become readily available in the market to fulfill the increased demand for fuel at any given time (MNRE 2009). An aspirational target of 20% blending of biofuels by 2017 was proposed, both for bio-diesel and bio-ethanol (Indian Express 2008; MNRE 2009). India achieved its highest ever ethanol market penetration at 3.3% in 2016 but will eventually settle below last year's level given tight supply through 2017.

The GOI and several state governments will promote the planting of Jatropha and other non-edible oilseeds by providing fiscal incentives to various public, private, and cooperative sectors (Prasad et al. 2017). The Planning Commission renamed as National Institution for Transforming India commission (NITI Ayog) had even set a grand target of planting Jatropha in11.2 to 13.4 million hectares (Mha) wasteland by the end of April 2012 but fell short (Aradhey 2017). By the year 2022, the GOI proposes to reduce its dependence on crude oil imports by 10% points in several ways: increasing domestic output; promoting energy efficiency and conservation; and encouraging the broader use of alternative fuels (Aradhey 2017).

India's biofuel policy continues to concentrate on the use of non-food resources; namely molasses for the production of bio-ethanol and non-edible oils for the production of biodiesel. Biodiesel blending targets will be regularly reviewed and adjusted as needed. The policy proposes establishing a National Registry of feedstock availability to help monitor production potential and set blending targets (MNRE 2009; Aradhey 2010). NBP drafted by the MNRE, assures that the biofuel programme would not compete with food security and the fertile farmlands would not be diverted for plantation of biofuel crops (MNRE 2009). The policy deals with many issues like Minimum Support Prices (MSPs) for biofuel crops. The policy also deals with subsidies for growers and farmers of biofuel crops, marketing, subsidies and fiscal concessions for the biofuel industry, research, and development, the obligatory blending of auto-fuel with biofuel, quality norms, testing and certification of biofuels (MNRE 2009, Prasad et al. 2012).

11.5 Biodiesel Production Process

"Biodiesel" refers to an oxygen-rich fuel prepared from biological resources, with characteristics similar to that of diesel (Planning commission 2003; Sukumar et al. 2010). Currently, soybean in the United States (US), rapeseed in Europe, and palm in SE Asia are the primary feedstock for biodiesel production. Biodiesel belongs to a non-petroleum-based diesel fuel containing mono-alkyl-esters of long chain fatty acids, produced by trans-esterification of oil/fat/lipid sources (Chisti 2007). Fig. 11.1. shows the transesterification reaction.

The edible/non-edible oil is composed of triglycerides of long chain fatty acids associated with glycerol. In the transesterification process, any natural oil (vegetable or animal) is mixed with any alcohol, in the presence of a catalyst leading to modification of one type of ester into another ester. The use of 1% sodium methylate as catalyst resulted in higher end-product yields of 98.6% over other same quantity of catalysts in a comparative study of trans-esterification conducted at 238 K with methanol to oil molar ratio of 6:1.

Petroleum based diesel comprises of several hydrocarbons of various lengths ranging mainly from 14–18 carbons, contaminated with aromatic hydrocarbons, sulfur and crude oil residues. Though similar in characteristics, biodiesel varies in its chemical composition from that of petroleum-based diesel fuel. With 16–20 carbon long hydrocarbon chains, biodiesel contains oxygen at one end, leading to 11% O_2 content by weight and no sulfur. The elevated O_2 content of the compound supports complete combustion of fuel (Kralova and Sjoblom 2010). These characters enhance fuel combustion efficiency and substantially reduce the pollutants emission from the vehicle exhaust. It can be utilized as an alternative fuel for vehicles in direct form or blended with fuel diesel without any modifications in existing diesel engines (Prasad et al. 2017).

The major success factor for biodiesel production is the availability of a sustainable supply of raw materials. Even though 9.3% of global oilseed production is from India and we occupy fourth position in the list of largest edible oil producing

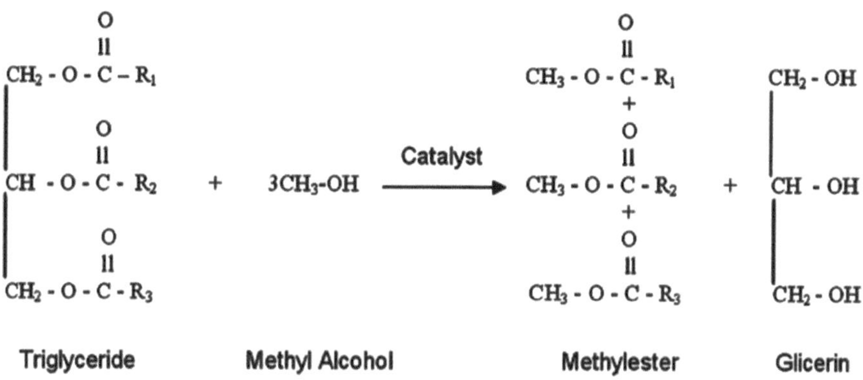

Fig. 11.1 Biodiesel production by transesterification reaction. (From Chisti 2007)

countries, presently up to 46% of edible oil for domestic needs is met through imports. As the main objective of production in India is meeting the domestic needs of people, it is not possible to divert the oilseeds towards biodiesel production. Hence, other feedstocks *viz.* non-edible and tree-borne oilseeds with higher yield potentials need to be sought out (Planning commission 2003; Sukumar et al. 2010; Prasad et al. 2012).

With a rich biodiversity of over 300 species of trees producing oilseeds, about 1 million tonne is the annual possible non-edible oil availability estimate. Though oil content varies with species, about one quarter of these trees have more than or at least 30% oil in their seeds/kernel (Planning commission 2003; Bandhopadhyay 2015). While soybean contains 20% and rapeseed ha 40% oil, the non-edible species also have comparative or even more content, for example, castor bean has 50%, neem seed has 30% and hemp contains 35% oil. *Jatropha curcas, Madhuca indica, Shorea robusta, Pongamia pinnata, Schleichera oleosa, Melia azadirachta, Garcinia indica, Diploknema butyracea, Sapindus mukorossi* and *Aleurites fordii* are some non-edible oilseed bearing trees with potential for biodiesel production.

Another feedstock of vital importance is the high oil algae. The major appeal for this group, particularly the aquatic unicellular green algae, is their particularly high growth rates owing to the higher photosynthetic efficiency and energy expenditure mainly for growth and high oil content (Chisti 2007). They are even known to double its biomass, under optimum conditions, in a day. In addition, with over 50% lipid contents, these high yielding green algae are good candidates for highly intensive cultivation and as a potential biodiesel feedstock (Schneider 2006; Chisti 2007).

In India, the railway department has taken a keen interest in biodiesel use as well as production, with the aim of reducing the use of high-speed diesel and thus reducing import expenses. On 31st December 2002, 5% biodiesel mix fuel was successfully tested in the Delhi-Amritsar Shatabdi Express. The south central railways also initiated a biodiesel mix fuel locomotive since 2015. Besides this, they have also initiated production of biodiesel from waste vegetable oils for their own use as well as have plans to utilize the land along the railway tracks for planting trees that provide oilseeds. GOI had set an ambitious goal of establishing Jatropha on 11.2–13.4 Mha of waste-land by 2012, to generate sufficient amount of biodiesel to blend at 20% with petrodiesel (MNRE 2009; Prasad and Dhanya 2011b).

11.6 Ethanol Production Process

Sugar containing resources, or those which may be modified into sugars, are suitable feedstocks for fermentation (Lynd et al. 2005; Prasad et al. 2012; Sheetal et al. 2017). The fermentable resources can be grouped into (i) direct fermentable sugar (ii) starchy materials, and (iii) lignocellulosic biomass. Commercially, production of ethanol by direct fermentation of molasses from sugarcane and beet, sugarcane, sugar beets and even sweet sorghum and fruit juices, has been reported through many studies (Prasad et al. 2007). Sugar-containing materials like molasses, sweet sorghum, and fruit juices do not require any pretreatment, whereas a starchy, lignocellulosic material needs pretreatment and hydrolysis to change these resources into fermentable substrates for ethanol production (Sun and Cheng 2002; Malav et al. 2017).

Fermentation substrates include substances which contain sugar and may be broken down to get monomers of glucose, which can be converted into the main product ethanol along with carbon dioxide, under anaerobic conditions, in the presence of microorganisms like fungi, bacteria, and yeast. *Saccharomyces cerevisiae* is one of the particular yeasts used commercially for this process. For every mole of glucose utilized, two moles each of ethanol and carbon dioxide are liberated by alcoholic fermentation (11.2). Theoretically, 51.4 g of ethanol and 48.8 g of carbon dioxide are produced for every 100 g of glucose consumed. Though, in practice, the actual yield is lesser than 51.4 g as some glucose is consumed by microorganisms for glucose (Badger 2002).

Production of alcohol from starchy materials like grains is a long process, since starch cannot be directly fermented by microbes and has to be broken down into fermentable sugars. Thus the process involves grain milling, physical/chemical/biological treatments and hydrolysis for release of sugar monomers from these materials, fermentation with yeasts and finally distillation to get ethanol. Mainly two enzymes are commonly used for hydrolysis, amylase and amyloglucosidase (Prasad et al. 2007). The biochemical reactions and aforementioned processes are shown in detail in Figs. 11.2 and 11.3 and in Eqs. (11.1) and (11.2).

$$\left(C_6H_{10}O_5\right)_n + nH_2O \rightarrow nC_6H_{12}O_6 \tag{11.1}$$

$$C_6H_{12}O_6 \rightarrow 2C_2H_5OH + 2CO_2 + \text{energy}\left(\text{stored as ATP}\right) \tag{11.2}$$

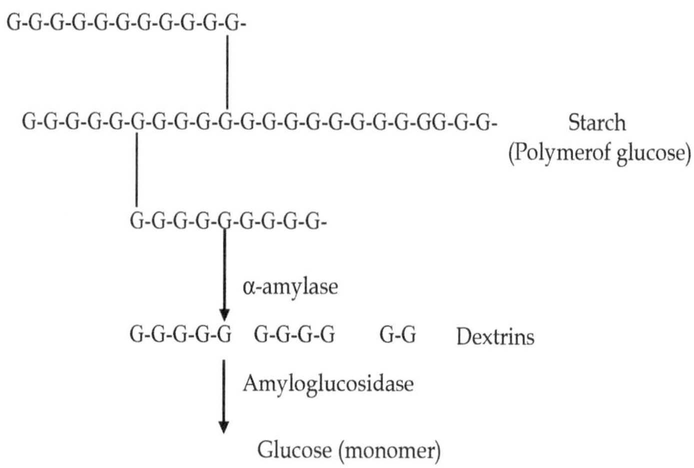

Fig. 11.2 Enzymatic hydrolysis of starch to glucose. (From Prasad et al. 2007)

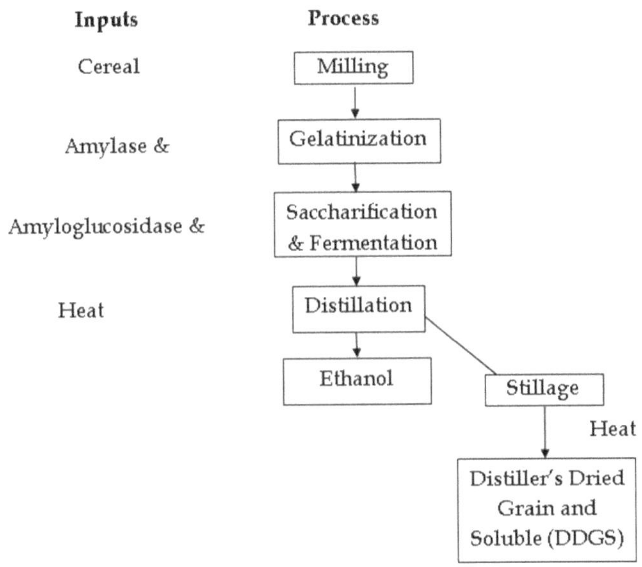

Fig. 11.3 Flowchart of ethanol production from cereal grains. (From Prasad et al. 2007)

Milling is the first step in ethanol production, and the purpose is to break up cereal grains into smaller particles that better adsorb water. The particles have to be reasonably small enough to yield high water access, but they should not be too small so that they cause problems in the recovery of co-products. The most common mill used in distilleries is a hammer mill, with hammers that rotate at high speed, crushing the grains in the grinding chamber. A retention screen is used to hold back the largest milling particles. Bran removal before milling and other fermentation processes begin has been reported to improve the ethanol production efficiency substantially (Prasad and Dhanya 2011b) in such production systems.

Each process based on cereal grains starts with a mixing stage, where grain meal and water are mixed under heat. The temperature and duration time varies but temperatures around 40–60 °C and duration time between a few minutes to 30 min are common. The mash is moved to a liquefaction tank where it is heated to 90 °C for 2 h. The enzyme α-amylase is added to break down starch into dextrins. At approximately 65 °C, starch granules have adsorbed so much water that they swell into large gel-filled sacs with lost crystalline structure. That phase is called gelatinization, where the mash has a very high viscosity. The swollen granules are now permeable for α-amylase, allowing it to penetrate and break down starch, which markedly reduces mash viscosity. The high liquefaction temperature is also used to minimize bacterial contamination.

The next process step is saccharification where dextrins are broken down into glucose by the enzyme glucoamylase (Fig. 11.3). The mash is kept at around 60 °C, and the pH is adjusted to 4.0–4.5. Fermentation is essentially accomplished with the addition of yeast that converts sugar produced during pretreatment to ethanol. After separation of fibers and yeast from the fermented mash, ethanol is extracted by distillation. The waste remaining after the fermentation process, such as alcohol from fermented broth and residual spillages are readily utilized as a part of animal feed after proper processing and is known as Distiller's Dried Grain and Soluble (DDGS) (Sheorain and Chavan 2000).

The presently available lignocellulosic materials could support the sustainable production of liquid transportation fuels, considering the vast availability and lower costs (Lynd et al. 2005; Joseph and Ronald 2010; Sheetal et al. 2017). Lignocellulosic materials can be transformed through pretreatment and hydrolysis process, into hexoses and pentoses, which under anaerobic conditions can be further converted to alcohol (Fig. 11.4). Hexose sugar (C6 sugar) utilization process has been studied and discussed in detail and is well understood (11.2). However, the concept of utilization of pentoses, which also contribute equally in fermentable sugar yield, is also important for higher efficiency of the process, and requires further study and attention. Microbes which can utilize both pentoses and hexoses are rare. As per the Eq. (11.3) given below, every three moles of pentose sugars consumed, liberates five moles each of ethanol and CO_2 (Prasad et al. 2007; Prasad and Dhanya 2011b).

$$3C_5H_{10}O_5 \rightarrow 5C_2H_5OH + 5CO_2 + \text{energy}\left(\text{stored as ATP}\right) \qquad (11.3)$$

On an industrial level, thermo-tolerant yeast could be more appropriate for fermentation. *S. cerevisiae,* the commercially used fermentation microbe, can utilize only hexose sugars. As mentioned formerly, there are few known microbes utilizing

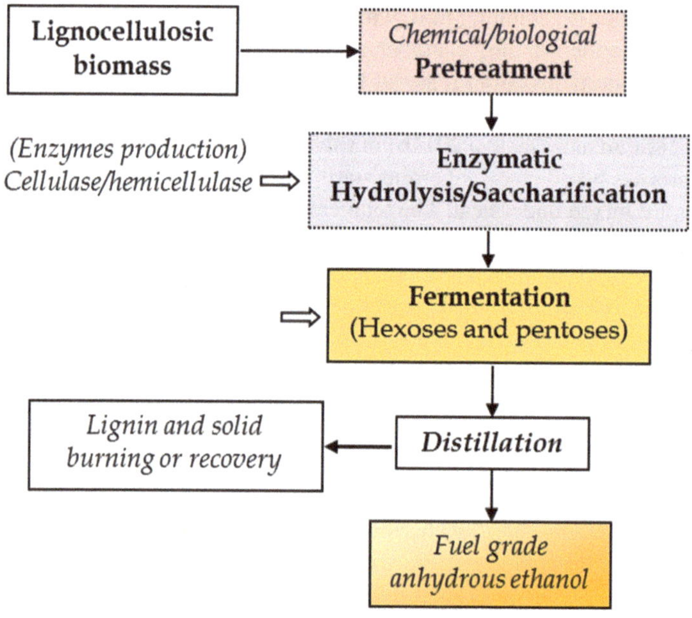

Fig. 11.4 Conversion of lignocellulosic biomass to ethanol. (From Prasad et al. 2012)

both C5 and C6 sugars, including *Pichia stipitis, Candida shehatae* and *Pachysolan tannophilus* (Steve et al. 2004). Two methods are suggested in general to overcome this problem through genetic engineering. Either ordinary hexose fermenting yeasts can be modified with addition of pentose metabolic pathways in their systems or ethanol fermenting efficiency may be improved in microbes which can utilize both hexoses and pentoses. Efficient utilization of both these major sugars found in lignocellulosic materials provides opportunity to make bioethanol production process economically feasible (Olsson and Hahn-Hagerdal 1996; Prasad et al. 2007; Malav et al. 2017).

As Fig. 11.4 shows, four necessary steps are involved in the lignocellulosic biomass conversion to ethanol: (1) pretreatment of raw biomass, (2) enzymatic hydrolysis for fermentable sugar production (3) ethanol fermentation and (4) distillation (Balat et al. 2008). The cellulosic biomass is first pretreated to begin breaking down the material and generating more surface area for the second step, hydrolysis. Hydrolysis is accomplished either by the use of enzymes or chemicals. In this step, the complex carbohydrate chains in the biomass are broken down into simple sugars. Finally, these sugars are fermented by microorganisms (yeast, or bacteria), which produce ethanol in a dilute form. In order to concentrate the fuel grade anhydrous ethanol, distillation techniques are used (Lynd et al. 2005; Prasad and Dhanya 2011a).

The agri-resources constitute a vast resource of biomass consisting of sugars, starch, cellulose and lignocellulosic residues. The enormous quantity of available biomass has a considerable potential for ethanol production using appropriate enzymes. The carbohydrate equivalent of straw of all crops is about 470–502 kg per ton of biomass. In the process of conversion, 1 kg of glucose polymer yields 1.1 × 0.511 kg ethanol (Garcha et al. 1987). The enzymes use the substrate energy from the biomass catalyzing the conversion for their sustenance.

The lesser efficiency of industrial processes also leads to energy losses of up to 10% during transformation as per previous studies. These losses were assumed as 10%, 5% and 20% for starchy cereals and tubers, sugarcane (sugars based conversion), and cellulose/hemicellulose based straw products respectively. The ethanol potential could be calculated for each product as follows, keeping in consideration the density of ethanol as 0.785 g/cc at 25 °C and concentration 95% (by weight) (Garcha et al. 1987):

$$\begin{aligned}&\text{Ethanol}\left(95\%\right)\text{liters / ton fresh biomass}\\&\quad=1000\,\text{kg fresh}\times\text{dry weight fraction}\times\text{carbohydrate fraction}\\&\qquad\times1.1\times0.511\times\left(0.9*\text{conversion efficiency}\right)/0.785/0.95\end{aligned}$$

The ethanol equivalent of major crops and their crop residues could be calculated assuming the appropriate dry weight fraction of different products. The results showed that the ethanol potential was highest for wheat being 403 liters per ton of biomass followed by paddy and maize (394 liters/ton), jowar (382 liters/ton) and bajra (366 liters/ton). Tapioca had higher ethanol potential of 157 liters/ton residue

due to relatively less water and high carbohydrate content as against these, straw of different crops had similar ethanol potential of 250–265 liters/ton of biomass (Prasad and Dhanya 2011b).

11.7 Greenhouse Gas (GHG) Management by Biofuels

Recently, managing energy and reducing GHG emissions are the primary drivers related to sustainability. Because most of the global warming issues are attributable to energy consumption, the use of biofuels helps address these challenges. Energy management involves adopting initiatives with two primary goals (i) to pick up fuel or energy efficiency, (ii) and to decrease potential adverse impacts resulting from energy consumption. Biofuels enhance the performance of fuel, support better fuel efficiency, and therefore help lower emissions. Supplementary use of biofuels with transport fuels contributes more to environmental sustainability than fossil-fuels alone, through lesser life cycle gaseous emissions. Advancements in the production process are expected to provide better environmental benefits (Bessou 2010; SRREN 2011). Even within biofuel resources, first-generation biofuels give higher emissions of 19 to 77 g CO_2eq/MJ compared to the second generation biofuels with −10 and 38 g CO_2eq/MJ lifecycle GHG emissions. In contrast, petroleum fuels were reported to release 85 to 109 g CO2eq/MJ which is much higher than any biofuel (Gaffney and Marley 2009). Figure 11.5 shows a range of reductions of GHG emissions per vehicle-km (v-km) obtained from various studies.

The impact of biofuels on the slowdown of climate change also stimulates its production. Theoretically, net emissions of GHGs from biofuels may reach zero because the carbon emitted while burning was sequestered during photosynthesis (Rajagopal and Zilberman 2017). Most researchers have found that the use of first-generation biofuels results in emission reductions of 20 to 60% of CO_2eq relative to fossil fuels. Expected cutbacks for future commercialized second-generation biofuels are in the range of 70–90% of CO_2eq relative to fossil fuels (Bessou 2010). The use of cellulosic biomass for energy production is expected to result in significantly higher carbon sequestration compared to starch and sugar-based biofuel (Tilman et al. 2006; Farrell et al. 2006).

Several investigations have revealed that using biodiesel produced from Jatropha seed oil, limits greenhouse gas (GHG) emissions by about 8–88% compared to the use of fossil diesel (Reinhardt et al. 2007; Achten et al. 2010). With plantation of 2 and 10 mha of wasteland with Jatropha, a CO_2 sequestration potential of 4.6 and 22.9 Mt. year^{-1} can be achieved in India as reported by Francis et al. (2005). The same study also predicted 2.25 CO_2 tons ha^{-1} year^{-1} of carbon sequestration with this plantation. Other biofuel sources like microalgae also help in this process, when grown in industrial sites utilizing nutrients from effluents and wastewater. Hence, the production and use of biomasses serve dual purposes of providing substrates for biofuel as well as helping in carbon sequestration and climate change mitigation (Sheetal et al. 2017).

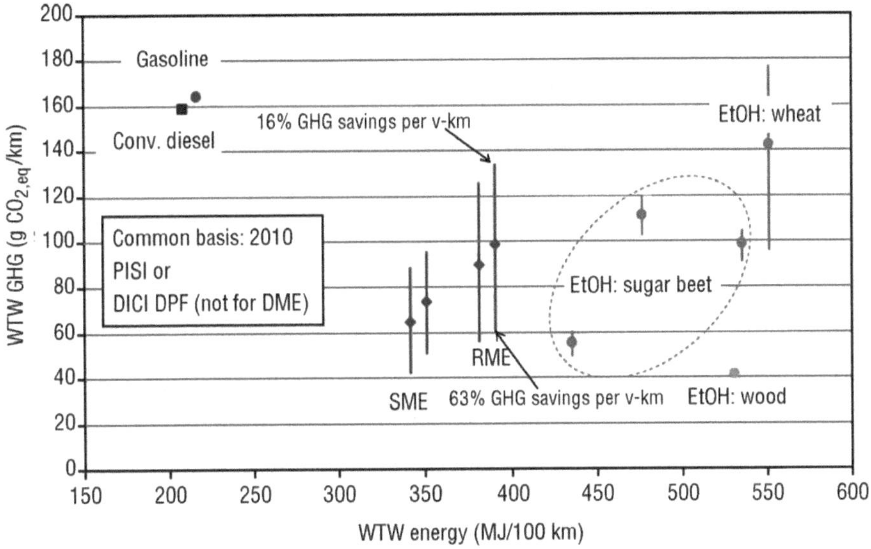

Fig. 11.5 A range of reductions of GHG emissions from biofuel vs. gasoline and diesel. (From Larson 2006)
Note: EtOH = ethyl alcohol (ethanol); SME = soy methyl ester; RME = rape methyl ester; PISI = port injection spark ignition; DICI DPF = direct injection compression ignition with diesel particulate filter

11.8 Air Quality Management by Biofuels

The gaseous discharges from engines using gasoline or petrol have various air pollutants such as sulfur dioxide (SO_2), nitrogen oxides (NOx), carbon monoxide (CO), unburned volatile organic compounds (VOC), and complex mixture of both organic and inorganic particles, which causes atmospheric pollution and responsible for deteriorating air quality. The pollutant gases particularly NOx and VOC react in the presence of sunlight by way of a series of photochemical reactions to form the secondary air pollutant ozone and released in the troposphere, the chief component of smog. CO is a deadly poison, and the inhaling of fine respirable particulate matter (PM2.5) is a severe health concern. Ethanol-gasoline blend is one of the best options for air quality management (Peter et al. 2003).

11.8.1 Air Quality Management by Ethanol

As a fuel additive, ethanol changes the emissions profile of gasoline, creating a cleaner and safer fuel. Emission tests performed with ethanol E10, E15, E22 and E1100 blends confirm that significant reductions in many air pollutants like NOx, CO, SO_2, and PM (Table 11.1). Comprehensively, a 41% decline in particulate

Table 11.1 Reduction in pollution emission with different percentages of Ethanol blending

	Emission (%)		Emission (g/km)	
Pollutant	10% ethanol	15% ethanol	22% ethanol	100% ethanol
Particulate matter	27	41	0.08	0.02
NOx	4	5	0.45	0.34
Carbon monoxide	20	27	0.76	0.65
Unburned hydrocarbons	–	–	0.004	0.02
Sulfur dioxide	–	–	0.064	0.0

From Prasad and Dhanya (2011a)

matter and 5% NOx and 27% CO emission has been observed with E15 blends (Prasad and Dhanya 2011a). An extensive evaluation of the feasibility and use of E 7.7 ethanol-blended fuel on the passenger bus fleet under the Karnataka State Road Transport Corporation was carried out to determine the reduction in air pollutants discharge and its influence on the air quality. The results revealed a significant decrease in PM, CO, CO2 and NOx (UNFCCC 2015). The blending of ethanol to gasoline has been found to reduce benzene, toluene, xylene, and 1,3-butadiene emissions, although may increase harmful acetaldehyde emission (Niven 2005; Yung-Chen et al. 2011).

Ethanol-gasoline blends may result in a moderation effect on exhaust emissions of the criteria pollutants, organic compounds and on ozone-forming potential. The high-mileage car also showed an emission drop while using ethanol-gasoline blends. In contrast, the toxicity-based emission ranking for the six air toxics of ethanol-gasoline blends exhibited high emission based on cancer and acute-effects. The Clean Air Act (CAA) of 1970 identified six common air pollutants of concern, called criteria pollutants. The reduction of criteria air pollutant emission and air quality improvement, the ethanol-gasoline blends are recommended for use as an alternative fuel in in-use passenger cars; ethanol content up to 20% in gasoline (E20) still suitable to be used in the car. However, in view of toxicity, the use of ethanol-gasoline blends needs more evaluation, especially in the case of high ethanol content (Yung-Chen et al. 2011).

11.8.2 Air Quality Management by Biodiesel

Biodiesel is a promising alternative fuel used for fueling diesel engines without necessary engine modifications. Use of biodiesel diminishes various toxic gaseous released from diesel engines. As biodiesel contains about 11% oxygen, the calorific value is lower than diesel, but it enhances the combustion process. Thus, the blend of biodiesel with diesel, reduce emissions of PM, CO, HC, SOx and smoke opacity (Prasad et al. 2012). The other environmental advantages of biodiesel include the fact that it is extremely bio-degradable. The privileges of 100% (B 100) and 20% (B 20) biodiesel blending, in terms of percent pollutants emission reduction (Planning

Table 11.2 Reduction in pollution emission with different percentages of biodiesel blending

	Emissions reduction (%)		Emission (g/km)		
Pollutant	B 100	B20	Diesel	B 10	B 15
Particulate matter	−30	−22	0.129	0.093	0.080
NOx	+13	+2	0.79	0.83	0.89
Carbon monoxide	−50	−20	0.77	0.65	0.62
Unburned hydrocarbons	−93	−30	0.37	0.22	0.16
Sulfur dioxide	−100	−20	--	---	---

From Prasad and Dhanya (2011a)
Less (−) and more (+) % of pollutant emission from biodiesel in comparison to 100% diesel

Commission 2003) and reduce emission in g/km for 10 and 15% (Vasudevan et al. 2005) is displayed in Table 11.2.

According to EPA's Renewable Fuel Standards Program, Regulatory Impact Analysis, published in 2010, biodiesel from soybean oil results on average in a 57% decrease in greenhouse gases (GHG) emission compared to fossil diesel, and biodiesel generated from waste grease results in an 86% decrease (Petracek 2011). The use of biodiesel results in a notable decrease in carbon monoxide, sulfur dioxide, particulate matter, and unburned hydrocarbons. However, a marginal increase in NOx (1–6%) is reported (Table 11.2). Nevertheless, the emission of nitrogen oxides increases to some extent with the use of biodiesel as fuel. The intensity of the particulate matter/smoke decreased up to 33% when the engine operated with100% biodiesel as fuel, compared to the 100% petroleum diesel (Zou and Atkinson 2003).

11.9 Carbon Cycle, Net Energy Balances and Biofuels

The real benefits which may be earned from biofuels depend on the energy supply and carbon emission, showing the extent of fossil fuel inputs and relative greenhouse gas emissions corresponding to fossil fuel savings through avoided greenhouse gas emissions, due to its use as alternative fuels (SRREN 2011). The Confederation of Indian Industry (CII) conducted a study to calculate the net energy and carbon balance for selected types of biofuels (Table 11.3). Among different biofuels, Jatropha-based biodiesel was noted to have the highest net energy and carbon balance annually. The great energy contribution from the co-products (seed husk, seed cake, and glycerol) received during biodiesel production; contribute approximately half of the total energy produced during the end use stage. Sweet sorghum-based ethanol was witnessed to have the best conversion efficiency in terms of converting input energy into output energy (CII 2010).

Table 11.3 Net energy balance and carbon balance for selected categories of biofuels

Biofuel type	Feedstock	Net energy ratio	Net energy balance (GJ/kl)	Net carbon balance (tCO2e/kl)	% Carbon emission reduction
Ethanol	Molasses	4.57	19.11	−1.1	75%
	Sweet sorghum	7.06	21.57	−1.4	86%
	Cellulosic (bagasse)	4.39	25.41	−1.7	70%
	Cellulosic (Rice straw)	3.32	22.79	−1.6	68%
Biodiesel	Jatropha-Transesterification	53.41	63.76	−4.0	30%
	Jatropha – SVO	4.38	66.73	−4.5	50%

From CII (2010)

11.9.1 Trading of Biofuels

Given its broad and extensive supply shortage resulting from strong market growth, India will remain to be a net importer of bioethanol. The small trade deficit that began in 2015 is anticipated to expand more through 2018, given the prediction of tight production this year and next. Aradhey (2017) predicted bioethanol imports to go up from 400 to 600 million liters from 2016 to 2018, under normal market situations. Currently, biofuel imports have no quantity restrictions, but traditionally India imports bioethanol only to satisfy the deficiency in demand during years of lower sugar production. Low import tariffs on bioethanol make imports attractive, convenient and economically viable, particularly when crude oil prices strengthen (MNRE 2014).

Demand is mainly fulfilled for consumption across the potable liquor and chemical industries and not for transport fuel. The United State was the largest ethanol supplier (80%) classified as un-denatured fuel use, followed by Brazil (18%) and Bhutan and Pakistan. In customary, imported ethanol is competitively high priced against local supplies. Usually, when domestic ethanol prices stay strong, industry users prefer to buy imported ethanol and sugar distilleries benefit from selling it to OMCs (Fig. 11.6).

Indian exports of ethanol have decreased by an annual average of 15%, since peak export trades in 2013 of 233.0 million liters, on tighter supply and strong local demand (Aradhey 2017). Ghana, Nigeria, Cameroon, Nepal, Sierra Leone, Tanzania, Jordan, Uganda, Rwanda, and Jamaica were the leading export destinations for Indian bioethanol in the last 5 years, although market share was lost to competition from the United States (Fig. 11.7).

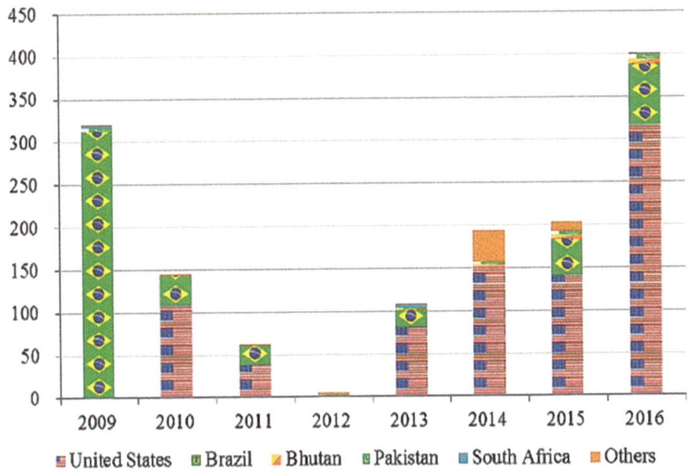

Fig. 11.6 India's ethanol imports (in million liters) (U.S. Census Bureau, GTA & Ministry of Commerce, GOI). (From Aradhey (2017), *GAIN* Report number IN7075)

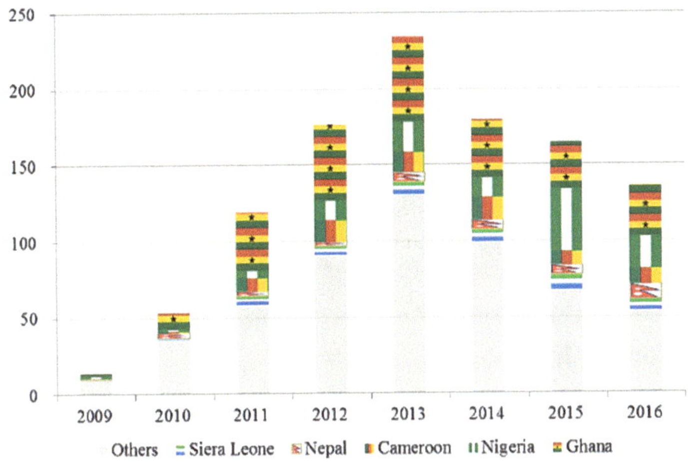

Fig. 11.7 India's ethanol exports (in million liters) (GTA & Ministry of Commerce, GOI). (From Aradhey (2017), *GAIN* Report number IN7075)

11.10 Future of Biofuels

India's biofuel production valued for only 1% of the global output in 2012. The biofuel industry is yet to reasonably mature in India, and it is challenging for the distillery industry to sustain without subsidies, raw material, fuel mandates, or additional government support (Aradhey 2017). As biofuels are usually considered as a cleaner, carbon-free and greener alternatives to fossil fuels, the plan of the subsidies and various policy supports to the sector is also usually done by keeping the potential positive benefits in view (Aradhey 2017; Prasad et al. 2017). The Indian approach towards the biofuel energy sector has been different to the current international approach as India's focus has been on non-feedstock which are raised on degraded or wasteland thus avoiding the conflict of fuel vs. food security. India's biofuels policy supports the utilization of domestic biomass feedstocks for the generation of biofuels (Sheetal et al. 2017).

As technology emerges, research and development improves, and policy responds, the competitiveness of biofuel with other renewable technologies will transform as a function of economic and environmental concerns beyond those considered now (Cotula et al. 2008). Second generation cellulosic ethanol, and third generation algal biodiesel may be able to provide massive benefits and can contribute significantly to the secure energy future for India (Prasad et al. 2017). The GOI has also conferred its support to remove obstacles to the growth of the biofuel industry and allowed the direct sale of biofuels by private manufacturers. Further, at the G20 summit 2014 held in Brisbane, Australia, the prime minister of India called for collective research and development effort and collaboration towards the path of clean energy (Sheetal et al. 2017).

India's dependence on oil imports can be decreased drastically, with small steps such as 5% biodiesel blending by bulk users such as Indian railways and defense establishments (Economic Times 2014). Although Ministry of New and Renewable Energy (MNRE), Govt. of India has actively executed and supported policy and budgetary provision for renewable energy over the years, there are still a few significant barriers that are limiting the substantial growth of renewable energy in India. There need to be considerable policy reformations to conceal the innovation induction, and strategies to up-scale deployment and provide investor's access to capital through technology development and adaptation.

11.11 Conclusion

Biofuels are considered eco-friendly and carbon-neutral fuel because entire CO2 released during the combustion of biofuels had been sequestered from the environment for the growth of crop plants. The GOI has initiated various policy measures to expand production and use of biofuels during the past few decades at the national level. Ethanol and biodiesel are the two biofuels essentially used in transport, either in pure or blended form, along with petrol and diesel. The agricultural residues which currently left unused or of less utilization can be a valuable and cheap resource to meet the entire demand of biomass for biofuels production provided the technology for its conversion is available and cost-effective. Since biofuels are produced solely from biomass, it does not carry any sulfur, and owning an oxygen content in it, increases the combustion efficiency of ignition engines and reduce the emissions. Increased use of biofuels could be an essential strategy and a viable option for managing India's energy security, air pollution abatement, and climate change mitigation.

Acknowledgments Authors wish to thank Indian Agricultural Research Institute, New Delhi, India for providing financial support.

References

Achten, W. M. J., Almeida, J., Fobelets, V., Bolle, E., Mathijs, E., Singh, V. P., Tewari, D. N., Verchot, L. V., & Muys, B. (2010). Life cycle assessment of Jatropha biodiesel as the transportation fuel in rural India. *Applied Energy, 87*(12), 3652–3660.

Aradhey, A. (2010). *India-biofuels annual Tech Rep Global Agricultural Information Network (GAIN) report*, USDA Foreign Agricultural Service, New Delhi, India, 2010.

Aradhey, A. (2014). *India-biofuels annual, IN4055 Tech Rep Global Agricultural Information Network (GAIN) report*, USDA Foreign Agricultural Service, New Delhi, India, 2014.

Aradhey, A. (2017). *India-biofuels annual Tech Rep Global Agricultural Information Network (GAIN) report*, USDA Foreign Agricultural Service, New Delhi, India, 2014.

Atsumi, S., & Liao, J. C. (2008). Metabolic engineering for advanced biofuels production from Escherichia coli. *Current Opinion in Biotechnology, 19*(5), 414–419.

Attal, M. M., Dave, M. C., Mahalle, A., Sunkavally, S., Shaik, M., & Babu, A. (2014). Experimental investigation on performance of a diesel engine using *Garciniaindica* (Kokam) and rice bran oil based biodiesels as fuels. *IOSR-Journal of Mechanical and Civil Engineering, 11*(5), 1–7.

Badger, P. C. (2002). Ethanol from cellulose: A general review. In J. Janick & A. Whipkey (Eds.), *Trends in new crops and new uses* (pp. 17–21). Alexandria: ASHS Press.

Balat, M., Balat, H., & Oz, C. (2008). Progress in bioethanol processing. *Progress in Energy & Combustion Science, 34*(5), 551–573.

Bandyopadhyay, K. R. (2015). Biofuel promotion in India for transport: Exploring the grey areas. The energy research institute. *Policy Brief, 16*, 1–11 Available at http://www.teriin.org/policy-brief/docs/biofuel.pdf.

Bessou, C. (2010). *Green house gas emissions of biofuels: Improving life cycle assessments by taking into account local production factors*. Earth Sciences. AgroParisTech, 2009. Available at http://unfccc.int/secretariat/momentum_for_change/items/7184.php

Biswas, S., Kaushik, N., & Srikanth, G. (2006). *Biodiesel: Fuel for the future?* Available online at: http://tifac.org.in/index.php?option=com_content&view=article&id=776:biodiesel-fuel-for-the-future&catid=120:publication-bioprocess-a-bioroducts&Itemid=1380

Chisti, Y. (2007). Biodiesel from microalgae. *Biotechnology Advances, 25*, 294–306.

CII. (2010). *Estimation of energy and carbon balance of biofuels in India*. Confederation of Indian Industry. Available at http://www.cii.in/webcms/Upload/Energy_Balance.pdf

Cotula, L., Dyer, N., & Vermeulen, S. (2008). *The implications of biofuels for land tenure and land policy*. Land Tenure Working Paper 1. Bioenergy and Land Tenure. Available at http://www.fao.org/nr/lten/abst/lten_080901_en.htm

Das, S., Priess, J. A., & Schweitzer, C. (2010). Biofuel options for India perspectives on land availability, land management, and land-use change. *Journal of Biobased Materials Bioenergy, 4*, 1–13.

Economic Times. (2014). *Oil ministry Cabinet note on allowing 5% blending of biofuels*. http://articles.economictimes.indiatimes.com/2014-11-06/news/55835735_1_bio-diesel-project-bio-diesel-use-bio-diesel-association

FAO. (2013). *Biofuels and sustainability challenge: A global assessment of sustainability issues, trends and policies for biofuels and related feedstocks*. Available at http://www.fao.org/docrep/017/i3126e/i3126e.pdf

Farrell, A. E., Plevin, R. J., Turner, B. T., Jones, A. D., O'Hare, M., & Kammen, D. M. (2006). Ethanol can contribute to energy and environmental goals. *Science, 311*, 506–508.

Francis, G., Edinger, R., & Becker, K. (2005). A concept for simultaneous wasteland reclamation, fuel production & socio-economic development in degraded areas in India: Need, potential & perspectives of Jatropha plantations. *Natural Resources Forum, 29*, 12–24.

Gaffney, J. S., & Marley, N. A. (2009). The impacts of combustion emissions on air quality and climate – From coal to biofuels and beyond. *Atmospheric Environment, 43*, 23–36.

Garcha, H. S., Sedha, R. K., & Singh, R. (1987). *Production of liquid fuels (ethanol) from biomass proceeding bioenergy society 4th convention & symposium '87* (pp. 121–131). New Delhi: Bioenergy Society India.

Indian Express. (2008). *Petrol with 20% biofuel to be mandatory by 2017*. Available at: http:/www.indianexpress.com/news/petrol-with-20-biofuel-tobemandatory-by/333541/

IPCC. (2007). Climate change 2007: impacts, adaptation and vulnerability. In M. L. Parry, O. F. Canziani, J. P. Palutikof, P. J. van der Linden, & C. E. Hanson (Eds.), *Contribution of working group II to the 4th Assessment report of the Intergovernmental Panel on Climate Change* (p. 976). Cambridge, UK: Cambridge University Press.

IPCC. (2014). *Summary for policymakers*. In Climate change 2014, mitigation of climate change: Contribution of working group III to the 5th Assessment report of the Intergovernmental Panel on Climate Change (IPCC). Available electronically at: http://www.climatechange2013.org/images/report/WG1AR5_SPM_FINAL.pdf

Joseph, B. B., & Ronald, T. R. (2010). Fermentable sugars by chemical hydrolysis of biomass. *PNAS, 107*(10), 4516–4521.

Kralova, I., & Sjoblom, J. (2010). Biofuels–renewable energy sources: A review. *Journal of Dispersion Science and Technology, 31*, 409–425.

Kumar, S., Prasad, S., Malav, M. K., & Gupta, N. (2016). Characterization and pretreatment of wheat straw using microwave assisted dilute acid to produce fermentable sugars for hydrogen production. *Green Farming, 7*(1), 95–98.

Larson, E. (2006). A review of life-cycle analysis studies on liquid biofuel systems for the transport sector. *Energy for Sustainable Development, 10*(2), 109–126.

Lynd, L. R., van Zyl, W. H., McBride, J. E., & Laser, M. (2005). Consolidated bioprocessing of cellulosic biomass: An update. *Current Opinion in Biotechnology, l6*, 577–583.

Malav, M. K., Prasad, S., Kharia, S. K., Kumar, S., Sheetal, K. R., & Kannojiya, S. (2017). Furfural and 5-HMF: Potent fermentation inhibitors and their removal techniques. *International Journal of Current Microbiology and Applied Sciences, 6*(3), 2060–2066.

MNRE. (2009). *National policy on biofuels (2009)*. Available at: http://mnre.gov.in/file-manager/UserFiles/biofuel_policy.pdf

MNRE. (2014). *Problems, prospects and economic analysis – MNRE*. Available at: www.mnre.gov.in/file-manager/akshay-urja/may-june-2014/EN/37-39.pdf

MoPNG. (2014). *Energizing the nation* (Annual report 2013–14), GOI. Available at http://petroleum.nic.in/docs/Annual_Report/AR13-14.pdf

Niven, R. K. (2005). Ethanol in gasoline: Environmental impacts and sustainability review article. *Renewable and Sustainable Energy Reviews, 9*, 535–555.

Olsson, L., & Hahn-Hagerdal, B. (1996). Fermentation of lignocellulosic hydrolysates for ethanol production. *Enzyme and Microbial Technology, 18*, 312–331.

Peter, J. G., David, J. G., & John, N. S. (2003). Wood-ethanol for climate change mitigation in Canada. *Applied Biochemistry and Biotechnology, 105*(1–3), 231–242.

Petracek, R. (2011). *The advantages of using biodiesel blends*. Available online at: http://www.buzzle.com/articles/the-advantages-of-using-biodiesel-blends.html.

Planning Commission, GOI. (2003). *Report of the committee on development of biofuels*. Available at: http://planningcommission.nic.in/reports/genrep/cmtt_bio.pdf

Prasad, S., & Dhanya, M. S. (2011a). *Air quality and biofuels, environmental impact of biofuels* (M. A. D. Santos Bernardes, Ed.). In Tech. ISBN: 978-953-307-479-5. doi:https://doi.org/10.5772/17889.

Prasad, S., & Dhanya, M. S. (2011b). Biofuel production potential, processing and utilization. In A. P. Haggerty (Ed.). *Biomass crops: Production, energy and the environment* (pp. 165–187). NOVA Publisher, ISBN: 978-1-61209-398-7.

Prasad, S., Singh, A., & Joshi, H. C. (2007). Ethanol as an alternative fuel from agricultural, industrial and urban residues. *Resources Conservation Recycling, 1854*, 1–39.

Prasad, S., Dhanya, M. S., Gupta, N., & Kumar, A. (2012). Biofuels from biomass: A sustainable alternative to energy and environment. *Biochem Cellular Archives, 12*(2), 255–260.

Prasad, S., Kumar, A., & Muralikrishna, K. S. (2014). Biofuels production: A sustainable solution to combat climate change. *Indian Journal of Agricultural Sciences, 84*(12), 1443–1452.

Prasad, S., Sheetal, K. R., Chandel, A. K., & Dhanya, M. S. (2017). Proximate technical and economic aspects and life cycle analysis of biodiesel production in India: An overview. In A. Chandel & R. Sukumaran (Eds.), *Sustainable biofuels development in India*. Cham: Springer.

Rajagopal, D., & Zilberman, D. (2017). *The use of environmental life-cycle analysis for evaluating biofuels*. https://s.giannini.ucop.edu/uploads/giannini_public/87/87/87878bc8-9b96-4575-9674-b8d8a4a380ef/v11n33.pdf. Accessed 9 Dec 2017.

Reinhardt, G., Gärtner, S. O., Rettenmaier, N., Munch, J., & Falkenstein, E. (2007). *Screening life cycle assessment of jatropha biodiesel* (Technical report). IFEU, Heidelberg,

Rothengatter, W. (2010). Climate change & the contribution of transport: Basic facts and the role of aviation. *Transportation Research Part D, 15*, 5–13.

Schneider, D. (2006). Grow your own? Would the widespread adoption of biomass derived transportation fuels really help the environment? *American Scientist, 94*, 408–409.

Sheetal, K. R., Prasad, S., & Chandel, A. K. (2017). Perspectives on climate change, water and land use, carbon accumulation, and effects on ecosystem after biofuel implementation in India. In A. Chandel & R. Sukumaran (Eds.), *Sustainable biofuels development in India*. Cham: Springer.

Sheorain, V. B., & Chavan, M. (2000). Ethanol production from sorghum. In A. J. Chandrasekhar, R. Bandyopadhyay, & A. J. Hall (Eds.), *Technical institutional option for Sorghum grain mold management: proceedings of an international consultation* (pp. 228–239).

SRREN. (2011). *PCC special report on renewable energy sources and climate change mitigation* (Summary for Policymakers, 11th Session of Working Group III of the IPCC). Abu Dhabi, United Arab Emirates. Available at https://www.ipcc.ch/pdf/special-reports/srren/SRREN_FD_SPM_final.pdf

Steve, S. H., Allison, M., Janet, L., Cameron, D. R., Sheldon, & Duff, J. B. (2004). Xylose fermentation by genetically modified Saccharomyces cerevisiae 259ST in spent sulfite liquor. *Bioresource Technology, 92*, 163–171.

Sudhakar, K., Trishna, A., Srivastava, T., & Premalatha, M. (2013). Assessment of carbon mitigation potential of various biofuels in Indian context. *International Journal of ChemTech Research, 5*(5), 2456–2461.

Sukumar, P., Saravanan, N., Nagarajan, G., & Vedaraman, N. (2010). Effect of biodiesel unsaturated fatty acid on combustion characteristics of a DI compression ignition engine. *Biomass and Bioenergy, 34*, 1079–1088.

Sun, Y., & Cheng, J. (2002). Hydrolysis of lignocellulosic material from ethanol production: A review. *Bioresource Technology, 83*, 1–11.

Tilman, D., Hill, J., & Lehman, C. (2006). Carbon-negative biofuels from low-input high-diversity grassland biomass. *Science, 314*, 1598–1600.

Tyagi, S., Tiwari, S., Mishra, A., Philip, K., Hopke, S. D., Attri, A. K., Srivastava, D., & Bisht, S. (2016). Spatial variability of concentrations of gaseous pollutants across the National Capital Region of Delhi, India. *Atmospheric Pollution Research, 7*(5), 808–816.

UNFCCC. (2015). *Use of ethanol blended diesel in Karnataka State Road Transport (KSRTC)*. Available at http://unfccc.int/secretariat/momentum_for_change/items/7184.php. Retrieved on 12.4.2015

USDoE. (2007). *US Department of energy on greenhouse gases*. 09.09.2007, Available http://www.eia.doe.gov/oiaf/1605/ggccebro/chapter1.html

Vasudevan, P., Sharma, S., & Kumar, A. (2005). Liquid fuel from biomass: An overview. *Journal of Scientific and Industrial Research, 64*, 822–831.

Vinish, K. (2005). *Vehicular pollution control–concept note, centre of excellence in environmental economics*. Madras School of Economics, Chennai. Available at: http://coe.mse.ac.in/dp/vinish_vehi.pdf

Yung-Chen, Y., Jiun-Horng, T., & Hsin-Hui, C. (2011). Air pollutant emission abatement using application of various ethanol-gasolines blends in high-mileage vehicles. *Aerosol Air Quality Research, 11*, 547–559.

Zou, L., & Atkinson, S. (2003). Characterizing vehicle emissions from the burning of biodiesel made from vegetable oil. *Environmental Technology, 24*(10), 1253–1260.

Chapter 12
Microbial Potential and Biofuel Production

Priyanku Teotia, Manoj Kumar, and Vivek Kumar

Abstract Exhaustion of fossil fuel has driven consideration and notion all over the world to ascertain an auxiliary and long-time supportable energy resources and sources of power to fulfil the requirement of human beings. The microbial originated fuels, known as micro-biofuels has immense potential to substantially reduce the transportation fuel crisis. For cost-effective biofuels production, the use of microbes using industrial, agricultural waste and renewable matters will sort out energy crisis, climate change apprehensiveness and food assurance. Quantum of plant biomass on our planet is remarkable and the biomass can be converted by the microbes and their enzymes into renewable energy sources. Currently, on a large scale, the bioethanol is produced in countries like United States of America using corn or other raw materials to meet the requirements of transportation sector. On the contrary, though methane gas is produced on a significant level, it is yet to gain currency for industrial and transportation purposes. As regards the biobutanol, it has huge potential to supplement the existing petroleum products.

Instead of producing bioethanol or biodiesel from microbes, researchers are trying to manufacture advanced microbiofuels, such as long chain isoprenoid, alcohols and fatty acids based fuels from *Saccharomyces cerevisiae* and *Escherichia coli* or production of hydrogen using the cyanobacteria. In this chapter, we analyse and discuss the present status of microbial based biofuel production, their constraints and challenges.

Keywords Biofuel · Biogas · *Saccharomyces cerevisiae*

P. Teotia
Department of Botany, CCS University, Meerut, India

M. Kumar
Centre for Life Sciences, Central University of Jharkhand, Ranchi, Jharkhand, India

V. Kumar (✉)
Himalayan School of Biosciences, Swami Rama Himalayan University, Dehradun, India

S. Shah et al. (eds.), *Sustainable Green Technologies for Environmental Management*, https://doi.org/10.1007/978-981-13-2772-8_12

12.1 Introduction

The upsurge in fossil fuel utilization has encouraged and provoked the scientists to exploit and explore unconventional energy resources to fulfil the existing and forthcoming energy demands. There is a wide range of biofuel production using organic matter, which chiefly includes the sugar cane, wheat, sugary and starch crop such as maize, oilseed crops for instance ground nut, mustard, vegetable oil, soya bean crop, Jatropha; wood, bagasse, fodder straw; other organic waste material. World's fuel energy is consumed as oil 33.5%, coal 26.8% and gas 20.9% (World Energy Resources 2016). During the eighteenth and nineteenth century, the industries were fuelled and developed by using fossil coal. Later on the growth in automobile industry used petro fuels to run engines or fuel for electricity generation, which resulted in the consumption of petroleum products as major source of energy in twentieth century.

To overcome the rising oil demand and associated fuel price hike, biofuel production using microbial sources is much sought after, chiefly in the form of sugar fermentation by *Saccharomyces* sp. for ethanol production (Balat and Balat 2009). Though there are several microorganisms which have been employed in production of ethanol commercially, the common alcohol producing fungal yeast *S. cerevisiae* is chiefly employed at industrial scale. For ethanol production, sugar and starch from the crop plants are used as principal beginning materials for this procedure (Bai et al. 2008). The utmost common feedstocks (commonly used carbon sources used by the microorganisms) are the agrarian waste commodities which can be easily handled to manufacture the simple sugar required for the process of fermentation. For production of bioethanol, in USA it is primarily the corn (maize, in the European Union it is wheat and in Brazil it is sugar cane (Hill et al. 2006; Balat and Balat 2009).

The bioethanol production usually simply exploits the sugary or starchy constituents of agrarian produce as ample energy is located in the biomass of the plant parts. Generally the unused plant parts, such as leaves, stems and root wooden part, are comprised of lignin and cellulosic materials, and owing to the complicated lignocellulosic constituents structure, it becomes very difficult for microbes to break down this structure. If by using any means (microbial and eco-friendly chemical) these cellulosic components could be manipulated, they could symbolise an enormous novel source of energy for production of bioenergy (Bokinsky et al. 2011). The lignocellulosic material is composed of various amounts of lignin, cellulose and hemicelluloses. Plant cell walls are composed of these lignocellulosic materials which provide structural support for the plant parts. In plant part or agricultural waste, the cellulosic part is homo-polysaccharide composed of simple units of glucose connected by β – (1,4) linkage between the monomers of glucose monomers. The unusual presence of β – linkage (in place of α – linkage, which is observed in the other biological system) is far more resilient for biodegradation by microbes. On the other hand, hemicellulose is a hetero-polysaccharide, which is present in variability and may be composed of several dissimilar sugar molecules, such as galactose, xylose and mannose. Owing to the diverse composition of hemicellulose, it becomes very problematic for several microorganisms to consume a varied variety and different sugar compositions. Lignin, the final constituent of lignocellulosic material, is not a polysaccharide, but it is a complicated units network which is composed of phenyl-propane. In the plant cellulosic material, the lignin is intertwined, which provides the structural backing and also protects it from microbial attack (Jorgensen et al. 2007).

The microbes are unbelievably varied with respect to their potential to exploit various raw materials, and because of this diversity the microbes have the chance to exploit various compounds for biofuels making along with using the other standard carbohydrates. To overcome the problem of complex plant materials, the fatty acids could provide fascinating alternative to the sugars. The reason is the fatty acids are much more reduced compared to the simple carbohydrate compounds. On the other hand, the fatty acids contain much more energy and carbon which can be helpful in producing higher yield of biofuel. Additionally, the fatty acid is easily converted into acetyl-CoA and do not have to pass via the intermediary formation of pyruvate. The fatty acids never undergoes the process of carboxylation, and therefore, not like the process of glycolysis, here every carbon particle from the material could be theoretically integrated into the final produce (Clomburg and Gonzalez 2010).

Microbiofuel is an alternate and unconventional source of energy and it may replace the conventional petro-hydrocarbon energy source in future. This fuel is manufactured from renewable resources such as microbial or plant biomass by the process of trans-esterification of triacyl glycerols from oil of plants. It yields mono alkyl esters, which are long chain fatty acid with short chain alcohol such as fatty acid ethyl esters and fatty acid methyl ester (FAMEs). In spite of several ecological welfares and advantages, a broad application of biological diesel is hindered owing to the wide land area needed for adequate and abundant manufacture of oilseeds crop. Consequently, novel techniques or manufacturing processes are immediately required to facilitate production of biofuels from more easily available massive plant materials such as cellulose or cheap and raw sugar sources. This innovative or unusual tactic might lead the way for industrialised biofuel production. Hopefully this biologically produced fuel may prove equivalent to the other renewable sources by using genetically engineered microbes which may enable the broad application of biological diesel like future fuel.

12.2 Microbes Based Systems for Biofuel Manufacture

The subsequent explanation of biological fuels and their process of manufacture may be different. People have employed algal cells to produce biodiesel, but the success rate is not appreciative. There are some process also where involvement of yeast cells and bacterial species have been done. Here things to be noted are that every microbial cell employed for biofuel production require an energy source. This energy source should be cost effective and it may be biological waste or cellulose based or some form of sugar, which can be easily metabolized by microbial cells.

12.2.1 Hydrogen as a Biofuel

Hydrogen gas is the purest form of biological fuels because this is oxidized to H_2O, without any emission of CO_2 in this procedure. As per the policy makers, this hydrogen fuel is very admired and accepted biological fuel. Keeping this in mind, hydrogen gas powered concept vehicles are presently being displayed and produced by various car companies to strengthen their conservational qualifications. The hydrogen is considered as a perfect and valuable future fuel for transport purpose, since it can be easily converted into electrical energy after burning, which is later changed into the mechanical energy. In the above said process, there is no production of carbon dioxide or any other pollution causing agents (Malhotra 2007). This concept has guided to a notion that an economy driven by hydrogen fuel would be a feasible possibility. Though, production of hydrogen is generally provoked by chemical/thermal ways and it is energy intensive process. Therefore, hydrogen gas produced in this manner might not be considered as a primarily renewable source of energy. On the other hand, production of hydrogen using biological means from different biomass resources would deliver an economic, energy-saving, and eco-friendly substitute. This concept should be considered comprehensively on the basis of above said points (Das and Verziroglu 2001; Esper et al. 2006).

Using biological means, hydrogen fuel can be generated by employing cyanobacterial and algal biological photolysis of water molecule or by the process of photolytic fermentation of organic substances using photosynthetic bacterial strains. Additionally, it can also be manufactured by "dark" fermentation reaction using anaerobic microorganisms from organic substrates such as acido-genic bacterial species. This technique has the extra benefit of reduction in waste organic material biomass (Wu et al. 2005). One can employ the thermophilic microbes, for instance *Thermotoga elfii* and/or *Caldicellulosiruptor saccharolyticus* for achieving higher hydrogen production (de Vrije et al. 2002; de Vrije and Claassen 2003; Claassen et al. 2004). These fermentation process may be carried out in aqua phase using immobilised microbial cells or by allowing the sludge or self-flocculated granular cells formation to inhibit wash out of hydrogen gas generating microbial cells. The process of hydrogen production is not at advanced stage using microbial means, therefore it is not a cost effective technique, moreover, it will take time to produce huge amount of hydrogen fuel using microbes. Further development and research is required to produce cost effective and huge amount of hydrogen using renewable biomass. If this approach is successful then we can have huge amount of hydrogen fuel. Ethanol along with hydrogen can be employed in fuel cells to generate electrical power for the purpose of transport or energy generation. The noted car manufacturing company BMW assert that the hydrogen fuel utilisation technology in combustion engines is at advanced stage, while on the other hand, other automobile companies do focus on the employment of fuel cells. Though vehicles with fuel cells are technologically possible, Volkswagen proclaimed that hydrogen fuel cars will not play a noticeable part until 2020 for mass production.

Hallenbeck and Ghosh (2009) reported that, in spite of the microbial technology, innovative engineering and improvement in design, the production of biological hydrogen, does not seems to be practical to solve the problem of huge amount of hydrogen required for transportation purpose. Owing to the hydrogen lower energy, huge compressed and strong storage tanks are required, which may prove hazardous and costly, moreover, such bulky tanks also acquire huge space in cars. These space can be utilized for luggage purposes. A huge set-up and substructure is also required for delivering and changing numerous energy devouring economic activities to a hydrogen fuel based economy. This will result in enormous drawback and shortcoming when hydrogen fuel is contrasted to biodiesel and alcohol. These two fuels (biodiesel and alcohol) could be utilized and transferred using prevailing set-up and structure for petro-hydrocarbon products.

12.2.2 Methane as a Biofuel

Biological gas plants generate methane (CH_4) gas along with some amount of carbon dioxide and hydrogen gas sustainably from the animal excreta and plant biomass. This waste may originate from any household, agricultural farms, waste from industry or from especially planted energy crops (Yadvika et al. 2004). The advantageous part of the biological gas process is the choice to employ the cellulosic part of the plant materials to manufacture energy. This energy may be heat, electrical power converted into mechanical energy, which could be easily managed and consumed by small scale industrial entities and also by rural households. On the other hand, the hydrogen gas can be compacted after the process of enrichment and purification and then later supply to the gas grid involved in electricity generation or may also be used as a green-fuel in vehicle combustion engine. Its supreme benefit is the eco-friendly feature of the future technology. This incorporates the possibility for comprehensive recycling of nutrients, minerals, and cellulosic fibre materials. All these things originate from the fields and revert back to the soil again, therefore, it plays a purposeful role by maintaining the soil vigour and sustainability for future crop plantation. This methane technology and knowledge is presently quite in mature stage, but there is amply of scope for improvement, which ultimately result in great technologically advanced plants for production with combined use of several by products. Substrates could be cow dung or any other animal dung such as goat, pigs, hoarse and chickens may be used, which is also convenient and suitable for inoculation purpose. Besides, household waste, industrial, slaughter waste, rotten food stuff, garden and agricultural organic wastes may also be used for inoculation purposes. Also the organic waste of municipal sewage sludge, non-infectious paper and cotton wastes of hospitals, waste from food production etc. can be utilized as consumable substrates. On the other hand, energy crops such as clover, young poplar tree, whole maize plant, grass, dried mustard, pearl millet plants and willow tree are specially grown for biological gas manufacture and are supplemented with the dung mixture. To guarantee a uniform substrates quality round the year, green plant materials are generally stored as fodder. This is done usually by a procedure which favours *Lactobacillus* sp. which are homofermentative in nature and minimises the carbon loss.

Though, on a small scale level, biological gas has huge possibility and is presently used widely. For instance, biological gas manufacturing amenities such as sewage waste water treatment plant and solid waste landfills can employ biological gas generated for operating the plants, therefore, it may become energy neutral. The application of biological gas on a native, inhabited small scale could be developed in rural and semi urban areas in developing countries like India. Interestingly, India had achieved huge success in consuming biological gas generated in pits connected with homes in rural areas with no services connected for biogas generation for cookery and electricity purposes (Singh et al. 2000). Cow dung has been employed as a source of inoculum for this purpose. Such tactic is presently being studied in Egypt for treating low nutrients agricultural waste, rice straw and other farm yard waste that cannot not be utilized as feedstock people use that waste as burning fuel. Such practices of burning agricultural and other wastes is partially accountable for the "Black Cloud" occurrence that has been occasionally watched for past many years in Egypt.

12.3 Microbial Products for Existing Fossil Fuel Recovery

Additional probable and possible employment of efficient microbes is to augment the fossil fuels production in existing natural gas and oil formation. For itself, we cannot say that the product is really a biofuel, because it is not manufactured from bioresources, but the approach is rather a biological used for taking out traditional petro-hydrocarbon fuels. The cost efficacy of such method is quite straight, simple and attractive. Basically, if the price of applying a particular procedure is lesser as compared to the income gained after additional recovered product selling, then the method or process is considered cost effective. Consequently, the charm of such procedures generally relies on fluctuating oil prices. These techniques are generally employed in petro-oil wells where day by day production is lessening, or stopped recently. As such, all substructure, transportation, and market related matters are generally in place for additional manufactured fossil fuel selling.

It is significant here to distinguish between two interconnected tactics which are discussed below; (1) microbial enhanced oil recovery (MEOR), and (2) microbial enhanced energy recovery (MEER). In microbial enhanced oil recovery, either microbes or their secondary products are added into oil formation to enhance the oil production from oil wells. These microbes are introduced in those wells, which are in their tertiary stage of oil production or where oil production level has reduced to that level that decide the oil extraction process not cost effective. For instances, this includes injecting biosurfactants or bacteria producing biosurfactants into the well to reduce water oil interface tension which results in improving recovery of oil

(Youssef et al. 2007), in addition to this, injecting microbes which produces acid and gas for recovering oil entrapped in formation of carbonates.

In microbial enhanced energy recovery, microbes accomplished in specific transformation processes are injected into the well. This is done to bring some *in situ* alteration in chemistry of fuel, which allows more effective recovery of energy. Common example includes injecting consortia of methanogenic bacteria which are efficient in anaerobic biological degradation of many petro-hydrocarbons into natural gas or oil reservoir to recover the unrecoverable products such as methane gas (Gieg et al. 2008), and also explore mechanism of motivating microbes present initially in petro-hydrocarbon formation to generate methane gas from native gases (Grigoryan and Voordouw 2008). The microbial enhanced energy recovery (MEER) research is in the process of developing a method to accelerate the process of methane gas generation in vast coal formation in the USA. Here, most part of the coal is either too dirty or unrecoverable to be consumed under existing US environmental regulation. To solve such critical issues, world's leading researches and huge amounts of private money has been involved. Though inspiring and positive reports on such issues has been published in the recent past (Volkwein et al. 1994; Strapo'c et al. 2008). But no known microbial culture or efficient microbial consortia that can reproducibly and persuasively transmute coal products to methane gas under anaerobic conditions has yet not been achieved.

12.4 Microbial Diesel as Biofuel

Biological diesel is defined as non-petro-hydrocarbon based diesel fuel which consists alkyl ester (mainly methyl group, but also ethyl as well as propyl group) of long chained fatty acid. Biological diesel can be manufactured from numerous plant and animal resources by the process of triglycerides esterification with methanol (Fucoda et al. 2001). Additionally, biological diesel can also be manufactured from numerous micro algal species (Chisti 2007). Investigation on biological diesel from algal species has been funded in national laboratory of USA, in the aquatic species program, sponsored by Department of Energy and started in 1978. The biodiesel production using micro algal species has manifold benefits and this technique has been called the third generation biological fuels (Tollefson 2008). Not like other oilseeds crop, micro algal species cultivate and raise very quickly and many algal species are very good at oil producing. The micro algal species generally make twice their biological mass within 24 h, and doubling time of biological mass during log phase growth are normally as short as 3.5 h. The content of oil in micro algal species can exceed up to 80% by weight of dry biological mass (Metting Jr. 1996; Spolaore et al. 2006), and the level of oil of 20–50% are pretty normal. Figure 12.1 shows the various energy sources produced by micro algal cells, such as ethanol, hydrogen, biogas and biofuel.

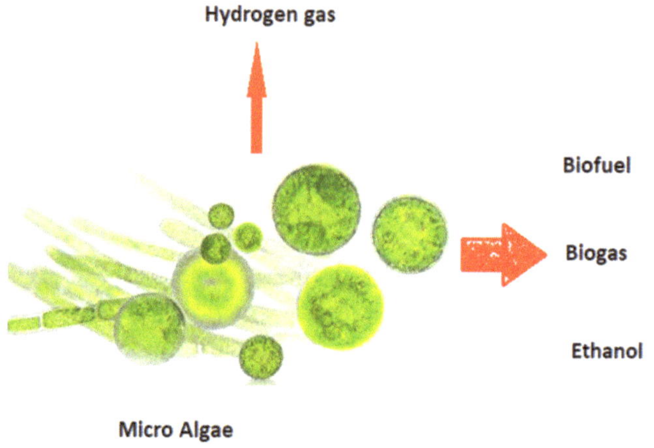

Fig. 12.1 Microalgae showing production of energy sources

Most significantly, owing to the photosynthetic behaviour, the autotrophic algal species do not struggle with starting plants material for production of biological fuel. In contrast, the algal species fixes and therefore lessen the CO_2 amount in atmosphere. Carbon dioxide gas is responsible for global warming process. On the other hand, the heterotrophic algal species have the benefit of accomplishing higher growth density, therefore there is more biological diesel concentration as compared to the phototrophic algal species. Additionally, growth of heterotrophic algal species under dark poses no production problem when comparing with phototrophic algal species. Though, this process necessitates starting plants material as substrate and the inclusive cost efficacy of the procedure is presently under investigation. It is presumed that algal species can be multiplied to produce biological diesel using specialized manmade aquatic ponds. Conversely, the cost efficacy of this progression is still ambiguous and unclear. It has been envisioned that the microbiological phases of the progression are exceptionally assuring, but the main problem is from the engineering point of view, owing to complexity of process. The chief engineering challenge presently is the algal collection and oil harvesting cost. Algal species multiply on the surface layer of water ponds, therefore, to get a huge amount of biodiesel, miles and miles of algal harvesting is required. To achieve this, enormously vast ponds are needed to cultivate micro algal species in huge quantities to make the biodiesel process commercially viable. Micro algal culture in natural ponds or shores of ocean has also been suggested. On the other hand, the intrusiveness of algal species might pose an environmental threat because the algae will devastate and engulf the environmental system.

12.5 Existing World Scenario

Already we are using the microbial manufactured biological fuel in the form of ethanol, which is being produced by efficient yeast strains using simple raw materials by the process of fermentation (Balat and Balat 2009). Though there several microbes which have been employed production of ethanol, but the species of yeast *Saccharomyces cerevisiae* is principally used at industrial level, uses sugars and starch of plant sources as the beginning material for fermentation progression (Bai et al. 2008). The commonest feedstock are in the form of carbon, which is supplied by agrarian products. These agricultural products can be simply treated to form the simple carbohydrates required for fermentation process. This approach is chiefly followed in USA by using corn crop and sugar cane in Brazil, while wheat is being used in European Union (Hill et al. 2006; Balat and Balat 2009).

Ethanol (alcohol) fermentation process by *Saccharomyces cerevisiae* is chiefly performed via glycolysis pathway (Fig. 12.2) (Bai et al. 2008).

In case of maize and other starch comprising plants, the simple necessary sugars are formed via hydrolysis process of starch to produce subunits of monosaccharides. In case of the sugarcane the sugar is hydrolysed only one time and then go rightly in the pathway (Clomburg and Gonzalez 2010). In this procedure, a single glucose molecule is oxidized into two pyruvate molecules. Here anaerobic conditions are essential, because molecular O_2 is not accessible for electron acceptor use, and therefore, instead of oxygen as terminal electron accepter pyruvate should be used. This process of fermentation involves the pyruvate decarboxylation to form CO_2 and acetaldehyde, and the successive acetaldehyde reduction to manufacture ethanol (Bai et al. 2008).

The application of fermented biological ethanol as a biofuel source provides some benefits as compared to conventional fossil fuels. The maize (corn) alcohol and sugarcane alcohol need a contribution of petro-hydrocarbon energy to produce

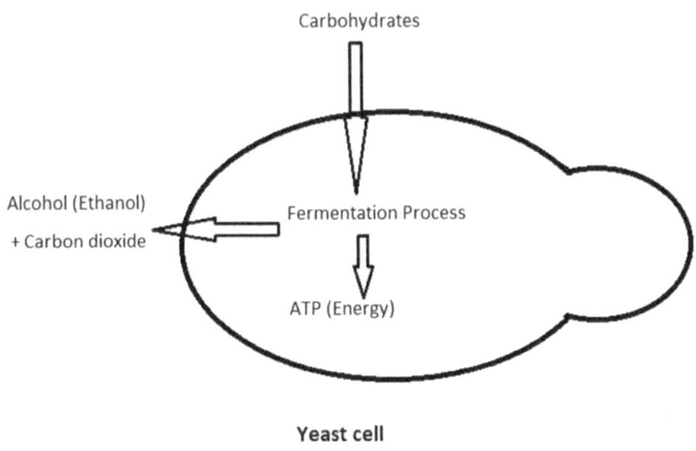

Fig. 12.2 A yeast cell showing biofuel production

them owing to the energy required for transportation, agricultural farming and industrial processing. Though, many research work reported that biological alcohol produces more energy as compared to the petro-hydrocarbon fuel energy. This gives them net energy balance (NEB), a positive aspect. It means, that biological fuels symbolise more effectual sources of energy. The net energy balance for corn (maize) alcohol is relatively small, at approximately 25% extra energy output than the required fossil fuel energy investment (Hill et al. 2006). In other words, we can say that, while the maize (corn) alcohol diminishes the quantity of essential petro-hydrocarbon fuel, but it does not provide a complete resolution of the drawback. Alcohol manufactured through fermentation of sugarcane by-products has a higher net energy balance as compared to maize alcohol, producing practically eight times biological fuel energy as compared to the necessitated input of petro-hydrocarbon fuel energy. Conversely, since the sugarcane crop is generally grown in particular parts of globe and require tropical climates, therefore, ethanol production from it does not provide a feasible answer for the universal problem (Goldemberg et al. 2008).

Fermentation of ethanol by yeast cells also provide some sort of help redress the problematic issue of emissions of green-house gases, though it does not provide a flawless answer from ecological point of view. All biological fuels with a positive net energy balance theoretically should discharge less CO_2. Since the carbon fixation process happening inside the cultivating plants should counter balance the CO_2 discharges of both the combustion of alcohol and invested petro-hydrocarbon fuels energy (Hill et al. 2006). Conversely, in real the fertilizer rich in nitrogen content employed in sustaining the crop plants and adding of extra plant biomass into agricultural soil encourages bacterial societies that produces N_2O, which is a stronger green-house gas as compared to CO_2. If we consider this whole system, ethanol production through corn (maize) fermentation process emits around 88% of the green-house gases content of petrol producing same energy amount (Von Blottnitz and Curran 2007). This average enhancement, combined with the other ecological insinuations such as chemical pesticides, make most present fermentation of ethanol technique of regulated usage, though biofuels are nonetheless are a positive substitute to the petro-hydrocarbon fuels.

12.6 Conclusion and Future Prospects

During the last few years have seen the advancement in technological development, also employment of genetically engineered efficient microbes and use of cheap and easily available non-food feedstocks to be consumed by microbes. On the other hand, creating microbes for better efficient enzyme producer or improved metabolic pathways may lead to advanced biofuels production. The practicability of these biological fuels manufacture pathways has been demonstrated by transformation of carbon dioxide, algal hydrolysates and switchgrasses into higher alcohol forms, fatty acid and isoprenoid originated biological fuels. Few years back, there was a simple question; which superior biological fuel can be manufactured from glucose in higher yield? On the other hand now scientists are also posing question that which feed stock can profitably, economically and efficiently manufacture advance type biological fuels in higher yield. The main problem lying in production of higher yield of advance biological fuels, requires the metabolic engineering of biological fuels producing pathway and substrates utilized. The process of developing easily consumable substrates, efficient metabolic pathways to produce potent enzymes will certainly lead in revolution of biological fuels industry.

The synthetic biology and data driven methods are dominant means for problem solving and improving genetically engineered microbial metabolic pathways. Improving the efficiency and production of biological fuels will further need more powerful tools. This will include methods and means for stable upholding of copy number of genes and more specific regulation of m RNA and proteins level. This has to be achieved in a vibrant fashion that will adjust automatically the pathways concurring to its own metabolic change. Additionally, enzymes activity and substrates specifications can also be transformed by means of engineering protein. Synthetic enzymes with novel performances can also be produced by incorporation of artificial amino acids and computational based proteins designing. Presumably, a lone answer methodologies or standardization of procedure for biological energy manufacture will not be the product of such investigation attempt. Considerably, a slow but careful advancement on numerous frontages will happen, and therefore, the concluding pattern for biological fuel manufacture will be an amalgamation of many tactics. The selection of the biological energy tactic to be employed in a particular place will ultimately rely on the energy requirements such as (electrical energy, fuel for transport, and gas for heating purpose), flora (agriculture aspects, grasslands, and forestry), and also considerations for politics. Undoubtedly, the global total yearly biological fuels production will last to rise progressively in coming future. But these increase will not be even, and annual alterations will be interrelated to oil process fluctuations, also will depend upon the policies of governing parties and results of elections, since fuel prices also play a significant role in election agenda especially in developing and developed countries. Nonetheless, the world-wide energy panorama will be considerably dissimilar in the coming two to three decades. Consequently, the microbial produced biological fuels will be a significant futuristic increase in petro-hydrocarbon fuel energy relatively than the solitary energy source within the nearby and in-between or transitional future. Therefore, significant developments (cost efficacy and technological advancement) in the fields of microbial biofuel energy will solve the issues fossil fuel dependency.

References

Bai, F. W., Anderson, W. A., & Moo-Young, M. (2008). Ethanol fermentation technologies from sugar and starch feedstocks. *Biotechnology Advances, 26*(1), 89–105.

Balat, M., & Balat, H. (2009). Recent trends in global production and utilization of bio-ethanol fuel. *Applied Energy, 86*(11), 2273–2282.

Bokinsky, G., Peralta-Yahya, P. P., George, A., Holmes, B. M., Steen, E. J., Dietrich, J., Lee, T. S., Tullman-Ercek, D., Voigt, C. A., Simmons, B. A., & Keasling, J. D. (2011). Synthesis of three advanced biofuels from ionic liquid-pretreated switchgrass using engineered *Escherichia coli. PNAS, 108*(50), 19949–19954.

Chisti, Y. (2007). Biodiesel from microalgae. *Biotechnology Advances, 25*(3), 294–306.

Claassen, P. A. M., de Vrije, T., & Budde, M. A. W. (2004). Biological hydrogen production from sweet sorghum by thermopilic bacteria. In *Proceedings 2nd world conference on biomass for energy* (pp 1522–1525), Rome.

Clomburg, J. M., & Gonzalez, R. (2010). Biofuel production in *Escherichia coli*: The role of metabolic engineering and synthetic biology. *Applied Microbiology and Biotechnology, 86*, 419–434.

Das, D., & Verziroglu, T. N. (2001). Hydrogen production by biological processes: A survey of literature. *International Journal of Hydrogen Energy, 26*, 13–28.

de Vrije, T., & Claassen, P. A. M. (2003). In J. H. Reith, R. H. Wijffels, & H. Barten (Eds.), *Dark hydrogen fermentations* (pp. 103–123). Petten: Dutch Biological Hydrogen Foundation.

de Vrije, T., de Haas, G. G., Tan, G. B., Keijsers, E. R. P., & Claassen, P. A. M. (2002). Pretreatment of Miscanthus for hydrogen production by *Thermotoga elfii. International Journal of Hydrogen Energy, 27*, 1381–1390.

Esper, B., Badura, A., & Rögner, M. (2006). Photosynthesis as a power supply for (bio-) hydrogen production. *Trends in Plant Science, 11*, 543–549.

Fukuda, H., Kondo, A., & Noda, H. (2001). Biodiesel fuel production by transesterification of oils. *Journal of Bioscience and Bioengineering, 92*(5), 405–416.

Gieg, L. M., Duncan, K. E., & Suflita, J. M. (2008). Bioenergy production via microbial conversion of residual oil to natural gas. *Applied and Environmental Microbiology, 74*(10), 3022–3029.

Goldemberg, J., Coelho, S. T., & Guardabassi, P. (2008). The sustainability of ethanol production from sugarcane. *Energy Policy, 36*(6), 2086–2097.

Grigoryan, A., & Voordouw, G. (2008). Microbiology to help solve our energy needs: Methanogenesis from oil and the impact of nitrate on the oil-field sulfur cycle. *Annals of the New York Academy of Sciences, 1125*, 345–352.

Hallenbeck, P. C., & Ghosh, D. (2009). Advances in fermentative biohydrogen production: The way forward? *Trends in Biotechnology, 27*(5), 287–297.

Hill, J., Nelson, E., Tilman, D., Polasky, S., & Tiffany, D. (2006). Environmental, economic, and energetic costs and benefits of biodiesel and ethanol biofuels. *PNAS, 03*(30), 11206–11210.

IEA (International Energy Association). (2012). http://www.ies.org/publications/freepublications/publication/English

Jorgensen, H., Kristensen, J. B., & Felby, C. (2007). Enzymatic conversion of lignocellulose into fermentable sugars: Challenges and opportunities. *Biofuels Bioproductions Biorefining, 1*, 119–134.

Malhotra, R. (2007). Road to emerging alternatives-biofuels and hydrogen. *Journal of Petrotech Society, 4*, 34–40.

Metting, F. B., Jr. (1996). Biodiversity and application of microalgae. *Journal of Industrial Microbiology and Biotechnology, 17*(5–6), 477–489.

Singh, B. P., Panigrahi, M. R., & Ray, H. S. (2000). Review of biomass as a source of energy for India. *Energy Sources, 22*(7), 649–658.

Spolaore, P., Joannis Cassan, C., Duran, E., & Isambert, A. (2006). Commercial applications of microalgae. *Journal of Bioscience and Bioengineering, 101*(2), 87–96.

Strapo'c, D., Picardal, F. W., Turich, C., Schaperdoth, I., Macalady, J. L., & Lipp, J. S. (2008). Methane-producing microbial community in a coal bed of the Illinois Basin. *Applied and Environmental Microbiology, 74*(8), 2424–2432.

Tollefson, J. (2008). Energy: Not your father's biofuels. *Nature, 451*, 880–883.

Volkwein, J. C., Schoeneman, A. L., Clausen, E. G., Gaddy, J. L., Johnson, E. R., & Basu, R. (1994). Biological production of methane from bituminous coal. *Fuel Processing Technology, 40*(2–3), 339–345.

Von Blottnitz, H., & Curran, M. A. (2007). A review of assessments conducted on bio-ethanol as a transportation fuel from a net energy, greenhouse gas, and environmental life cycle perspective. *Journal of Cleaner Production, 15*(7), 607–619.

World Energy Resources. (2016). https://www.worldenergy.org/wp-content/uploads/2016/10/World-Energy-Resources-Full-report-2016.10.03.pdf

Wu, S. Y., Hung, C. H., Lin, C. N., Chen, H. W., Lee, A. S., & Chang, J. S. (2005). Fermentative hydrogen production and bacterial community structure in high-rate anaerobic bioreactors containing silicone immobilized and self-flocculated sludge. *Biotechnology and Bioengineering, 93*, 934–946.

Yadvika, S. S., Sreekrishnan, T. R., Kohli, S., & Rana, V. (2004). Enhancement of biogas production from solid substrates using different techniques-a review. *Bioresource Technology, 95*, 1–10.

Youssef, N., Simpson, D. R., Duncan, K. E., McInerney, M. J., Folmsbee, M., & Fincher, T. (2007). *In situ* biosurfactant production by *Bacillus* strains injected into a limestone petroleum reservoir. *Applied and Environmental Microbiology, 73*(4), 1239–1247.

Chapter 13
Microbial Biofortification: A Green Technology Through Plant Growth Promoting Microorganisms

Amir Khan, Jyoti Singh, Viabhav Kumar Upadhayay, Ajay Veer Singh, and Shachi Shah

Abstract The hidden hunger or malnutrition is considered to be the most dignified global challenge to human kind. Malnutrition afflicts approximately more than one billion of world's population in both developed and developing countries. Malnutrition includes diet related chronic diseases as well as overt nutrient deficiencies which leads to morbidity, reduced physical and mental growth. However, strategies to enhance supplementation of mineral elements and food fortification have not always been successful. Plant growth promoting microorganisms are known to fortify micro- and macro-nutrient contents in staple food crops through various mechanisms such as siderophore production, zinc solubilization, nitrogen fixation, phosphate solubilization, etc. Inoculation of potential microorganisms along with mineral fertilizers can increase the uptake of mineral elements, yield and growth. Therefore, biofortification of staple food crops by the implications of plant growth promoting microorganisms has an ability to attain mineral elements, is advocated as novel strategy not only to increase concentration of micronutrient in edible food crops but also to improve yields on less fertile soils.

Keywords Microbial biofortification · Plant growth promoting microorganisms · Malnutrition · Zinc · Micronutrient

---

A. Khan · J. Singh · V. K. Upadhayay · A. V. Singh (✉)
Department of Microbiology, College of Basic Sciences and Humanities, GBPUA&T, Pantnagar, Uttarakhand, India

S. Shah
School of Interdisciplinary and Transdisciplinary Studies, Indira Gandhi National Open University, New Delhi, India

S. Shah et al. (eds.), *Sustainable Green Technologies for Environmental Management*, https://doi.org/10.1007/978-981-13-2772-8_13

13.1 Introduction

For sustainable agriculture, the use of microbial based biofertilizers has prestigious role in enhancing level of crop productivity and in food safety. Microorganisms as invisible soil engineers maintain soil health, construct a hub for different biogeochemical cycles (Gadd 2010) and many soil microorganisms such as bacteria, actinomycetes, cyanobacteria and mycorrhiza present an eco-friendly approach for improved uptake of nutrients and enhanced plant growth. Microorganisms particularly plant growth promoting rhizobacteria (PGPR) are dwelled in rhizospheric region and efficiently colonize the roots of plants and confer tolerance in plants against several abiotic and biotic stresses (Prasad et al. 2015). Plant growth promoting microorganisms make nutrients available to plants by numerous mechanisms such as atmospheric nitrogen fixation, solubilizing the nutrients fixed in the soil matrix, and production of phytohormones. Such microorganisms make sure for further enhancement of micronutrients in plants, as they play a key role in organic material mineralization and as well as transforming inorganic nutrients. Microorganisms can also influence nutrient availability through presenting different characteristics such as chelation, solubilization and oxidation or reduction (Khan 2005) and also conferred resistance from pathogens causing diseases to the host plant by the secretions of antibiotics (Bonfante and Genre 2015). The aim of modern agriculture system, besides augmented crop yield, is also to produce nutritious safe food crops with improved level of micronutrients in the edible portion of crop plants. Human population is mainly dependent on crop based foods for the basic diet, and having foods with poor level of essential micronutrients creates serious health issues in humans. Deficiency of micronutrients (zinc, iron, selenium, copper, manganese and vitamins) in both humans and plants is narrated as 'hidden hunger' (Sharma et al. 2016), and bestows threat of malnutrition among world population. Therefore, implementation of biofortification strategy is an important mode for providing the preeminent solution for producing food crops with elevated level of necessary micronutrients. 'Biofortification' and 'standard fortification' are two different terms, where biofortification is related with consigning the nutrients aggregation inside plant cells whereas latter involves use of additives with the foods. Biofortification process deals with several approaches for enhancing bioavailability of key nutrients in crops.

13.2 Micronutrient Associated Malnutrition

Micronutrient deficiency occurs in humans, where populations of developing countries intake diet in the form of staple foods characterized by reduced bioavailability of essential micronutrients. Prevalence of micronutrient deficiency increase the risk of extensive disease burden in low and middle income countries (Black 2014), where populations of impecunious people cannot afford costlier nutrient rich foods and other nutrient supplements and suffers from wide varieties of micronutrients malnutrition associated ailments. According to the United Nations System Standing Committee on Nutrition (UNSSCN) (2004) micronutrients starvation are associated with more than 50% of all child mortality and also present the foremost risk factor for maternal mortality. Zinc (Zn), iron (Fe), and selenium (Se) are considered as important micronutrients and these are required in appropriate amount through routine diet for maintaining several life processes. Deficiency of one or more micronutrients creates negative impact on human health express in wide arrays of diseases (Fig. 13.1). The micronutrient zinc is most essential for all organisms including humans, and also has important structural roles in several proteins. Zinc deficiency is most prevalent micronutrient dearth and is associated with numerous human health related issues such as impairments of physical growth, greater risk of various infections, retarded growth, deferred wound healing, diarrhea, skeletal abnormalities and increased risk of abortion (Salgueiro et al. 2000).

Deficiency of iron causes chlorosis in plants and results in reduced crop yield, and eventually affects human health through food-chain, specifically to people whose diets generally rely on plant resources. Iron deficiency engenders nutritional anemia and also associated with impaired immune functions in children and as well impaired neurocognitive development (Murray-Kolb 2013). Selenium is another instance of essential micronutrient possessing role in wide range of metabolic pathways such as antioxidant defense and thyroid hormone metabolism. Selenium (Se) deficiency may linked with numerous ailments including heart diseases, reduced male fertility, hypothyroidism, weakened immune system and high risk of infections, cancer, oxidative stress-related conditions and epilepsy (Hatfield et al. 2014). Deficiency of vitamins (Vitamin B6, Vitamin B12, Vitamin C, Vitamin E and folic acid), zinc and iron also

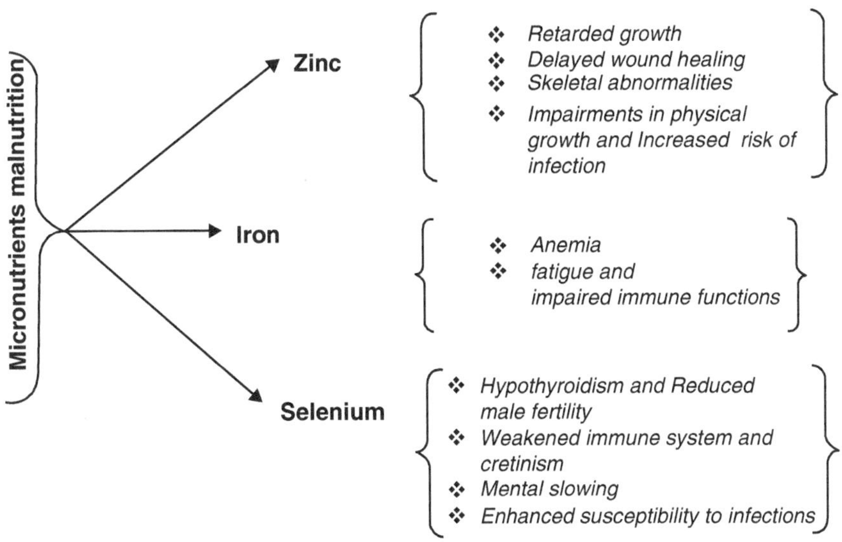

Fig. 13.1 Schematic representation of health effects of micronutrient (Zn, Fe and Se) malnutrition in humans

results in DNA damage through using similar strategies as radiation and various chemicals, and therefore considered a major factor to cause cancer and other disabilities. To circumvent problems associated with micronutrients malnutrition, investigation of those strategies are required which can improve the nutrient assimilation in plants.

13.3 Approaches for Biofortification

13.3.1 Biofortification Through Genetic Modification

Biofortification of vital food crops through genetic amendment and various biotechnological techniques is a sustainable solution for alleviating the micronutrient malnutrition. The techniques of genetic modification are being optimized for the development and production of healthy foods in addition to step up in the levels and activity of biologically active components in food crop system. Techniques of genetic modification have typically been targeted at escalating yields of staple food crops in developing countries. Though, the food crops with improved nutritional quality have gathered less consideration. An excellent example of a genetically modified biofortified crop is golden rice. Ordinary rice is not able to synthesise beta-carotene; however, due to genetic modification, golden rice can produce rice with beta-carotene in it. Moreover, stearidonic acid assimilation in soybean crop is also reported through genetic transformation (Singh et al. 2017).

13.3.2 Transgenic Approaches

Biotechnological techniques accredit the screening and selection of flourishing genotypes, isolation as well as cloning of favorable traits and formation of transgenic crops for sustainable agriculture system. Transgenic approaches can be used to increase the micronutrient content of staple food crops such as legumes and cereals, which can be achieved by insertion of specific genetic trait with the ability to produce the desired nutrients that are typically deficient in recipients. It involves the characterization, insertion or deletion of specific gene to improve the desired trait like nutritional quality from donor organisms. This may be achieved by the introduction of genes that code for trace element binding proteins, over expression of storage proteins already present or the expression of other proteins that are responsible for micronutrient uptake in plants. Furthermore, metabolic pathways from any microorganism and other organisms can also be applied into crops to utilize alternative pathways for metabolic engineering. Thus, these technologies provide a powerful tool that is unconstrained by the gene pool of the host. In addition, the transgenic approaches can be targeted to the edible portions of commercial crops (Hirschi 2009). As shown in Table 13.1, several crops have been genetically modified with traits of macronutrient and micronutrient that may provide reimbursement to consumers (Newell 2008).

Table 13.1 Genetically modified crops with description of macronutrient and micronutrient assimilation

Characteristics	Crop (details of characteristics)
Protein and amino acids	
Quality of protein and level	Maize (amino acid composition; protein↑)
	Potato (amino acid composition; protein↑)
	Rice (amino acid composition; protein↑)
	Soybean (amino acid balance)
	Sweet potato (protein↑)
Essential amino acid	
	Maize (Lys↑, Met↑, Trp↑)
	Potato (Met ↑)
	Sorghum (Lys↑)
	Soybean (Lys↑, Trp↑, Cys↑, Met↑)
Oils and fatty acids	
	Canola (lauric acid↑; + ω-3 fatty acids; 8:0 and 10:0 fatty acids↑; lauric and myristic acids↑; oleic acid↑; γ-Linolenic acid)
	Cotton (oleic acid↑, oleic + stearic acids↑)
	Grass, legumes (↓trans-fatty acids)
	Linseed (+ ω-3 and ω-6 fatty acids)
	Maize (oil↑)
	Oil palm (oleic acid↑ or stearic acids↑, oleic acid↑,+palmitic acid↓)
	Rice (α-linolenic acid↑)
	Soybean (oleic acid↑, α-linolenic acid↑, stearidonic acid↑, Arachidonic acid↑)
Carbohydrates	
Fructans	Maize (fructan↑)
	Potato (fructan↑)
Starch	Rice (amylase↑)
	Wheat (amylose↑)
Micronutrients and functional metabolites (Vitamins and carotenoids)	
	Canola (vitamin E↑)
	Maize (vitamin E↑, vitamin C↑, provitamin A)
	Mustard (+β-carotene)
	Soyabean (Vitamin E)
	Potato (β-carotene and lutein↑)
	Rice (+β-carotene, Vitamin B9↑)
	Wheat (provitamin A↑)
	Strawberry (vitamin C↑)
	Tomato (folate↑, phytoene and β-carotene↑, lycopene↑)

(continued)

Table 13.1 (continued)

Characteristics	Crop (details of characteristics)
Mineral availabilities	
	Alfalfa (phytase↑)
	Carrot (calcium↑)
	Lettuce (iron↑)
	Rice (iron↑, zinc↑)
	Maize (phytase↑, ferritin↑)
	Soybean (phytase↑)
	Wheat (phytase↑, iron↑, zinc↑)
	Alfalfa (phytase↑)
	Carrot (calcium↑)
	Lettuce (iron↑)
	Barley (zinc↑, phytase↑)

Modified from Singh et al. (2017)

13.3.3 Agronomic Biofortification

Biofortification through agronomical approach can be achieved through the implication of nutrient-rich fertilizers to foliage or soil to increase the micronutrient concentration in edible crop parts and thus increase the intake of essential micronutrients by consumers. Interaction between micronutrient and macronutrient can influence the efficiency of agronomic biofortification. Good quantity of macro elements (N, P and K) in crop has a positive effect on development of root architecture and transportation of nutrients from vegetative tissues to the seeds. Consequently, there is increased concentration of micronutrients in edible parts of the food crop (Prasad et al. 2014). However, when food crops are grown where mineral elements become straight away unavailable in the soil system, targeted application of soluble chemical fertilizers to foliar parts and roots are practised.

13.3.4 Chemical Fertilizer

Effectiveness of chemical fertilizer on crop performance is influenced by type of fertilizer. The interaction between chemical fertilizer and different forms of nutrients can have positive, negative or neutral effect on food crop in yield and nutrient bioavailability. The implication of fertilizer with soil is often the most efficient manner. However, soils often contain huge amounts of iron, but only little amount of iron is phytoavailable. The implication of inorganic Fe fertilizers to such soils is more often futile as it rapidly becomes unavailable to plant root system through adsorption, oxidation reactions and precipitation. For this reason, Fe chelators are often used as soil Fe fertilizers (Rengel et al. 1999). Zinc is commonly applied to crops as $ZnSO_4$ or as synthetic chelators (Shuman 1998). The application of Zn fertilizers to the soil system is effective for increasing Zn concentrations in cereal grains, growing mostly in soils and foliar applications of either $ZnSO_4$ or Zn chelators can increase Zn concentrations in plant via ample Zn mobility in the phloem. Similarly, applications of Zn fertilizers in soil and foliar can increase Zn concentrations in leaf, tuber and fruit (Shuman 1998; Rengel et al. 1999).

13.3.5 Biofortification Through Plant Growth Promoting Microorganisms (PGPM)

Biofortification of crops through implications of PGPMs can be considered as a promising accompanying measure, which along with transgenic varieties, can lead to augmented micronutrient concentrations in food crop system, besides improving yield and soil fertility. Plant growth promoting microorganism's have been reported to biofortify the micronutrient contents in food crops besides improving the soil fertility and crop yield (Rana et al. 2012). In addition, plant growth promoting microorganisms also facilitate the plant growth through N_2 fixation, insoluble phosphorus solubilization, production of phytohormones, lowering of ethylene concentration, antibiotics and antifungal metabolites synthesis and induced systemic resistance. In this way, PGPM are also known to boost the soil fertility in return, the plant acquiesce by supplying essential nutrients, growth regulators and enhancing the ethylene mediated stress by 1-aminocyclopropane-1-carboxylate (ACC) deaminase production along with improved plant stress tolerance to drought, salinity, metal and toxicity of pesticide (Singh and Prasad 2014; Singh and Singh 2017). Moreover, the potentiality of PGPM in agriculture is progressively increased as it provides an attractive approach to replace the exploitation of chemical fertilizers, pesticides and other supplements. Subsequently, biofortification of crops through application of PGPMs can be therefore considered as a potential supplementary approach, which along with breeding varieties, can escort to augment the concentrations of micronutrient in wheat crop, besides improving yield and soil fertility (Singh et al. 2017).

13.4 Mechanisms of Plant Growth Promoting Microorganisms

A variety of plant growth promoting microorganisms (PGPM) has been reported to enhance plant growth and productivity by means of various mechanisms. Illustration of some of the mechanisms is as follows.

13.4.1 Iron Chelation

Iron is a vital component for all forms of life including prokaryotes as well as eukaryotes. It is the component of electron transport carrier, cofactor of various enzymes and important part of various constituent such as hemoglobin. Due to the aerobic environment conditions, iron is present in its oxidized form (Fe^{3+}, insoluble at neutral pH) instead of reduced form (Fe^{2+}, soluble at neutral pH) which are taken up by plants. To sequester the iron, many fungi, bacteria and some plants have an unusual adaptation to produce low molecular weight compounds called as siderophores, a group of low molecular weight compounds (<10 KD) those have immense affinity towards Fe^{3+} ions. Siderophores are PGPM secreted compounds that are ultimately taken up by plants therefore transporting molecule of iron to the plants cells. Plant roots might be able of take up siderophore and use them as sources of iron. Therefore, microbial siderophore can enhance plant growth by improving iron uptake as well as by inhibiting the plant pathogen by means of competition ultimately leads to the iron biofortification in plants and their grains (Srivastava et al. 2013). Many researchers has been reported the siderophore production in wide range of bacterial species *viz. Bacillus, Pseudomonas, Azotobacter, Arthrobacter, Burkholderia, Enterobacter, Rhodospirrilum, Serratia, Azospirillum* and *Rhizobium* and fungal species *viz. Aspergillus Penicillium Rhizopus*, *Syncephalastrum* (Leong and Neilands 1982; Das et al. 2007; Duran et al. 2016; Srivastava et al. 2013). There are many types of siderophore such as hydroxamate, catacholate and carboxalate that are secreted by microbes varies from species to species. Furthermore, mixed type of siderophore has been secreted by many baceterial species (Wandersman and Delepelaire 2004). Hence we can say that use of siderophore producing PGPM is better approach over other conventional methods such as chemical fertilizers to enhance iron content in plants and grains.

13.4.2 Zinc Solubilizer

Zinc is one of the essential nutrients required for growth and metabolic activities. Zinc ions takes part in many physiological activities, it act as cofactor in various enzymes; take part in defense; play role in cell division and growth in prokaryotes as well as in eukaryotes. Zinc ions are highly reactive in nature and present in close interaction with soil constituents therefore soluble zinc is very low in soil. Generally, it is found in the form of oxides, phosphates and carbonates. Plant associated microorganisms adopt several mechanisms to solubilize zinc such as chelation (Whiting et al. 2001), reduction in soil pH (Subramanian et al. 2009), or through improving root growth and root absorptive area (Burkert and Robson 1994). Zinc chelation by microbes and making them available for plants roots is a well known phenomenon. Microbes produce chealating compounds, which forms complex upon binding with zinc. In addition, they releases chealated zinc at the root surface and enhance the zinc availability ultimately lead to the zinc biofotification in plants. Whiting et al. (2001) reported production of metallophores as the possible strategies used by bacteria to chelate Zn. Reduction in soil pH also enhance availability of zinc. Decline in pH has been reported, when *Pseudomonas* and *Bacillus* spp. solubilized zinc complex compounds (ZnS, ZnO and $ZnCO_3$) into zinc ions in a broth culture (Saravanan et al. 2004). Zinc solubilization methods differ from one microorganism to another to improve Zn availability in soil system and plant tissues. Many microbes including many bacterial and fungal species (*Pseudomonas, Microbacterium, Enterobacter, Bacillus, Arbuscular mycorrhizae*) have the incredible capability to solubilize Zn from complex compounds (Whiting et al. 2001; Fasim et al. 2002; Subramanian et al. 2009) and consequently take part in improvement of food quality and nutrient status of plants and grains.

13.4.3 Biofertilizers

In the present era, use of chemical fertilizers to enhance plant growth, productivity and to replenish soil nutrient status is very common. But several problems coincide with the use of chemical fertilizers such as high cost, unavailability of large portion of nutrients, toxic and non-degradable nature, leading to enhancement of environmental pollution and making land unsuitable for cultivation. Therefore as an alternative strategy, application of biofertilizers can be used to enhance crop productivity and biofortification of nutrients in grains. Biofertilizers which are the fertilizers based on source of biological origin such as microbes (including bacteria and fungi) are used instead of synthetic compounds. Use of biofertilizers is an ecofriendly approach as they are of biological origin and keeps environment healthy due to its low persistence. The objective of biofertilizers is to increase the soil organic contents improve soil structure and reduce loss of nutrients such as nitrogen, iron, zinc, calcium and phosphorus (Lal and Greenland 1979). Biofertilizers serves as source of all nutrients due to their ability to solubilize complex form of nutrients into soluble form (Singh et al. 2010, 2013, 2018). Plants utilize phosphorus in the form of orthophosphate (Pi). Plant growth promoting microorganisms possess various mechanisms to enhance phosphate solubilization (Fig. 13.2). Jones et al. (1998) reported 3.1–4.7 times more efficacy in mycorrhizal associated plants for phosphorus uptake than nonmycorrhizal plants. Commercially biofertilizers are developed by coating of various bacteria (*Azotobacter, Azospirillum, Rhizobium, Pseudomonas* and *Bacillus*) on seeds, a process called bacterization. *Azotobacter chrococcum* secretes azotobacterin, *Bacillus megaterium* secretes phosphobacterin, which is used for preparation of biofertilizers (Kumar and Bohra 2006). These bacteria may or may not form symbiotic association but enhances lateral root hairs of plants to increase mineral and water absorption, also increases nitrogen availability, secretes plant growth stimulating substances such as vitamin, auxins, gibberellic acid, cytokinins which leads to increase in photosynthesis capacity ultimately enhancing nutrient status in plants. Rhizobial biofertilizers are able to fix 50–150 N/ha/annum.

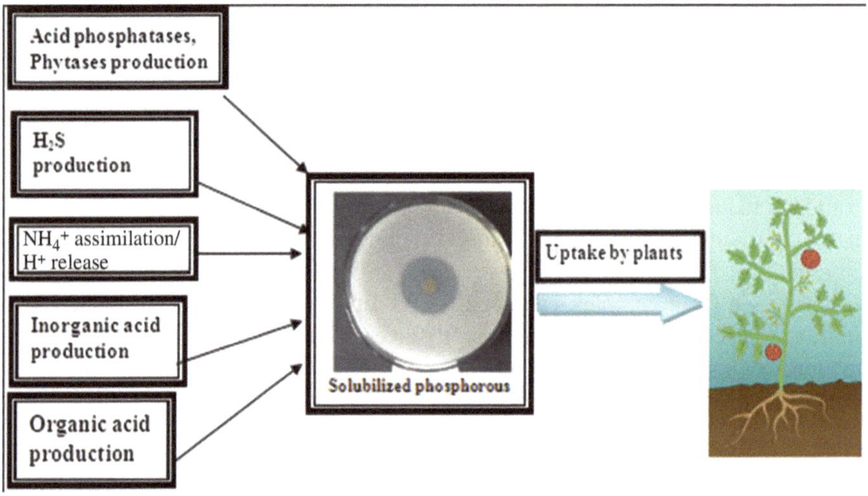

Fig. 13.2 Phosphate solubilization mechanisms of plant growth promoting microorganisms

It has been well known, that application of biofertilizers with plants, significantly increases plant growth, high nutrient status, low level of pathogen attack (Gupta et al. 2003; Yadav et al. 2016). Therefore, such free living and symbiotic microorganisms are promoted to reduce the dependence on chemical fertilizers. Some biofertilizers are listed in Table 13.2.

13.4.4 Biocontrol Agents

Under natural environmental conditions, plants are continuously exposed to various pathogenic bacteria and fungi, which causes disease in plants leading to the reduction in crop production or death of the plants. Therefore, plant diseases need to be controlled to maintain the quality and nutritional status of plants. Different approaches may be used to prevent or control plant diseases. In present circumstances, many pesticides are used to prevent disease but the disadvantage is that the pest may adapted towards pesticide and it is also cost intensive as well as act as environmental pollutant due to its persistence. Therefore, there is a novel approach i.e. application of biocontrol agent to suppress or demise pathogen growth. Biocontrol agents are microorganisms, which controls the fungi, insect, pest and any other pathogen (Beattie 2006). Generally bacteria, fungi, virus and protozoans are used as biocontrol agents. Generally plant growth promoting microorganisms produces various substances that protect plants against pathogens by direct interactions i.e. antagonistic activity or indirectly by inducing host resistance (Induce systemic resistance or systemic acquired resistance). Plant growth promoting microorganisms that indirectly act on pathogens may have some mechanisms to control plant pathogen such as the following (Table 13.3).

Table 13.2 Biofertilizers, their mode of action and crops benefitted

Organisms		Action	Crop	References
Bacteria	*Rhizobium leguminosarum*	Symbiotic nitrogen fixation	All leguminous crops	Bagali (2012)
	Bacillus megaterium	Phosphate solubilization	Mustard	Kang et al. (2014)
	Bacillus subtilis	Micronutrient solubilizer	Cotton	Yao et al. (2006)
	Pseudomonas fluorescens	Micronutrient solubilizer	Bean	Alemu (2013)
	Azotobacter sp.	Free living nitrogen fixation	Leguminous crops	Bagali (2012)
	Azospirillum sp.	Associated Symbiotic nitrogen fixation	Leguminous crops	Bagali (2012)
Fungi	*Penicillium bilaiae*	Phosphate solubilization	Coffee, Casurina	Malhi et al. (2013)
	Glomex sp.	Phosphate solubilization	Coffee, Casurina	Malhi et al. (2013)
Cyanobacteria	*Nostoc*	Free living nitrogen fixation	Rice	Vaishampayan et al. (2001)
	Anabana	Free living nitrogen fixation	Rice	Vaishampayan et al. (2001)
	Anabana- Azolla	Symbiotic nitrogen fixation	Rice	Vaishampayan et al. (2001)

Table 13.3 Major biocontrol agents, their target pathogen and mechanism of action

Biocontrol agent	Target pathogen	Crop	Action	References
Bacillus thuringiensis	*R. solani,* All phytopathogen	Cotton	Lytic enzymes	Shaikh and Sayyed (2015)
Pseudomonas fluorescence	*Erwinia carotovora,Puccinia ultimum, Fusarium glycinia*	Potato, wheat, Sugar beat	Siderophor, Antibiotics production	Shaikh and Sayyed (2015)
Strepromyces sp.	*S. sclerotiorum*	Potato, Tomato	Lytic enzymes	Shaikh and Sayyed (2015)
Trichoderma harzianum	*Botrytis cinerea, Meloidogyne javanica*	Bean, Tomato	Lytic enzymes, Competition	Woo et al. (1999), Sahebani and Hadavi (2008) and Puyam (2016)
Trichoderma viride	*Sclerotium rolfsii*	Groundnut	Lytic enzymes, Competition	Hirpara et al. (2017) and Puyam (2016)
Pseudomonas cepacia	*Bipolaris maydis*	Maize	Antibiotics production	Shaikh and Sayyed (2015)

1. The PGPR may have ability to produce siderophore that chelates iron, which makes iron unavailable for plant pathogens (Singh et al. 2017)
2. The PGPR may possess capacity to secrete some anti-pathogenic metabolites such as antibiotics, cell wall degrading hydrolytic enzyme (Glucanases, chitinases, proteases, lipase, pectinases) or hydrogen cyanide (HCN), which suppresses pathogen growth (Maksimov et al. 2011).
3. The PGPR may compete for nutrients and niche with pathogen (Kamilov et al. 2005).
4. The PGPR may stimulate Induced Systemic Resistance (ISR) or Systemic Acquired Resistance (SAR) (Van loon et al. 1998).

A wide range of microorganisms are reported to act as biocontrol agents, as they possess one or more than one mechanisms to suppress pathogen attack or growth. Due to less persistence in environment, specificity for target pest, cost effectiveness and ecofriendly nature, biocontrol agents are good alternative to pesticides.

13.5 Conclusion

Development of crops with elevated concentration of micronutrients is immensely and urgently needed to combat the problem of micronutrient deficiency. Plant growth promoting microorganisms (PGPMs) make interaction with plants and exert plant growth promoting activities and enhance the capability of the plant for uptake of micronutrients from surrounding soils. Zinc solubilization and siderophore secreting microorganisms enhance the level of zinc and iron in the various edible portions of crop plants and provide an alternative strategy to fortify micronutrients and produce micronutrients rich foods. Application of such microorganisms inoculants reduces the dependency on costly biofortification approaches i.e. agronomic and genetic approaches. In future, formulation of microorganisms possessing multiple beneficial traits can be applied for biofortification strategies to tackle the problem of hidden hunger. Using plant growth promoting microorganisms for biofortification as part of green technology approach can be a better strategy to achieve environmental sustainability.

References

Alemu, F. (2013). Isolation of *Pseudomonas fluorescens* from rhizospheric soil of faba bean and assessment of their phosphate solubility: In vitro study, Ethiopia. *Scholars Academic Journal of Biosciences, 1*(7), 346–351.

Bagali, S. S. (2012). Review: Nitrogen fixing microorganisms. *International Journal of Microbiology, 3*(1), 46–52.

Beattie, G. A. (2006). Plant-associated bacteria: Survey, molecular phylogeny, genomics and recent advances. In S. S. Gnanamanickam (Ed.), *Plant-associated bacteria* (pp. 1–56). Dordrecht: Springer.

Black, R. E. (2014). Global distribution and disease burden related to micronutrient deficiencies. *Nestle Nutrition Institute Workshop Series, 78*, 21–28.

Bonfante, P., & Genre, A. (2015). Arbuscular mycorrhizal dialogues: Do you speak 'plantish' or 'fungish'? *Trends in Plant Science, 20*, 150–154.

Burkert, B., & Robson, A. (1994). Zn uptake in subterranean clover (*Trifolium subterraneum* L.) by three vesicular-arbuscular mycorrhizal fungi in a root-free sandy soil. *Soil Biology and Biochemistry, 26*, 1117–1124.

Das, A., Prasad, R., Srivastava, A., Giang, P. H., Bhatnagar, K., & Varma, A. (2007). Fungal siderophores: Structure, functions and regulations. In A. Varma & S. B. Chincholkar (Eds.), *Microbial siderophores* (pp. 1–42). Berlin/Heidelberg: Springer.

Durán, P., Acuña, J. J., Armada, E., López-Castillo, O. M., Cornejo, P., Mora, M. L., & Azcón, R. (2016). Inoculation with selenobacteria and arbuscular mycorrhizal fungi to enhance selenium content in lettuce plants and improve tolerance against drought stress. *Journal of Soil Science and Plant Nutrition, 16*(1), 211–225.

Fasim, F., Ahmed, N., Parsons, R., & Gadd, G. M. (2002). Solubilization of zinc salts by a bacterium isolated from the air environment of a tannery. *FEMS Microbiology Letters, 213*(1), 1–6.

Gadd, G. M. (2010). Metals, minerals and microbes: Geo-microbiology and bioremediation. *Microbiology, 156*(Pt 3), 609–643.

Gupta, R. K., Kaushik, S., Sharma, P., & Jain, V. K. (2003). Biofertilizers: An ecofriendly alternative to chemical fertilizers. In A. Kumar (Ed.), *Environmental challenge of 21th century* (pp. 275–287). New Delhi: APH Publication Corporation.

Hatfield, D. L., Tsuji, P. A., Carlson, B. A., & Gladyshev, V. N. (2014). Selenium and selenocysteine: Roles in cancer, health, and development. *Trends in Biochemical Sciences, 39*(3), 112–120.

Hirpara, D. G., Gajera, H. P., Hirpara, H. Z., & Golakiya, B. A. (2017). Antipathy of *Trichoderma* against *Sclerotium rolfsii* Sacc.: Evaluation of cell wall-degrading enzymatic activities and molecular diversity analysis of antagonists. *Journal of Molecular Microbiology and Biotechnology, 27*, 22–28.

Hirschi, K. D. (2009). Nutrient biofortification of food crops. *Annual Review of Nutrition, 29*, 401–421.

Jones, M. D., Durall, D. M., & Tinker, P. B. (1998). A comparison of arbuscular and ectomycorrhizal *Eucalyptus coccifera*: Growth response, phosphorus uptake efficiency and external hyphal production. *New Phytologist, 140*(1), 125–134.

Kamilova, F., Validov, S., Azarova, T., Mulders, I., & Lugtenberg, B. (2005). Enrichment for enhanced competitive plant root tip colonizers selects for a new class of biocontrol bacteria. *Environmental Microbiology, 7*, 1809–1817.

Kang, S. M., Radhakrishnan, R., You, Y. H., Joo, G. J., Lee, I. J., Lee, K. E., & Kim, J. H. (2014). Phosphate solubilizing Bacillus megaterium mj1212 regulates endogenous plant carbohydrates and amino acids contents to promote mustard plant growth. *Indian Journal of Microbiology, 54*(4), 427–433.

Khan, A. G. (2005). Role of soil microbes in the rhizospheres of plants growing on trace metal contaminated soils in phytoremediation. *Journal of Trace Elements in Medicine and Biology, 18*(4), 355–364.

Kumar, A., & Bohra, C. (2006). Green technology in relation to sustainable agriculture. In *Green technologies for sustainable agriculture*. New Delhi: Daya Publishing House.

Lal, R., & Greenland, D. J. (1979). *Soil physical properties and crop production*. Chichester: Wiley.

Leong, S. A., & Neilands, J. B. (1982). Siderophore production by phytopathogenic microbial species. *Archives of Biochemistry and Biophysics, 281*, 351–359.

Maksimov, I. V., Abizgil'dina, R. R., & Pusenkova, L. I. (2011). Plant growth promoting rhizobacteria as alternative to chemical crop protectors from pathogens. *Applied Biochemistry and Microbiology, 47*, 333–345.

Malhi, S. S., Vera, C. L., & Brandt, S. A. (2013). Relative effectiveness of organic and inorganic nutrient sources in improving yield, seed quality and nutrient uptake of canola. *Agricultural Sciences, 4*(12), 1–18.

Murray-Kolb, L. E. (2013). Iron and brain functions. *Current Opinion in Clinical Nutrition and Metabolic Care, 16*(6), 703–707.

Newell, M. M. (2008). Nutritionally improved agricultural crops. *Plant Physiology, 147*, 939–953.

Prasad, R., Shivay, Y. S., & Kumar, D. (2014). Agronomic biofortification in cereal of cereal grains with iron and zinc. *Advances in Agronomy, 125*, 55–91.

Prasad, R., Kumar, M., & Varma, A. (2015). Role of PGPR in soil fertility and plant health. In D. Egamberdieva, S. Shrivastava, & A. Varma (Eds.), *Plant growth-promoting rhizobacteria (PGPR) and medicinal plants* (pp. 247–260). Cham: Springer.

Puyam, A. (2016). Advent of *Trichoderma* as a bio-control agent – A review. *Journal of Applied and Natural Sciences, 8*(2), 1100–1109.

Rana, A., Joshi, M., Prasanna, R., Shivay, Y. S., & Nain, L. (2012). Biofortification of wheat through inoculation of plant growth promoting rhizobacteria and cyanobacteria. *The European Journal of Soil Biology, 50*, 118–126.

Rengel, Z., Batten, G. D., & Crowley, D. E. (1999). Agronomic approaches for improving the micronutrient density in edible portions of field crops. *Field Crops Research, 60*, 27–40.

Sahebani, N., & Hadavi, N. (2008). Biological control of the root-knot nematode *Meloidogyne javanica* by *Trichoderma harzianum. Soil Biology and Biochemistry, 40*, 2016–2020.

Salgueiro, M. J., Zubillaga, M., Lysionek, A., Cremaschi, G., Goldman, C. G., Caro, R., De Paoli, T., Hager, A., Weill, R., & Boccio, J. (2000). Zinc status and immune system relationship: A review. *Biological Trace Element Research, 76*(3), 193–205.

Saravanan, V. S., Subramoniam, S. R., & Raj, S. A. (2004). Assessing in vitro solubilization potential of different zinc solubilizing bacterial (zsb) isolates. *Brazilian Journal of Microbiology, 35*(1–2), 121–125.

Shaikh, S. S., & Sayyed, R. Z. (2015). Role of plant growth-promoting rhizobacteria and their formulation in biocontrol of plant diseases. In *Plant microbes symbiosis: Applied facets* (pp. 337–351). New Delhi: Springer.

Sharma, P., Aggarwal, P., & Kaur, A. (2016). Biofortification: A new approach to eradicate hidden hunger. *Food Reviews International, 33*(1), 1–21.

Shuman, L. M. (1998). Micronutrient fertilizers. *Journal of Crop Production, 1*, 165–195.

Singh, A. V., & Prasad, B. (2014). Enhancement of plant growth, nodulation and seed yield through plant growth promoting rhizobacteria in lentil (*Lens culinaris Medik* cv. VL125). *International Journal of Current Microbiology and Applied Sciences, 3*(6), 614–622.

Singh, J., & Singh, A. V. (2017). Microbial strategies for enhanced phytoremediation of heavy metals contaminated soils. In R. N. Bharagava (Ed.), *Environmental pollutants and their bioremediation approaches* (Vol. 9, pp. 249–264). Boca Raton: Taylor & Francis.

Singh, A. V., Shah, S., & Prasad, B. (2010). Effect of phosphate solubilizing bacteria on plant growth promotion and nodulation in soybean (*Glycine max* (L.) Merr). *Journal of Hill Agriculture, 1*(1), 35–39.

Singh, A. V., Chandra, R., & Reeta, G. (2013). Phosphate solubilization by *Chryseobacterium* sp. and their combined effect with N and P fertilizers on plant growth promotion. *Archives of Agronomy and Soil Science, 59*(5), 641–651.

Singh, J., Singh, A. V., Prasad, & Shah, S. (2017). Sustainable agriculture strategies of wheat biofortification through microorganisms. In A. Kumar, A. Kumar, & B. Prasad (Eds.), *Wheat a premier food crop*. New Delhi: Kalyani Publishers.

Singh, A. V., Prasad, B., & Goel, R. (2018). Plant growth promoting efficiency of phosphate solubilizing *Chryseobacterium* sp. PSR 10 with different doses of N and P fertilizers on Lentil (*Lens culinaris* var. PL-5) growth and yield. *International Journal of Current Microbiology and Applied Sciences, 7*(05), 2280–2289.

Srivastava, M. P., Tewari, R., & Sharma, N. (2013). Effect of different cultural variables on siderophores produced by *Trichoderma* spp. *International Journal of Advanced Research, 1*, 1–6.

Subramanian, K. S., Tenshia, V., Jayalakshmi, K., & Ramachandran, V. (2009). Role of arbuscular mycorrhizal fungus (*Glomus intraradices*)–(fungus aided) in zinc nutrition of maize. *Journal of Agricultural Biotechnology and Sustainable Development, 1*, 29–38.

United Nations System Standing Committee on Nutrition (UNSSCN). (2004). *5th report on the world nutrition situation nutrition for improved development outcomes*. Geneva: SCN.

Vaishampayan, A., Sinha, R. P., Hader, D. P., Dey, T., Gupta, A. K., Bhan, U., & Rao, A. L. (2001). Cyanobacterial biofertilizers in rice agriculture. *The Botanical Review, 67*(4), 453–516.

Van Loon, L. C., Bakker, P. A. H. M., & Pieterse, C. M. J. (1998). Systemic resistance induced by rhizosphere bacteria. *Annual Review of Phytopathology, 36*, 453–483.

Wandersman, C., & Delepelaire, P. (2004). Bacterial iron sources: From siderophores to hemophores. *Annual Review of Microbiology, 58*, 611–647.

Whiting, S. N., de Souza, M. P., & Terry, N. (2001). Rhizosphere bacteria mobilize Zn for hyperaccumulation by Thlaspi caerulescens. *Environmental Science & Technology, 35*, 3144–3150.

Woo, S. L., Donzelli, B., Scala, F., Mach, R. L., Harman, G. E., Kubicek, C. P., Del Sorbo, G., & Lorito, M. (1999). Disruption of the ech42 (endochitinase-encoding) gene affects biocontrol activity in *Trichoderma harzianum* P1. *Molecular Plant-Microbe Interactions, 12*, 419–429.

Yadav, R., Singh, A. V., Kumar, M., & Yadav, S. (2016). Phytochemical analysis andplant growth promoting properties of endophytic fungi isolated from tulsi and *Aloe vera*. *International Journal of Agricultural and Statistics Sciences, 12*(1), 239–248.

Yao, A. V., Bochow, H., Karimov, S., Boturov, U., Sanginboy, S., & Sharipov, A. K. (2006). Effect of FZB 24® *Bacillus subtilis* as a biofertilizer on cotton yields in field tests. *Archives of Phytopathology and Plant Protection, 39*(4), 323–328.

Chapter 14
Harnessing Microbial Potential for Wastewater Treatment in Constructed Wetlands

Manoj Kaushal, Suhas P. Wani, and Mukund D. Patil

Abstract Microbial community constitute a major component of constructed wetlands (CWs), playing a major role in these systems capacities for treating wastewater. Constructed wetland system has a hydraulic regime, although the volume of inflow in the wetland is never the same as the outflow. Wetland are either of Free Water Surface (FWS) or Subsurface Flow (SF). Nitrogen, the most important component in constructed wetlands undergoes transformation by various processes converting N into one to another form and by plant uptake. For instance, nitrification is more impactful for ammonia reduction and its removal relies on the configuration of the wetland and the dissolved oxygen (DO). The chapter discusses the types of wetlands and their physical, chemical and biological processes in the removal of various contaminants. It also gives an overview of different microbial processes and their mechanisms involved during the treatment of wastewater inside constructed wetland systems.

Keywords Waste water treatment · Constructed wetlands · Biological transformation · Nutrient removal

14.1 Introduction

"Constructed wetlands are engineered ecosystems that have been made to optimize the natural processes such as wetland flora, media, and associated microbial community for rejuvenating wastewater (Vymazal 2007)". So wetlands created for treating effluents like domestic (rural or urban), industrial wastewater, stormwater runoff are termed as constructed wetlands (CW). These engineered systems are natural alternative that acts as biofilters and use natural functions of vegetation,

M. Kaushal (✉)
International Institute of Tropical Agriculture, Ibadan, Oyo State, Nigeria

S. P. Wani · M. D. Patil
Research Program – Asia, International Crops Research Institute for the Semi-Arid Tropics, Patancheru, India

S. Shah et al. (eds.), *Sustainable Green Technologies for Environmental Management*, https://doi.org/10.1007/978-981-13-2772-8_14

microorganisms and media to treat different water streams. Wastewater purification can be governed by a number of factors including phytoremediation, adsorption or nitrification, filtration by gravel and gravitational sedimentation. Dominant factor among all is biological filtration through biofilm formation by aerobic or facultative bacteria.

14.2 Constructed Wetlands

Constructed Wetlands are designed primarily for wastewater treatment with the aim to maximise pollutant storage and transformation. Components of wetland include water, media (gravel, sand), plants and microbes. Plants selected for use in CWs must have following characteristics:

- Adaptation to local climatic conditions;
- Adaptation to pollutants and waterlogged conditions;
- Ecological acceptability;
- Contaminant removal capacity, either through direct or indirect mechanisms.
- Ability to establish, spread and growth

The constructed wetlands are of following types:

- Horizontal surface flow systems;
- Horizontal subsurface flow systems;
- Vertical flow systems with upstream or downstream and undefined loading.

CWs functions are highly under the control of microorganisms (bacteria, fungi, algae, yeasts, protozoa) and their metabolism (Wetzel 1993). Wastewater treatment within CWs occurs as it passes through the wetland media (gravel/sand) and the plant rhizosphere. Major microbial process involved in wastewater treatment is transformation (which may be aerobic or anaerobic) of organic and inorganic substances into insoluble substances and alteration of redox potential of media. In general, microbes adapt to alternations with characteristics of the media and the water reached to them, but when environmental conditions are unsuitable, they remain dormant for years (Hilton 1993). However, toxic substances, such as pesticides and heavy metals may affect microbial community of a constructed wetland.

Metabolic activity of microorganisms in the root zone gets affected by the supply of oxygen. Rhizosphere, the most active reaction zone of constructed wetlands is responsible for physicochemical and biological processes induced plant-microbe interactions as well as with the soil and pollutants. In general, marsh plants give efficient results for wastewater treatment as they possess specific characteristics of growth physiology and able to survive even under extreme rhizosphere conditions such as acidic or alkaline pH, toxicity, salinity, etc. Many physical, chemical and biological mechanisms play important role in the treatment of wastewater.

14.2.1 Sedimentation

Sedimentation plays lead role in removal of contaminants and biological degradation pathways. Plant and leaf litter in the marsh slows down the water flow and allow sediments in the wastewater to be deposited into the bed of the marsh. While predominates are the anaerobic processes in subsurface flow systems, surface flow systems usually prevail with aerobic processes. Pollutant/contaminants accumulation is a prolonged process for phosphorus sink that only possible in a FWS system. Retention rates due to these accretion goes up to 75 g m^{-2} $year.^{-1}$ (Vymazal 2007). Sedimentation plays lead role in the reduction of some microbes within the wetland (Gray 2004). Protozoan cysts and helminth eggs can automatically settle through gravity flow in a free-flowing surface water constructed wetland. Bacteria and viruses will not be removed unless they adhere to larger particles. Only some of viruses are known adhere to particles in stabilized ponds that get attached and observed to be too small to settle down on the surface of media (Characklis et al. 2005; Symonds et al. 2014). Thus, removal of pathogens can be directly correlated with particle removal in constructed wetlands (Chouinard et al. 2014; Wu et al. 2016).

14.2.2 Adsorption

Pollutants in the wastewater adhere to plant material and media colloids because of attractive forces, allowing them to settle to the base of the wetland. The movement of inorganic P from the soil to mineral surfaces and thereafter its accumulation at the media surface is known as adsorption. High presence of Al, Fe and Ca levels in media can be directly related with P adsorption.

14.2.3 Precipitation

Insoluble substances (heavy metals) can become insoluble settle onto media and plant material. Formation of phosphate ions with metallic cations, resulting in amorphous solids when both are at peak level in terms of their concentrations is regarded as precipitation reaction.

14.2.4 Nutrient Uptake

Plants growing in CWs use pollutants for growth due to abundance of nutrient availability in the wastewater. Ammonia and nitrate are utilized in the process of plant uptake or assimilation, which are regarded as spring process in temperate zones. In this process, wetland plants behave for long cycle of aboveground biomass like Typha spp. or *Phragmites australis*. Presence of nutrients and growth rate of plant tissues tapped the potential rate of uptake of nutrients by the plants. Likely, it impacts plant tissue in terms of nutrient storage and their concentrations and finally with accumulation of biomass which is maximum for standing vegetation.

14.2.5 Decomposition and Volatilisation

Various elements such as nitrogen and sulphur existing in gaseous form are released to the atmosphere which is important pathway for their removal in CWs. Water mass distinct the soluble organic matter, by the process of sorption to solid surfaces inside the constructed wetlands or through volatilization for volatile organic compounds (VOC) (US EPA 2000).

This chapter will dwell on the mechanisms of important processes carried out by microorganisms in the root zone of constructed wetlands which helps in removal of contaminants from wastewater.

14.3 Role of Microbes in Constructed Wetlands

A diverse array of microbial communities performs and impact the removal efficiency of contaminants in constructed wetlands. Also, microbial communities are mostly responsible for the improvement in water quality from effluents of constructed wetlands (Ibekwe et al. 2003; Dong and Reddy 2010; Long et al. 2016). Archaeal, bacterial, and fungal OTUs are always greater in the influent than in the effluent of the any constructed wetland (Ibekwe et al. 2017). Rhizosphere is the region where complex reactions take place due to close interaction between roots and the wetland media. Rhizosphere is also identified as zone of diverse substances as root exudates contains sugars, minerals, vitamins, polysaccharides, organic acids, phenol and other organic compounds (Miersch et al. 2001). Bacteria, fungi, algae and protozoa are the most common microorganisms responsible for the proper functioning of CWs. Rhizosphere with root exudates stimulates the microbial communities for the degradation of organic pollutants. Rhizodeposition products perform the mobilizing of nutrients. Organic acid excretion increased during nutrient limiting conditions which enhancing iron and phosphate solubility, thus the plant's nutrient supply is boosted (Hoffland et al. 1992). Also, microorganisms use organic compounds as substrates (sugars and amino acids) and excreted vitamins stimulate microbial growth which is termed as rhizosphere effect. Microbial cells adheres to each other and to wetland surface to form biofilms. They are frequently embedded within a self-produced matrix of extracellular polymeric substance (EPS). However, wetland media is the main supporting material for the formation of microbial film and hence, responsible for removal of contaminants. In addition to contaminant/

pathogen removal, there are other four major microbial reactions (fermentation, nitrification, denitrification and phosphorus removal) mainly responsible for the performance of constructed wetlands (Mitchell 1996). Nitrogen removal in constructed wetlands has been documented to be the result of the activities of microbial communities, which impacts Anammox (Oehl et al. 2004) and nitrification-denitrification (Kroger et al. 2012) pathways. In addition, microbial activities also effect P removal partially through mineralization (Truu et al. 2005) and immobilization.

14.3.1 Fermentation

With respect to climate change, constructed wetlands are also a major area of concern and need to focussed as they are the major significant producer of atmospheric methane. Designated by diverse microbial communities and water-logged surface constructed wetlands are emerged and acclimatized to continuous water presence. Oxygen poor environments favours methanogenesis and microbial communities living in moist and hot environments consume oxygen at much faster rate than it diffuses from air. Thus, wetlands are the most ideal candidates for fermentation processes with anoxic environments enabling easier decomposition of organic carbon and production of high energy compounds (e.g., alcohol, volatile fatty acids, methane). In a process of methanogenesis, microorganisms produce methane by fermenting acetate into methane and carbon dioxide.

$$H_3C-COOH \xrightarrow{\text{Halobacterium, Ferroglobus}} CH_4 + CO_2 \tag{14.1}$$

Depending on the wetland type, in another process archaea oxidize hydrogen with carbon dioxide to produce methane and water.

$$4H_2 + CO_2 \xrightarrow{\text{Pyrolobus, Pyrococcus, Methanosarcinales}} CH_4 + 2H_2O \tag{14.2}$$

14.4 Microbial Mediated Transformations in CWs

Microbial communities in wetland cells are the main cause for perturbation, concentration of dissolved organic matter, and stress related to chemical compounds (Wassel and Mills 1983; Hirayama et al. 2005; Bodtker et al. 2008; Nelson 2009). It has also been shown that, shifts in the bacterial communities structure can be related with alteration in many physico-chemical soil properties such as texture and availability of nitrogen (Frey et al. 2004; Lauber et al. 2008). Moreover, inside the constructed wetland, diverse microbial communities are the significant drivers greatly impacting the final quality of wetland effluents (Calheiros et al. 2009).

Biogeochemical cycling of nitrogen involves both biotic and abiotic transformation resulting in a wide variety of inorganic and organic nitrogen compounds. Most of the forms of the nitrogen are necessary for the existence of life as the nitrogen forms the structural constituent of cell. Nitrogen transformation in wetlands are necessary for functioning of wetland ecosystem and the transformation are enabled by a diverse group of micro-organisms. The important processes that leads to the transformation of nitrogen are ammonification, volatilization, nitrification, denitrification, nitrogen fixation, assimilation, Ammonia absorption and ANAMMOX (Vymazal 2007) occurring in the constructed wetlands.

14.4.1 Ammonification

Ammonification is a catabolic process involving biological conversion of organic nitrogen compounds like amino acids into ammonia. This is essentially a multi-step energy releasing biochemical process. The energy so released is utilised by the microbes involved in the transformations. The rate of ammonification is a function of pH, temperature, C/N ratio and soil physical conditions. However, the optimum temperature (40–60 °C) and pH (6.5–8.5) are reported for ammonification process. The important mechanism for the reduction of nitrogen content from constructed wetlands is by the ANAMMOX process. When anaerobic ammonium oxidized, ammonium and nitrite are converted to dinitrogen gas (Kuenen 2008). The ANAMMOX reaction is as follows:

$$NH_4^+ + NO_{2^-} \rightarrow N_2 + 2H_2O \tag{14.3}$$

Partial-nitrification in constructed wetlands with Anammox having huge deportation efficacy of total nitrogen has been investigated over conventional methods (Dong and Sun 2007), thus explaining microbial role in altering wastewater quality in engineered wetlands. It has been reported that Anammox utilize carbon dioxide as source of carbon to yield biomass and electron acceptor (nitrite) for oxidation of ammonium, and electron donor for the discharge of carbon dioxide (Kuenen 2008).

Ammonia formation from organic compounds during organic matter degradation is called ammonification, creating energy for growth, and then the ammonia is directly incorporated into cell biomass (Vymazal 2005). Various factors including temperature, pH, available nutrients, C/N ratio, such as texture and structure influencing ammonification rates (Han and Lee 2005; Gustavsson and Engwall 2012) and the values fluctuates between 0.004 and 0.53 g N/m^2. d (Reddy et al. 2001; Scholz and Lee 2005). After formation, ammonium can also be absorbed by plants with their complex root systems (Forbes et al. 2010). The process occurs both under aerobic as well as anoxic conditions, but has been moderately slow with anaerobic conditions (Mitsch and Gosselink 2007). Ammonium (NH_4^+) gets absorbed by plants with root and root hairs after the formation and immobilized in the sediments by ion exchange, volatilized as gas, anoxically changed back to organic matter by different microbial community (*Bacillus*, *Clostridium*, *Proteus*, *Pseudomonas*, and *Streptomyces*) called ammonifying bacteria. Then, process of deamination takes place which results in the removal of the amino groups and producing ammonia (NH_3). In most soils, the ammonia dissolves in water and get converted into ammonium ions (NH_4^+).

14.4.2 Nitrification

Nitrification refers to the biological oxidation of ammonium to nitrate. Nitrification takes place in the occurrence of dissolved oxygen (DO), when microbes change ammonium to nitrate nitrogen, with an intermediate (nitrite) in the reaction sequence (Oopkaup et al. 2016). Soil organisms like *Nistrospira, Nitrosomonas, Nitrosococcus, Nitrovibrio, Nitrosolobus* are found to oxidize ammonia to nitrite and in the process generate energy for its growth (Oehl et al. 2004; Ipsilantis and Sylvia 2007). These organisms are chemolithotrophic aerobes. Subsequently, the facultative chemolitrotrophic bacteria (*Nitrobacter*) oxidize nitrite to nitrate. Two step nitrification process is as follows:

$$NH_4^+ + 1.5O_2 \rightarrow NO_{2^-} + 2H^+ + H_2O \tag{14.4}$$

$$NO_{2^-} + 0.5O_2 \rightarrow NO_{3^-} \tag{14.5}$$

$$NH_4^+ + 2O_2 \rightarrow NO_{3^-} + 2H^+ + H_2O \tag{14.6}$$

Nitrification process is influenced by both physico-chemical and environmental factors which includes temperature, pH, alkalinity, inorganic carbon source, moisture, microbial population, and ammonium-N and dissolved oxygen concentrations of (Vymazal 2007; Sundberg et al. 2007). Nitrate is either converted to biomass after subsequently absorbed by plants or microbes or be decreased through denitrification (Truu et al. 2005). Both the rhizosphere and biofilms (aerobic process) are the major spots of biofilms. Nitrogen removal by nitrification mediated by microorganisms in the presence of dissolved oxygen is a two-step process. Ammonium is converted to nitrite and nitrate nitrogen by microbial community either in the water columns or within the biofilms. The chemoautotrophic bacteria that accomplish the process are called nitrifying bacteria. The process occurs in the presence of oxygen in water column by suspended microbes and air within any aerobic biofilms. The water column or sediments pore water still remains with nitrate as it is not immobilized by wetland media which can further be utilized by plants or microbial cell factories in assimilatory nitrate reduction using them to produce additional plant biomass.

14.4.3 Denitrification

Denitrification occurs after nitrification as nitrate is one of the prerequisites for it. In d*enitrification* nitrate get reduced in anoxic conditions to a gaseous dinitrogen. Denitrification is overruled by bacteria in the absence of oxygen with a terminal electron acceptor (nitrate) and electron donor in the form of organic carbon (Button et al. 2015). Nitrate is converted into N_2 through intermediates nitrite, nitric oxide and nitrous oxide (Weber et al. 2011; Nolvak et al. 2013; Button et al. 2016). Biochemical conversion from nitrate to gaseous dinitrogen is as follows:

$$2NO_{3^-} \rightarrow 2NO_{2^-} \rightarrow 2NO \rightarrow N_2O \rightarrow N_2 \quad (14.7)$$

Also, the absence of O_2, redox potential, temperature, pH, presence of denitrifiers, nitrate concentration impacts the denitrification rates (Vymazal 2007; Wang et al. 2012; Herbst et al. 2016). Denitrification reaction occurs mostly in the sediments of constructed wetlands and in the periphyton and phytoplankton films of water column where available carbon is very high and dissolved oxygen is very low. Denitrification or dissimilatory nitrate reduction again carried out by denitrifying bacteria under these anoxic conditions with the products (N_2 and N_2O gases) that will readily exit in the CWs.

Plant root exudates are act as potential sources of biodegradable organic carbon for the constructed wetland plants after they get decomposed. Further, microorganisms (bacteria and fungi) coagulate colloidal material, remove soluble and stabilize OM, and convert it into various gases and new cell tissue (Mitsch and Gosselink 1986). Dissimilatory nitrate reduction then supplies N_2 for fixation by bacteria and for uptake by plants in the vicinity of roots of sediments (where system is nitrogen-lacking) with the remain as N_2 in the water column.

14.4.4 Nitrogen Fixation

Atmospheric nitrogen is assimilated into organic compounds, especially by certain microorganisms (diazotrophs) that contain the enzyme nitrogenase. Diazotrophs are a diverse group of prokaryotes that includes cyanobacteria, green sulfur bacteria, and diazotrophs Azotobacteraceae, rhizobia and *Frankia*. Either aerobic or anaerobic bacteria and blue-green algae carried out nitrogen fixation in the overlying water in free water surface zones, sediment, oxidized rhizosphere, and on the surfaces of plants i.e. on the leaf and stems (Reddy and Graetz 1988).

14.5 Phosphates

Phosphorus is an important nutrient of CWs ecosystem which can have secondary effects by influencing eutrophication process that leads to algal blooms and other problems related to quality of water in constructed wetlands. Phosphorus in engineered wetlands occurs as phosphates in organic and inorganic compounds. Phosphorus elimination from water in CWs occurs takes place through plant use (assimilation/uptake) and microbes; aluminium, iron oxides and hydroxides adsorption; aluminum, iron, and calcium phosphates complexation; and removal of phosphorus adsorbed with sand/media sediments or organic matter (Richardson 1985; Johnston 1991; Walbridge and Struthers 1993). Also, until the transformation into a soluble inorganic form, dissolved organic phosphate remains unavailable to plants. Again, suspended microbes and biofilms in sediments are responsible for these transformations in the water column (Kaushal et al. 2016). Microbial removal (uptake by biofilm) of phosphorus from CWs is very fast and highly effective; however, following cell death, the phosphorus is discharged again. Polyphosphate-accumulating organisms are *Acinetobacter* spp., *Lampropedia* spp., *Microlunatus phosphovorus* and *Tetrasphaera* spp. Phosphate accumulating organisms (PAO) take up volatile fatty acids (VFAs), convert them to Polyhydroxybutyrate (PHB) and store as soluble organics. PAO break energy-rich Poly-P bonds to produce energy needed to produce PHB and thus Ortho-P is released into sediments. Accretion processes are also responsible for the loss of phosphate within the sediments.

14.6 Heavy Metals

CWs receiving industrial wastewater are high in heavy metal content which can be removed by the plant material uptake, adsorption and precipitation through microbial behaviour. Fe (II) is oxidized to Fe (III) in surface flow CW by abiotic process and microbial oxidation; precipitation of other elements are also occur such as arsenic (Kaushal et al. 2017). Under anoxic conditions, by microbial dissimilatory sulfate reduction some of the heavy metals are immobilized to form H_2S.

14.7 Conclusion

Constructed wetlands are low-rate biological treatment systems to for the treatment of variety of wastewaters including rural, urban, industrial and agricultural. CWs are less sophisticated in operation and maintenance. Although such treatment technologies tend to be land intensive, they are often more effective in removing pathogens and continuous to work with proper maintenance. Also in any CW, microbial processes are particularly needed in the conversion of nitrogen into different biologically appropriate forms, which are made available for plant metabolism. Microbes are also responsible for phosphorus uptake by plants which converts insoluble into soluble forms that again become available for the plants. In addition, microbes decompose the organic compounds, and release carbon dioxide in the oxygenic zones of CW and a variety of gases (CO_2, H_2S, and CH_4) in anoxic zones. Microbial activity may be concentrated at solid surfaces provided by plants, biomass, and sediments. Microbial activities vary through seasonal influences, with the least during winters.

References

Bodtker, G., Thorstenson, T., Lillebo, B. L. P., Thorbjornsen, B. E., Ulvoen, R. H., et al. (2008). The effect of long-term nitrate treatment on SRB activity, corrosion rate and bacterial community composition in offshore water injection systems. *Journal of Industrial Microbiology & Biotechnology, 35*, 1625–1636.

Button, M., Weber, K. P., Nivala, J., Aubron, T., & Muller, R. A. (2015). Community-level physiological profiling of constructed wetland microbial communities: Effects of sample preparation. *Applied Biochemistry and Biotechnology, 178*, 960–973.

Button, M., Auvinen, H., Van Koetsem, F., Hosseinkhani, B., Rousseau, D., Weber, K. P., & Du Laing, G. (2016). Susceptibility of constructed wetland microbial communities to silver nanoparticles: A microcosm study. *Ecological Engineering, 97*, 476–485.

Calheiros, C. S. C., Duque, A. F., Moura, A., Henriques, I. S., Correia, A., et al. (2009). Substrate effect on bacterial communities from constructed wetlands planted with *Typha latifolia* treating industrial wastewater. *Ecological Engineering, 35*, 744–753.

Characklis, G. W., Dilts, M. J., Simmons, O. D., Likirdopulos, C. A., Krometis, L.-A. H., & Sobsey, M. D. (2005). Microbial partitioning to settleable particles in stormwater. *Water Research, 39*, 1773–1782.

Chouinard, A., Balch, G. C., Jorgensen, S. E., Yates, C. N., & Wootton, B. C. (2014). *Tundra wetlands: The treatment of municipal wastewaters*. RBC blue water project: performance & operational tools. CWAT, Fleming College pp. 380.

Dong, X., & Reddy, G. B. (2010). Soil bacterial communities in constructed wetlands treated with swine wastewater using PCR-DGGE technique. *Bioresource Technology, 101*, 1175–1182.

Dong, Z., & Sun, T. (2007). A potential new process for improving nitrogen removal in constructed wetlands-promoting coexistence of partial-nitrification and ANAMMOX. *Ecological Engineering, 31*, 69–78.

Forbes, D. A., Reddy, G. B., Hunt, P. G., Poach, M. E., Ro, K. S., et al. (2010). Comparison of aerated marsh-pond-marsh and continuous marsh constructed wetlands for treating swine wastewater. *Journal of Environmental Science and Health, 45*, 803–809.

Frey, S. D., Knorr, M., Parrent, J. L., & Simpson, R. T. (2004). Chronic nitrogen enrichment affects thee structure and function of the soil microbial community in temperate hardwood and pine forests. *Forest Ecology and Management, 196*, 159–171.

Gray, N. F. (2004). *Biology of wastewater treatment, Series in Environmental Science and Management*. London: Imperial College Press.

Gustavsson, L., & Engwall, M. (2012). Treatment of sludge containing nitro-aromatic compounds in reed-bed mesocosms- water, BOD, carbon and nutrient removal. *Waste Management, 32*, 104–109.

Han, H. S., & Lee, K. D. (2005). Phosphate and potassium solubilizing bacteria effect on mineral uptake, soil availability and growth of eggplant. *Research Journal of Agriculture and Biological Sciences, 1*, 176–180.

Herbst, F. A., Lunsmann, V., Kjeldal, H., Jehmlich, N., Tholey, A., von Bergen, M., Nielsen, J. L., Hettich, R. L., Seifert, J., & Nielsen, P. H. (2016). Enhancing metaproteomics-the value of models and defined environmental microbial systems. *Proteomics, 16*, 783–798.

Hilton, B. L. (1993). Performance evaluation of a closed ecological life support system (CELSS) employing constructed wetlands. In G. A. Moshiri (Ed.), *Constructed wetlands for water quality improvement* (pp. 117–125). Boca Raton: CRC Press.

Hirayama, H., Takai, K., Inagaki, F., Yamato, Y., Suzuki, M., et al. (2005). Bacterial community shift along a subsurface geothermal water stream in a Japanese gold mine. *Extremophiles, 9*, 169–184.

Hoffland, E., Van Den Boogaard, R., Nelemans, J., & Findenegg, G. (1992). Biosynthesis and root exudation of citric and malic acids in phosphate-starved rape plants. *The New Phytologist, 122*, 675–680.

Ibekwe, A. M., Grieve, C. M., & Lyon, S. R. (2003). Characterization of microbial communities and composition in constructed dairy wetland wastewater effluent. *Applied and Environmental Microbiology, 69*, 5060–5069.

Ibekwe, A. M., Ma, J., Murinda, S., & Reddy, G. B. (2017). Microbial diversity in continuous flow constructed a wetland for the treatment of swine waste. *Hydrology Current Research, 8*, 277.

Ipsilantis, I., & Sylvia, D. M. (2007). Abundance of fungi and bacteria in a nutrient impacted Florida wetland. *Applied Soil Ecology, 35*, 272–280.

Johnston, C. A. (1991). Sediment and nutrient retention by freshwater wetlands: Effects on surface water quality. *Critical Reviews in Environmental Control, 21*(5), 491–565.

Kaushal, M., Wani, S. P., Patil, M. D., & Datta, A. (2016). Monitoring efficacy of constructed wetland for treating domestic effluent-microbiological approach. *Current Science, 110*, 1710–1715.

Kaushal, M., Patil, M. D., & Wani, S. P. (2017). Potency of constructed wetlands for deportation of pathogens index from rural, urban and industrial wastewater. *International Journal of Environmental Science and Technology, 15*, 637–648. https://doi.org/10.1007/s13762-017-1423-y.

Kroger, R., Pierce, S. C., Littlejohn, K. A., Moore, M. T., & Farris, J. L. (2012). Decreasing nitrate-N loads to coastal ecosystems with innovative drainage management strategies in agricultural landscapes: An experimental approach. *Agricultural Water Management, 103*, 162–166.

Kuenen, J. G. (2008). Anammox bacteria: From discovery to application. *Nature Reviews. Microbiology, 6*(4), 320–326.

Lauber, C. L., Strickland, M. S., Bradford, M. A., & Fierer, N. (2008). The influence of soil properties on the structure of bacterial and fungal communities across land-use types. *Soil Biology and Biochemistry, 40*, 2407–2415.

Long, Y., Yi, H., Chen, S., Zhang, Z., Cui, K., et al. (2016). Influences of plant type on bacterial and archaeal communities in constructed wetland treating polluted river water. *Environmental Science and Pollution Research, 23*, 19570–19579.

Miersch, J., Tschimedbalshir, M., Barlocher, F., Grams, Y., Pierau, B., Schierhorn, A., & Kraus, G. J. (2001). Heavy metals and thiol compounds in Mucor racemosus and Articulospora tetracladia. *Mycological Research, 105*, 883–889.

Mitchell, C. (1996). *Pollutant removal mechanisms in artificial wetlands: Course notes for the IWES 96*. Gold Coast: International Winter Environmental School.

Mitsch, W. J., & Gosselink, J. G. (1986). *Wetlands*. New York: Van Nostrand Reinhold.

Mitsch, W. J., & Gosselink, J. G. (2007). *Wetlands* (4th ed.). Hoboken: Wiley p 582.

Nelson, C. E. (2009). Phenology of high-elevation pelagic bacteria: The roles of meteorologic variability, catchment inputs and thermal stratification in structuring communities. *The ISME Journal, 3*, 13–30.

Nolvak, H., Truu, M., Tiirik, K., Oopkaup, K., Sildvee, T., Kaasik, A., Mander, U., & Truu, J. (2013). Dynamics of antibiotic resistance genes and their relationships with system treatment efficiency in a horizontal subsurface flow constructed wetland. *Science of the Total Environment, 1*, 636–644.

Oehl, F., Frossard, E., Fliessbach, A., Dubois, D., & Oberson, A. (2004). Basal organic phosphorus mineralization in soils under different farming systems. *Soil Biology and Biochemistry, 36*, 667–675.
Oopkaup, K., Truu, M., Nõlvak, H., Ligi, T., & Preem, J. K. (2016). Dynamics of bacterial community abundance and structure in horizontal subsurface flow wetland mesocosms treating municipal wastewater. *Water, 8*, 457.
Reddy, K. R., & Graetz, D. A. (1988). Carbon and nitrogen dynamics in wetland soils. In D. D. Hook (Ed.), *Ecology and management of wetlands. Ecology of Wetlands Portland* (pp. 307–318). Portland: Timber Press.
Reddy, G. B., Hunt, P. G., Phillips, R., Stone, K., & Grubbs, A. (2001). Treatment of swine wastewater in marsh-pond-marsh constructed wetlands. *Water Science and Technology, 44*, 545–550.
Richardson, C. J. (1985). Mechanisms controlling phosphorus retention capacity in freshwater wetlands. *Science, 228*, 1424–1427.
Scholz, M., & Lee, B. H. (2005). Constructed wetlands: A review. *International Journal of Environmental Studies, 62*, 1256–1261.
Sundberg, C., Tonderski, K., & Lindgren, P. E. (2007). Potential nitrification and denitrification and the corresponding composition of the bacterial communities in a compact constructed wetland treating landfill leachates. *Water Science and Technology, 56*, 159–166.
Symonds, E. M., Verbyla, M. E., Lukasik, J. O., Kafle, R. C., Breitbart, M., & Mihelcic, J. R. (2014). A case study of enteric virus removal and insights into the associated risk of water reuse for two wastewater treatment pond systems in Bolivia. *Water Research, 65*, 257–270.
Truu, J., Nurk, K., Juhanson, J., & Mander, U. (2005). Variation of microbiological parameters within planted soil filter for domestic wastewater treatment. *Journal of Environmental Science and Health. Part A, Toxic/Hazardous Substances & Environmental Engineering, 40*, 1191–1200.
US EPA. (2000). *Constructed wetlands treatment of municipal wastewaters* (1st ed.). Cincinnati: United States Environmental Protection Agency.
Vymazal, J. (2005). Horizontal sub-surface flow and hybrid constructed wetlands systems for wastewater treatment. *Ecological Engineering, 25*(1), 478–490.
Vymazal, J. (2007). Removal of nutrients in various types of constructed wetlands. *Science of the Total Environment, 380*(1–3), 48–65.
Walbridge, M. R., & Struthers, J. P. (1993). Phosphorus retention in non-tidal palustrine forested wetlands of the Mid-Atlantic region. *Wetlands, 13*(2), 84–94.
Wang, Y., Hayatsu, M., & Fujii, T. (2012). Extraction of bacterial RNA from soil: Challenges and solutions. *Microbes and Environments, 27*, 111–121.
Wassel, R. A., & Mills, A. L. (1983). Changes in water and sediment bacterial community structure in a lake receiving acid-mine drainage. *Microbial Ecology, 9*, 155–169.
Weber, K. P., Mitzel, M. R., Slawson, R. M., & Legge, R. L. (2011). Effect of ciprofloxacin on microbiological development in wetland mesocosms. *Water Research, 45*, 3185–3196.
Wetzel, R. G. (1993). Constructed wetlands: Scientific foundations are critical. In G. A. Moshiri (Ed.), *Constructed wetlands for water quality improvement* (pp. 3–7). Boca Raton: CRC Press.
Wu, S., Carvalho, P. N., Muller, J. A., Manoj, V. R., & Dong, R. (2016). Sanitation in constructed wetlands: A review on the removal of human pathogens and fecal indicators. *Science of the Total Environment, 541*, 8–22.

Chapter 15
Role of Vermicomposting in Agricultural Waste Management

Charu Gupta, Dhan Prakash, Sneh Gupta, and Monica Azucena Nazareno

Abstract Agricultural wastes including food processing wastes are the by-products of various food industries that have not been recycled or utilized for other purposes. These agri-horticultural wastes constitute a big problem in municipal landfills due to their high rate of biodegradability. In other words, they are actually the unutilized raw materials whose industrial applications are less than their cost of collection and recovery; and therefore they are generally considered as wastes. The major agricultural sources are livestock, crop residues, tree wastes, aquatic weeds, agro-industrial byproducts, marine wastes and tank silt. The advancement of agricultural biotechnology has led to the development of high yielding variety crops and their subsequent crop residues such as straw, leaves twigs, stubbles along with huge amounts of grasses and weeds. During vermicomposting, stabilization of organic waste occurs through the joint activity of earthworms and aerobic micro-organisms. Vermicomposting is ecofriendly and an economic technique for management of agricultural waste. The earthworm *Eisenia foetida* is one of the most common species used in vermicomposting. The temperature of the earthworm feed should be in the range of 20–35 °C along with relative humidity between 60–80%. Commonly known as farmer's friends, the earthworms improves the fertility of the soil by decomposition of organic matter. In this process, the earthworms leave behind their castings that are exceptionally a rich source of bio-fertilizer. Physico-chemical analysis had shown that vermicomposting decreases total organic carbon (TOC) and carbon-nitrogen (C/N) ratio but increases nitrogen-phosphorus-potassium (NPK) content when compared to compost and other agricultural wastes. The other areas of its application are for crop improvement through pathogen destruction (biocon-

C. Gupta (✉) · D. Prakash
Amity Institute of Herbal Research and Studies, Amity University-UP, Noida, Uttar Pradesh, India

S. Gupta
Department of Zoology, RGPG College, Meerut, Uttar Pradesh, India

M. A. Nazareno
Laboratorio de Antioxidantes Procesos Oxidativos, Centro de Investigacin Transferencia de Santiago del Estero (CITSE-CONICET), Universidad Nacional de Santiago del Estero (UNSE), Santiago del Estero, Argentina

S. Shah et al. (eds.), *Sustainable Green Technologies for Environmental Management*, https://doi.org/10.1007/978-981-13-2772-8_15

trol), improving the water holding capacity of soil and production of plant growth regulators. All these factors will ultimately lead to improved crop growth and yield, and better soil physical, chemical and biological properties.

Keywords Earthworms · Vermicompost · Waste-management · Agricultural waste · Organic manure

15.1 Introduction

In a study in 1994, it was estimated nearly 700 million tonnes organic waste is generated annually in cities and rural areas of India, which is either burned or land filled (Bhiday 1994). Food processing industries produces large volumes of wastes, comprising of both solids and liquid. These food wastes pose though rich in nutrients and biomass poses threat to the environment thereby increasing pollution problems. Many studies have shown that these organic wastes have potential to be converted into useful products or even as raw material for other industries. They can also be used as animal feed after little biological treatment.

The problem in environment management usually arises due to the farmers that bring their agriculture produce to the market without grading and cleaning it. Thus the organic waste in the market area increases, that in turn pressurizes and disturbs the ecosystem of agricultural-waste management. Most of the uncollected waste starts decaying at the market site. This in turn creates sanitation problems and hygiene hazards to the common people (Mane and Rasker 2012). Thus it can be observed that inadequate management of organic wastes can lead to the disease outbreak and also have harmful effects on environment.

In the view of above, vermicomposting offers an attractive alternative in environment management. It also generates viable animal feed protein in the form of worm biomass, while alleviating the negative effects of poor organic waste management. Vermicompost develops due to the biological activity of earthworms that consumes mainly organic materials, such as food preparation residuals and leftovers, scrap paper, animal manure, crop residues, including organic by-products from industries, and yard trimmings. Thus wastes are converted into a valuable biofertilizers that can be used in soil amendment for plants and crops.

The other advantages of vermicomposts are it boosts the soil nutrients, increases the availability of nutrients to plants, improves soil structure and drainage, increases the plant growth and suppresses plant disease and insect pest attacks (biocontrol). Thus vermicompost is one of the efficient means to mitigate environmental-pollution problems and for its management (Waleed 2016).

15.2 Vermicomposting

There is a vast microflora that includes both micro- and macro-organisms that have the ability to decompose organic wastes into valuable resources containing both plant nutrients and organic matter, which plays a vital role in maintaining soil productivity. Vermicomposting mainly involves the biodegradation activity of earthworms to maintain nutrient flow from one system to another. It is observed that the earthworm population decreases with soil degradation and are used as a sensitive indicator of soil degradation.

Charles Darwin described earthworms as the 'unheralded soldiers of mankind', and Aristotle called them as the 'intestine of earth', as they could digest a wide variety of organic materials (Darwin and Seward 1903; Martin 1976). Earthworms participate in cellulose degradation, soil formation and humus accumulation. Due to the biological activity of earthworms, the physical, chemical and biological properties of soil are adversely affected. Earthworms feed on organic wastes and are unique as they consume only a small portion from these wastes for their growth and excrete a major proportion of wastes in a partially-digested form (Jambhekar 1992). This is because the intestine of earthworms contain an array of micro-organisms, hydrolytic enzymes and hormones which helps in rapid decomposition of partially-digested material thereby transforming the complex organic matter into vermi-compost in a relatively smaller duration of 1–2 months (Nagavallemma et al. 2004) as compared to traditional composting process which takes the longer duration of nearly 5 months (Sánchez-Monedero et al. 2001).

The mechanism of action of digestion and formation of vermicompost by earthworms occurs in many steps. The organic matter passes through the gizzard of the earthworm where it is grounded into fine powder. Thereafter the hydrolytic enzymes such as cellulase, amylase, lipase, protease, urease and chitinase (Munnoli et al. 2010), micro-organisms and other fermenting substances further helps in their breakdown within the gut, that finally passes out in the form of "casts". These are finally known as the "vermicomposts" (Dominguez and Edwards 2004).

Such earthworms are known as "epigeic" earthworms. Their characteristic features are efficient bio-degradation and nutrient release, tolerant to disturbances, and helps in early decomposition of litter. Epigeic earthworms include *Lumbricus rubellus, Eisenia foetida, Eiseniella tetraedra, Lumbricus castaneus, L. festivus, Bimastus minusculus, B. eiseni, Dendrobaena veneta, Dendrodrilus rubidus,* and *D. octaedra.* As a result of earthworm's activity, the soil loosens and becomes porous. The porosity increases aeration, water absorption, drainage and easy root penetration. The soil aggregates thus formed by earthworms and associated microbes, helps in maintaining soil ecosystem. These aggregates are mineral granules bonded in a way to resist erosion and to avoid soil compaction both in wet and dry condition. Earthworms speed up soil reclamation and make them productive by restoring beneficial microflora (Nakamura 1996). Thus degraded unproductive soils and land degraded by mining could be engineered physically, chemically and biologically and made productive by earthworms. Hence earthworms are termed as ecosystem engineers (Munnoli et al. 2010).

Vermicomposting does not lead to rise in temperature and so can be aptly defined as a non-thermophilic, bio-oxidative process that produces highly fertile compost with the help of biological activity of earthworms and other associated microbes present in the soil. The final product thus obtained is called vermicompost. The bio-physico-chemical properties of vermicomposts include finely divided, peat like, porous, good aeration and drainage, high water holding capacity, with greater microbial activity and buffering capacity. During vermicomposting, many plant growth-regulating hormones and enzymes are produced that enhances soil biodiversity by promoting the beneficial microbes which in turn enhances plant growth. They also help to control plant pathogens, nematodes and other pests, thereby enhancing plant health and minimize the yield loss. Thus vermi-compost offers an attractive alternate to promote sustainable agriculture and safe management of agricultural, industrial, domestic and hospital wastes (Pathma and Sakthive 2012).

The various types of raw materials and earthworm species used in vermicomposting are presented in Table 15.1.

15.2.1 Precautions During Vermicomposting

Various research studies have shown that only the African and not the Indian species of earthworms, *Eisenia foetida* and *Eudrilus eugenae* are ideal for the preparation of vermicompost. The other consideration is that select only plant-based materials such as grass, leaves or vegetable peelings in preparing vermicompost. Materials of animal origin including bone, meat, eggshells, chicken droppings, etc. should be avoided for preparing vermicompost. *Gliricidia* loppings and tobacco leaves as earthworm feed are also not desirable for their rearing. Earthworms are susceptible to be attacked by birds, termites, ants and rats hence they should be protected from them; adequate moisture should also be maintained during their rearing because both stagnant water or lack of moisture are lethal for them. As the process gets completed, the vermicompost should be removed from the bed and replaced by fresh waste materials.

15.3 Management of Flower Waste by Vermicomposting

Management of flower wastes through vermicomposting is another extension to its application. In a recent study, the management of flower waste in the temples of Indore city, India was assessed. The waste from the temples are usually rich in organic matter and comprises of mainly vegetable material such as flowers, leaves, fruits, sugar, jaggery, grains, milk and milk products, and water. These wastes are normally biodegradable and are readily available for the growth of microbes. These

Table 15.1 Various types of Raw Materials and Earthworm Species Used in Vermicomposting

S. No.	Raw material	Earthworm species
1.	Agricultural residues	*Eudrilus eugeniae*
2.	Agriculture waste and sugar cane thrash	*Eudrilus eugeniae, Perionyx excavates*
3.	Board mill sludge	*Lumbricus terrestris*
4.	Canteen waste & vegetable waste	*Eisenia foetida*
5.	Cattle manure	*Eudrilus eugeniae*
6.	Deciduous forest waste, cow-dung	*Eisenia foetida, Perionyx excavatus* and *Dicogaster bolaui*
7.	Different mammalian animal waste	*Eisenia foetida*
8.	Domestic waste + cow-dung	*Perionyx excavates, Perionyx sansibaricus*
9.	Fly ash + cow dung	*Eisenia foetida*
10.	Gaur gum	*Eudrilus eugeniae*
11.	Grass clippings, cow dung	*Eisenia foetida,*
12.	Green waste	*Eisenia andrei*
13.	*Imperata cylindrica* grass	*Perionyx excavatus, Eisenia foetida*
14.	Municipal solid waste	*Eisenia foetida, Eudrilus eugeniae*
15.	Municipal, agricultural and mixed solid waste	*Eudrilus eugeniae*, Perionyx excavates
16.	Onion residue/waste	*Eisenia foetida, Eudrilus eugeniae*
17.	Organic matter, moistened peat moss, crushed leaves, dried yard waste	*Eisenia foetida, Lumbricus rubellis*
18.	Organic wastes	*Lumbricus rubellus, Eisenia jetida, Eisenia andrei, Dendrobdena rubida, Eudrilus eugeniae, Perionyx excavatus* and *Eiseniella tetraedra.*
19.	Paper mill sludge	*Lumbricus rubellus, Eisenia foetida*
20.	Pig manure, food wastes, leaf wastes, yard wastes, bark wastes, chicken manure	*Eisenia foetida*
21.	Potato peels	*Pheretima elongate*
22.	Press mud	*Pheretima elongate, Eudrilus eugeniae, Eisenia foetida, Megascolex megascolex, Perionyx ceylanensis, Drawida willsi*
23.	Bagasse, sugar cane trash	*Drawida willsi*
24.	Sago waste	*Lampito mauritii, Eisenia foetida*
25.	Sericulture waste	*Perionyx excavates*
26.	Sheep manure + cotton industrial waste	*Eisenia foetida*
27.	Shredded paper or newspaper, coir (coconut husk fiber)	*Perionyx excavatus*
28.	Source separated from human faeces	*Eisenia foetida*
29.	Sugar cane residues	*Pheretima elongate*
30.	Vegetable waste + floral waste	*Eudrilus eugeniae, Eisenia foetida, Perionyx excavates*
31.	Wooden or plastic	*Eisenia foetida, Eudrilus eugeniae, Perionyx excavates*

From Sobana et al. (2016)

wastes are either directly released in the water bodies or dumped at the available places of land as such without giving any further treatment. This creates severe environmental pollution and health hazards. So vermicomposting method was adopted to manage temple waste. The steps include the mixing of cow dung in temple waste-solids and are allowed to decompose for 45 days at 30 °C. After partial decomposition, *Eudrilus eugeniae* earth worm species are introduced into the waste. The process is subjected to optimization of parameters like pH of material, electrical conductivity, C/N ratio and temperature. The optimum parameters for vermicomposting of flower waste are temperature (25 °C), pH 8.0, Electric conductivity (200 microSiemens/cm). Vermicompost obtained by this method was rich in C/N ratio 12.3 after 45 day of vermicomposting. The cost production of vermicomposting of flower waste was worked out and its viability for the Indore city was also justified (Kohli and Hussain 2016).

15.4 Management of Bagasse by Vermicomposting

Agro-industrial wastes such as bagasse are pre-decomposed for 40 days, and subsequently vermicomposting is performed for 30 days through the introduction of *Eisenia foetida* at 26 °C temperature and 62–82% moisture content. The earthworm species decomposes the organic matter through their digestive enzymes and during this process leaves behind castings that function as a valuable fertilizer. Physico-chemical analysis revealed that there is a decrease in total carbon (TC), C/N ratio while increase in NPK content in vermicompost when compared to compost and initial agro-industrial waste (bagasse). The other advantages of vermicomposting are in crop improvement, pathogen destruction (biocontrol), improves water holding capacity of soil and production of plant growth regulators. All these factors lead to improved crop growth and yield, improved soil physical, chemical and biological properties (Jaybhaye and Bhalerao 2016).

15.5 Management of Banana Agro-Waste by Vermicomposting

The waste generated from banana is vast as it is a major crash crop in India. A study was conducted to investigate the vermicomposting potential of banana agro-waste (dried leaves and pseudo-stem biomass). The organic waste material was mixed with cow-dung using earthworm *Eudrilus eugeniae,* it was found that the optimal growth and reproduction was obtained in the ratio of 200 g banana waste and 800 g cow dung mixture. Favourable growth and reproduction was also observed in combination of 200 g banana waste and 600 g cow dung mixture. It was also observed that the earthworm could not survive in the treatments of 1000 g banana waste alone; mixture of 800 g banana waste and 200 g cow dung; and also in mixture of 600 g banana waste and 400 g cow dung. Thus the above research showed that higher amounts of banana waste in feed mixtures is injurious for the growth and reproduction of earthworms. Besides this, irrespective of proportion of banana waste and cow dung in the vermicomposting treatment mixture, a decrease in pH, organic carbon (OC), C/N ratio, and an increase in N, P and K was observed. Thus banana waste can be easily managed through vermicomposting if mixed with cow dung (CD) in suitable amounts (Kavitha et al. 2010).

15.6 Management of Agro-Industrial Waste Water Through Vermicomposting

There are several studies that reported the management of agro-industrial waste water through vermicomposting. A research showed that when highly polluted agro-industrial wastewater of a palm oil mill was treated using earthworms into vermicompost, the process significantly reduced the C/N ratio (0.69–79%), soluble chemical oxygen demand ranging from (20–88%) and volatile solids ranging from (0.7–53%). The palm oil mill effluent (POME), which is a waste water, was absorbed into amendments (soil or rice straw), and was used as feed-stocks for the earthworm *Eudrilus eugeniae.* During vermicomposting, there was a significant increase in pH, electrical conductivity and nutrient content. In this study, it was found that rice straw was a better amendment and absorbent as compared to soil, due to its higher nutrient content and greater reduction in soluble chemical oxygen demand (COD) and a lower C/N ratio. In addition, the growth of the earthworms was also reduced in all treatments. It was concluded that the treatment involving mixture in the ratio of 1 part rice straw and 3 parts palm oil mill effluent (w/v) produced the best quality vermicompost with high nutritional status (Lim et al. 2014).

In yet another study, the chemical characteristics of vermi-composts obtained from cattle manure (CM), orange peel (OP) and filter cake (FC) was evaluated. Three compost piles were set up in the sequence 2:1 OP and CM, 3:1 FC plus CM and CM. The piles were initially composted for 60 days and thereafter, earthworms were added to initiate the vermicomposting process. It was found that the pH and organic carbon contents were above the minimum recommended values for organic fertilizers but the C/N ratio was in the required range. However, the N content was low. So it was concluded that co-vermicomposting of filter cake and orange peel with cattle manure can be applied in sustainable agriculture (Pigatin et al. 2016).

15.7 Status of Vermicomposting of Agro-Waste

It is now well known that vermicomposting is the most promising bio-fertilizer which besides increasing the plant growth and productivity by nutrient supply, is economic and ecofriendly. As a result of degradation activity of earthworms, the mineralization of nutrients is enhanced, that increases crop productivity. Vermicompost produced from the farm wastes improves soil health and growth, enhances quality and crop yield and helps in pollution control. The technique can also be used to generate additional revenue (economic benefits) e.g. as for Baramati Agriculture Produce Marketing Committees (APMC). Nowadays, vermicomposting production units are being constructed to solve the problem of disposing the agro-wastes as no hazardous effluents are generated during the process such as no pesticide residues, weed seeds, heavy metals, and, termite or wax, plant root diseases, etc. Moreover, vermicompost can be used for all types of crops of agricultural, horticultural, ornamental and vegetables and at any stage. Practicing vermicompost for disposal of fruits and vegetable wastes will thus reduce the requirement of more land in near future thereby creating better environments, and reducing ecological risk (Mane and Raskar 2012).

In a study by Nithya and Lekeshmanaswamy (2010) vermicompost showed higher percentage of biomass production in the vermicompost medium as compared to the garden soil.

In yet another study, scientists have devised a low-maintenance vermicomposting system for processing manure and food waste for small-holder farmers. This system was first set up for treating cow manure and food waste in Kampala, Uganda, and monitored for approx. 6 months. The rate of biomass degradation and protein production were observed and calculated after every 2 months and finally at the end of the experiment. The organic biomass was reduced by around 46% and waste-to-biomass conversion rate was around 4% on a total solid basis. However, the conversion rate can be increased by increasing the frequency of worm harvesting. Thus it can be safely concluded from the above study that vermicomposting is an effective and economically viable manure management method for a small-scale agriculture. It was also found during the study that the return of investment is 280% for treating the cow manure of a 450 kg. Although the vermicompost is not sanitized, but the hygiene quality can be improved by including a post-stabilization step in which no fresh material is added. The animal feed protein generated in the process can be used as an incentive to improve current manure management strategies (Lalander et al. 2015).

In another study, vermicompost was also shown to play the role in reducing the heavy metal content in polluted soils in Thai region. The experiment was conducted using vermicompost at various concentrations of cadmium as cadmium chloride. The change/decrease in cadmium content was calculated by analyzing the physico-chemical properties of soil before and after the treatment with compost and vermicompost. The promising results were obtained showing that vermicompost absorbed more cadmium in sludge waste and subsequently reduced the cadmium concentration as compared to compost alone. The earthworms in vermicompost increased the pH of soil, the availability of P, K, Na, Mg, Ca while the content of organic carbon and cadmium contamination in soil were decreased (Nuntawut et al. 2010).

In yet another study, scientists studied the various integrated approaches available for different composting methods for the management of solid waste. Composting is not only a method for waste disposal but also includes waste recycling that can be used for agricultural purposes. The experiment was carried out in the following steps. The solid waste was first composted for 22 days that was further subjected to vermicomposting. Samples were routinely analyzed for the change in content of carbon, nitrogen, moisture, pH and temperature in order to determine the quality of composting. Besides this, decrease in moisture content to around 32%, and relative decrease in carbon and nitrogen content were also observed. Among the different types of treatment studied, only the mixture of municipal solid waste and activated sludge integration showed promising results that was followed by vermicomposting of mixture of municipal solid waste and activated sludge combination. All the results were compared to the other combinations used such as mixture of dried activated sludge, municipal solid waste plus activated sludge semisolid and municipal solid waste plus sewage water. All these results proved that windrow composting method followed by vermicomposting is an effective alternative as compared to other methods (Kumaresan et al. 2016).

Conventionally, vermicomposting is done manually by the farmers in their fields but research has improved and mechanized the production methods of earthworms and their castings. A mechanized process through a continuous flow reactor was developed that employs an elevated bed that allows the feedstock addition at the top level up to a height with 1–2 m thick bed of earthworms. These earthworms degrade the organic matter into 'castings' which are collected from below the bed, thus allowing the earthworms to work continuously with-out being disturbed. This in-vessel technology is more efficient than the customized windrow technology that is used in an outdoor environment and is subjected to variation in weather, predation, and moisture.

In yet another study, the comparison of effect of vermicompost with commercially available growth media) on plant growth was determined on an artificial synthetic media and soil. The research conclusion was that earthworms in the concentration of 10–20% produces optimum castings in media that resulted in improved root and shoot development, increase in leaf size, formation of flowers, increase in crop yield, and overall health of plants. They also produced certain antagonistic substances that helped from plant pathogens. The other recent findings on the functions of casting research are their insect-repellent properties, suggesting their potential to be used as an organic, non-toxic bio-pest repellent.

Vermicomposting can be used in treatment of wastewater residuals (bio-solids). Scientific studies have shown the nearly complete removal of four indicator species of human pathogens (*E. coli, Salmonella*, enteric viruses, helminth ova). It is still yet to be approved by USEPA or USDA as a safe and effective means for treatment of bio-solids for reducing pathogens (Vermico-http://www.vermico.com/vermicomposting-technology-for-waste-management-agriculture-an-executive-summary/).

Anwar et al. (2015) reported that application of compost prepared from a mixture of dairy manure with wheat straw and sawdust yielded higher plant biomass. However, compost prepared from cattle manure and rice straw contained high levels of total nitrogen and C/N ratio which is suitable to be used as soil amendment. Zhen et al. (2014) tried to reclaim degraded soils by applying manure compost and bacteria fertilizers alone or in combination on maize growth. They found that the number of microorganisms increased by the application of compost manure due to improved microbial activity and diversity of degraded irrigated lands. Ewulo et al. (2007) determined the effect of cow dung on soil, leaf mineral composition and pepper yield. The results showed that plant height, yield and fruit weight increased when cow dung was added up to 7.5 t ha^{-1}. Wani et al. (2013) observed that cow dung based compost, prepared by using the epigeic earthworm *Eisenia fetida*, contained high concentration of nitrogen, phosphorus and potassium nutrients compared to other waste materials. Ngakou et al. (2014) observed that the compost prepared from cow dung was higher in N, P and K contents as compared to kitchen manure.

Suthar (2008) studied the potential of the epigeic earthworm *Eisenia fetida* commonly used in vermicomposting for stabilization of sludge after being mixed with cow dung under laboratory conditions. It is found that all the vermicompost ponds showed a decline in organic carbon by 8–26% and pH by 8–19%; however and an increase in total nitrogen by 130–171%, available phosphorus by 22–121%, exchangeable potassium by 105–160%, exchangeable calcium by 49–118% and exchangeable magnesium by 14–51% content was observed. Thus it was concluded that *Eisenia fetida* maximized the degradation and mineralization efficacy in vermibeds showing it as a useful method for organic manure management. Garg and Kaushik (2005) found that earthworm population mortality was more in textile mill sludge vermibeds but it can be minimized by adding sufficient amount of cow dung or plant residues (Suthar 2007a). The other factors that affect the growth of earthworms are related to physiochemical and nutrient characteristics of waste feed stocks (Suthar 2007b).

Some studies have also shown that these earthworm species also secretes phosphatase enzyme in the soil during their decomposition activity and excreted through their cast deposition (Le Bayon and Binet 2006).

Previous studies indicated that vermicompost earthworms can help in the bioremediation of heavy metals (Gupta et al. 2005). Yamada et al. (2007) developed another method of composting cattle dung wastes that utilizes hyper-thermophilic pre-treatment reactor along with a general windrow post-treatment system.

15.8 Vermicomposting: International Appeal

Vermicomposting has a great international appeal. It is especially useful in areas where temperate weather conditions allow for implementation of outdoor systems. Besides India, where vermicomposting has been used for waste management and for the production of marketable castings; China utilizes these earthworms in their traditional medicines and also as pharmaceutical agents. Some of their clinical applications are in the treatment for diseases related to nervous, blood, cardiovascular, and respiratory systems. Earthworm treatments have been used for the treatment of conditions such as asthma, epilepsy, high blood pressure, schizophrenia, mumps, eczema, burns, ulcers and cancer.

In Cuba, vermicomposting animal manures has been practiced from the time as early as after the break-up from USSR and the loss of chemical fertilizers from the Soviet Union. In Australia, some researchers reported that these earthworms increased the grapes yield in vine yards by up to 35%. Similarly, increase in yield and fruit size of cherries by up to 25% was also observed with the use of earthworms for up to at least two annual harvests without further additions of vermicompost. Vermicomposting is practiced on large scale in countries like Mexico where more than 40 companies or individual farmers operate vermicomposting plants in 13 states. Their production capacities range from 0.3 to 4 tons/day of castings, chiefly from coffee pulp. Similarly, vermicompost can be used for damaged agriculture particularly in poor regions of Mexico. Its regular use can slowly improve the soil health and flora and at the same time can generate some income to rural farmers. Thus vermicomposting can be used as a social support and an ecological defense tool in developing countries (Gonzales and Morales 2002).

15.9 Conclusion

Vermicomposting technology is an old age practice in India and a well-known technology throughout the world. It represents an attractive, efficient and ecofriendly approach in treatment and management of solid wastes generated from all sources such as industrial, agricultural and domestic. The other added advantage is that in vermicomposting the material is neither landfilled nor burned but recycled. Thus, vermicomposting is a technology that focuses on conservation of resources and their sustainable utilization. Vermicomposting can also be used for the treatment of food-waste, paper, cardboard, manures, and bio-solids. It can be used in soil amendment. In addition, vermicomposting may also help in employment generation. Vermicompost can also be used in greenhouse application, in establishing new plants such as rootstock in vineyards. Vermicomposts can be used for both agricultural and horticultural production. However, there are still many research gaps that need to be addressed such as insufficient scientific study on enhancing the growth rate of earthworms.

Acknowledgements We duly acknowledge Department of Science & Technology (DST), Delhi for the financial support. The authors feel grateful to Dr. A.K. Chauhan, Founder President and Mr. Atul Chauhan, Chancellor, Amity University UP, India for their research facilities and moral support.

References

Anwar, Z., Irshad, M., Fareed, I., & Saleem, A. (2015). Characterization and recycling of organic waste after co-composting-a review. *Journal of Agricultural Science, 7*, 68–79.

Bhiday, M. R. (1994). Earthworms in agriculture. *Indian Farming, 43*(12), 31–34.

Darwin, F., & Seward, A. C. (Eds.). (1903). *More letters of Charles Darwin. A record of his work in series of hitherto unpublished letters* (p. 508). London: John Murray.

Dominguez, J., & Edwards, C. A. (2004). Vermicomposting organic wastes: A review. In S. H. Shakir Hanna & W. Z. A. Mikhail (Eds.), *Soil zoology for sustainable development in the 21st century* (pp. 369–395). Cairo: Safwat H. Shakir Hanna.

Ewulo, B. S., Hassan, K. O., & Ojeniyi, S. O. (2007). Comparative effect of cowdung manure on soil and leaf nutrient and yield of pepper. *International Journal of Agricultural Research, 2*, 1043–1048.

Garg, V. K., & Kaushik, P. (2005). Vermi-stabilization of textile mill sludge spiked with poultry droppings by an epigeic earthworm *Eisenia foetida*. *Bioresource Technology, 96*, 1063–1071.

Gonzales, C., & Morales, J. (2002). Vermicomposting fits needs of a developing country. *BioCycle: Journal of Composting & Organics Recycling, 43*, 64–69.

Gupta, S. K., Tewari, A., Srivastava, R., Murthy, R. C., & Chandra, S. (2005). Potential of *Eisenia foetida* for sustainable and efficient vermicomposting of fly ash. *Water, Air & Soil Pollution, 163*, 293–302.

Jambhekar, H. (1992). *Use of earthworm as a potential source of decomposing organic wastes* (pp. 52–53). Coimbatore: Proceedings of National Seminar on Organic Farming.

Jaybhaye, M. M., & Bhalerao, S. A. (2016). Vermicomposting- a sustainable approach for management of agro-industrial waste (Bagasse). *Galaxy International Interdisciplinary Research Journal, 4*(4), 27–34.

Kavitha, P., Ravikumar, G., & Manivannan, S. (2010). Vermicomposting of banana agro-waste using an epigeic earthworm *Eudrilus eugeniae* (kinberg). *International Journal of Recent Scientific Research, 1*, 032–035.

Kohli, R., & Hussain, M. (2016). *Management of flower waste by vermicomposting*. International Conference on Global Trends in Engineering, Technology and Management (ICGTETM-2016). pp. 34–38.

Kumaresan, K., Balan, R., Sridhar, A., Aravind, J., & Kanmani, P. (2016). An integrated approach of composting methodologies for solid waste management. *Global Journal of Environment Science and Management, 2*(2), 157–162.

Lalander, C. H., Komakech, A. J., & Vinnerås, B. (2015). Vermicomposting as manure management strategy for urban small-holder animal farms- Kampala case study. *Waste Management, 39*, 96–103.

Le Bayon, R. C., & Binet, F. (2006). Earthworms change the distribution and availability of phosphorus in organic substrates. *Soil Biology and Biochemistry, 38*, 235–246.

Lim, S. L., Wu, T. Y., & Clarke, C. (2014). Treatment and Biotransformation of highly polluted agro-industrial wastewater from a palm oil mill into vermicompost using earthworms. *Journal of Agricultural and Food Chemistry, 62*(3), 691–698.

Mane, T. T., & Raskar, S. S. (2012). Management of agriculture waste from market yard through vermicomposting. *Research Journal of Recent Sciences, 1*(ISC-2011), 289–296.

Martin, J. P. (1976). *Darwin on earthworms: The formation of vegetable moulds*. Ontario: Bookworm Publishing.

Munnoli, P. M., Da Silva, J. A. T., & Saroj, B. (2010). Dynamic soil, dynamic plant. In *Dynamics of the soil-earthworm-plant relationship: A review* (pp. 1–21).

Nagavallemma, K. P., Wani, S. P., Stephane, L., Padmaja, V. V., Vineela, C., Babu Rao, M., & Sahrawat, K. L. (2004). *Vermicomposting: Recycling wastes into valuable organic fertilizer* (p. 20). Global Theme on Agro-ecosystems Report no. 8. Patancheru 502324. Andhra Pradesh: International Crops Research Institute for the Semi-Arid Tropics.

Nakamura, Y. (1996). Interactions between earthworms and microorganisms in biological control of plant root pathogens. *Farming Japan, 30*, 37–43.
Ngakou, A., Koehler, H., & Ngueliaha, H. C. (2014). The role of cow dung and kitchen manure composts and their non-aerated compost teas in reducing the incidence of foliar diseases of *Lycopersicon esculentum* (Mill). *International Journal of Agricultural Research, Innovation and Technology, 4*, 88–97.
Nithya, G., & Lekeshmanaswamy, M. (2010). Production and utilization of Vermicompost in agro industry. *Journal of Phytology, 2*(2), 68–72.
Nuntawut, C.-N., Chuleemas, B. I., & Mongkon, T.-O. (2010). Vermicompost: Tool for agro-industrial waste management and sustainable agriculture. *International Journal of Environmental and Rural Development, 1*(2), 38–43.
Pathma, J., & Sakthivel, N. (2012). Microbial diversity of vermicompost bacteria that exhibit useful agricultural traits and waste management potential. *Springer Plus, 1*, 26.
Pigatin, L. B. F., Atoloye, I. A., Obikoya, O. A., Borsato, A. V., & Rezende, M. O. O. (2016). Chemical study of vermicomposted agro industrial wastes. *International Journal of Recycling of Organic Waste in Agriculture, 5*(1), 55–63.
Sánchez-Monedero, M. A., Roig, A., Paredes, C., & Bernal, M. P. (2001). Nitrogen transformation during organic waste composting by the Rutgers system and its effects on pH, EC and maturity of the composting mixtures. *Bioresource Technology, 78*, 301–308.
Sobana, K., Agnes Sharmila, M., & Jegadeesan, M. (2016). A review on vermicomposting of bio waste using different earthworm species. *Journal of Environmental Science,Computer Science and Engineering & Technology Sec. A, 5*(1), 193–200.
Suthar, S. (2007a). Vermicomposting potential of *Perionyx sansibaricus* (Perrier) in different waste materials. *Bioresource Technology, 98*, 1231–1237.
Suthar, S. (2007b). Production of vermifertilizer from guar gum industrial wastes by using composting earthworm *Perionyx sansibaricus* (Perrier). *Environmentalist, 27*, 329–335.
Suthar, S. (2008). Bioconversion of post-harvest crop residues and cattle shed manure into value-added products using earthworm *Eudrilus eugeniae* Kinberg. *Ecological Engineering, 32*, 206–214.
Waleed, S. A. (2016). Cow manure composting by microbial treatment for using as potting material: An overview. *Pakistan Journal of Biological Sciences, 19*, 1–10.
Wani, K. A., Mamta, W. K. A., & Rao, R. J. (2013). Bioconversion of garden waste, kitchen waste and cow dung into value-added products using earthworm *Eisenia fetida*. *Saudi Journal of Biological Sciences, 20*, 149–153.
Yamada, T., Miyauchi, K., Ueda, H., Ueda, Y., Sugawara, H., Nakai, Y., & Endo, G. (2007). Composting cattle dung wastes by using a hyperthermophilic pre-treatment process: Characterization by physicochemical and molecular biological analysis. *Journal of Bioscience and Bioengineering, 104*, 408–415.
Zhen, Z., Liu, H., Wang, N., Guo, L., Meng, J., et al. (2014). Effects of manure compost application on soil microbial community diversity and soil microenvironments in a temperate cropland in China. *PLoS One, 9*, e108555. https://doi.org/10.1371/journal.pone.0108555.

Additional Readings

Vermico-http://www.vermico.com/vermicomposting-technology-for-waste-management-agriculture-an-executive-summary/

Index

S. Shah et al. (eds.), *Sustainable Green Technologies for Environmental Management*, https://doi.org/10.1007/978-981-13-2772-8

R

S

FSC
www.fsc.org
MIX
Papier | Fördert
gute Waldnutzung
FSC® C083411

Kolibri 360 GmbH
Ferdinand-Jühlke-Straße 7
99095 Erfurt, Deutschland
produktsicherheit@kolibri360.de